CBS PROBLEMS and SOLUTIONS Series

Problems and Solutions in POWER SYSTEMS

- This book is suitable for examinations of all Indian universities.
- Solutions of **215 problems** are given in most simplified manner.
- All types of problems are included and the solutions given in detail.
- To explain the basics lucidly, computerised figures have been liberally added in the solutions. This volume contains more than **200 line diagrams**.

CBS Problems and Solutions *Series*

- Engineering Thermodynamics
- Strength of Materials
- Integrated Electronics
- Communication Systems
- Signals and Systems
- Power Systems
- Electrical Machines and Transformers
- Network Analysis
- Power Electronics
- Control Systems
- Analog Systems

plus many ... more new topics

CBS PROBLEMS and SOLUTIONS *Series*

Problems and Solutions in POWER SYSTEMS

R. GOPAL
Engineer

CBS

CBS Publishers & Distributors Pvt. Ltd.

New Delhi • Bengaluru • Chennai • Kochi • Kolkata • Mumbai
Hyderabad • Nagpur • Patna • Pune • Vijayawada

Problems and Solutions in
POWER SYSTEMS

ISBN: 81-239-1197-1

First Edition: 2005
Reprint: 2008, 2011, 2017

Published by **Satish Kumar Jain** and produced by **Varun Jain** for
CBS Publishers & Distributors Pvt. Ltd.,
4819/XI Prahlad Street, 24 Ansari Road, Daryaganj, New Delhi - 110002
delhi@cbspd.com, cbspubs@airtelmail.in • www.cbspd.com
Ph.: 23289259, 23266861, 23266867 • Fax: 011-23243014

Corporate Office: 204 FIE, Industrial Area, Patparganj, Delhi - 110 092
Ph: 49344934 • Fax: 011-49344935
E-mail: publishing@cbspd.com • publicity@cbspd.com

Branches:
- ***Bengaluru:*** 2975, 17th Cross, K.R. Road, Bansankari 2nd Stage, Bengaluru - 70 • Ph: +91-80-26771678/79 • Fax: +91-80-26771680 E-mail: cbsbng@gmail.com, bangalore@cbspd.com
- ***Chennai:*** No. 7, Subbaraya Street, Shenoy Nagar, Chennai - 600030 Ph: +91-44-26681266, 26680620 • Fax: +91-44-42032115 E-mail: chennai@cbspd.com
- ***Kochi:*** Ashana House, 39/1904, A.M. Thomas Road, Valanjambalam, Ernakulum, Kochi • Ph: +91-484-4059061-65 Fax: +91-484-4059065 • E-mail: cochin@cbspd.com
- ***Kolkata:*** 6-B, Ground Floor, Rameshwar Shaw Road, Kolkata - 700014 Ph: +91-33-22891126/7/8 • E-mail: kolkata@cbspd.com
- ***Mumbai:*** 83-C, Dr. E. Moses Road, Worli, Mumbai - 400018 Ph: +91-9833017933, 022-24902340/41 • E-mail: mumbai@cbspd.com

Representatives:

- Hyderabad: 0-9885175004
- Nagpur: 0-9021734563
- Patna: 0-9334159340
- Pune: 0-9623451994
- Vijayawada: 0-9000660880

Printed at:
J.S. Offset Printers, Delhi (India)

Preface

Power Systems is one of the essential subjects for the students of electronics, electronics and telecommunication and electrical engineering. The author feels it a great pleasure in presenting this very complex subject through typical problems and their solutions.

This book covers all important and major topics taught at the undergraduate level in an engineering college. All the **215 problems** have been discussed at length, and detailed **200 line diagrams** given extensively to help the reader understand the *basics* and *grasp the logic*. This book is suitable for preparing for all university examinations and also to face the competitions. This feature is expected to help the reader in visualising the patterns of problems and the extent and level of difficulty in solving the questions likely to be encountered in the examinations, and thus tackling the problems with confidence and success.

Although every care has been taken to ensure accuracy, yet some errors might have crept in. The author will be grateful if these are brought to his notice so that these could be rectified in the subsequent printings and editions of the book. Readers are requested to send their suggestions for further improvement and revision to me through the publishers at the following e-mail address: cbspubs@del3.vsnl.net.in.

R. GOPAL

Contents

Chapter 1

Basic Concepts of Power Systems

Chapter 1

Basic Concepts of Power Systems

Problem 1.1. If $v = 141.4 \sin(\omega t + 30°)$ V and $i = 11.31 \cos(\omega t - 30°)$ A, find for each (a) the maximum value, (b) the rms value, and (c) the phasor expression in polar and rectangular form if voltage is the reference. Is the circuit inductive or capacitive?

Solution: Given

$$v = 141.4 \sin(\omega t + 30°) \text{ V}$$
$$i = 11.31 \cos(\omega t - 30°) \text{ A}$$

(a) To calculate maximum value of voltage and current. Compare the given equation with

$$v = V_m \sin(\omega t + \phi) \text{ and } i = I_m \cos(\omega t + \phi').$$

where V_m = maximum value of voltage

I_m = maximum value of current

Hence, Maximum value of voltage (V_m) = 141.4 V **Ans.**

Maximum value of current (I_m) = 11.31 A **Ans.**

(b) We know that

$$\text{R.M.S. Value} = \frac{\text{maximum value}}{\sqrt{2}}$$

$$\therefore \text{ R.M.S. value of voltage } (V_{rms}) = \frac{V_m}{\sqrt{2}} = \frac{141.4}{\sqrt{2}}$$

$$= 100 \text{ V} \quad \textbf{Ans.}$$

$$\text{And} \quad \text{R.M.S. Value of current } (I_{rms}) = \frac{I_m}{\sqrt{2}} = \frac{11.31}{\sqrt{2}}$$

$$= 7.997 \text{ A}$$

$$\cong 8 \text{ A} \quad \textbf{Ans.}$$

(2) Since the phasor rotation of a quantity is expressed in terms of rms value of the quantity and the phase angle with respect to certain reference.

Assuming voltage as reference phasor

Phase angle of the voltage $= 0°$

rms value of the voltage $= 100\ V$

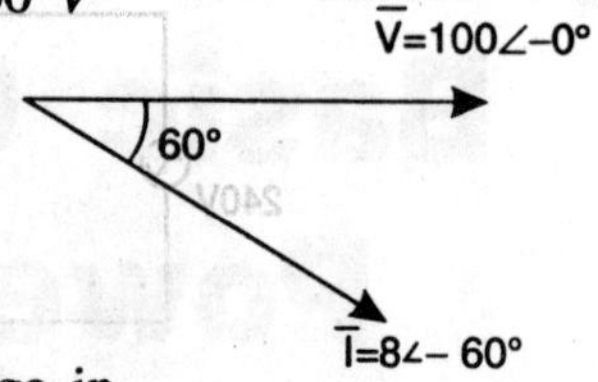

Fig. P. 1.1

Hence, phasor expression of the voltage in polar form $= V_{rms} \angle \phi$

$$\overline{V} = 100 \angle 0°\ \text{V} \quad \textbf{Ans.}$$

Now, phase angle between voltage and current $= 30° - (-30°)$

$= 60°$

RMS value of the current $= 8\ A$

Since current is lagging, the phase angle of current with respect to voltage is $-60°$.

Hence, phasor expression of the voltage in polar form

$$= I_{rms} \angle \text{phase angle}$$

$$= 8 \angle -60°\ \text{A} \quad \textbf{Ans.}$$

In rectangular form $\overline{V} = 100 \cos 0° + j\, 100 \sin 0°$

$$= 100 + j\, 0°\ \text{V} \quad \textbf{Ans.}$$

And $\overline{I} = 8 \angle -60°$

$$= 8 \cos(-60°) + j\, 8 \sin(-60°)$$

$$= 4 - j\, 6.928\ \text{A} \quad \textbf{Ans.}$$

To analyse whether the circuit is inductive or capacitive, we have to calculate complex power.

Now, Complex power $= \overline{V}\, \overline{I}$

$$= 100 \angle 0° \times 8 \angle 60°$$

$$= 800 \angle 60°$$

$$= \underset{\text{Active}}{400} + j\, \underset{\text{Reactive}}{692.82}$$

Since reactive power is positive, this shows that the circuit is inductive. **Ans.**

> **Problem 1.2.** A single-phase ac voltage of 240 V is applied to a series circuit whose impedance is 10 ∠60° Ω. Find *R, X, P, Q* and the power factor of the circuit.

Solution: Given

$$V_{rms} = 240\ \text{V},\ \overline{Z}_L = 10 \angle 60°\ \Omega.$$

Let *R* be the resistance and *X* be the reactance of the series circuit. The circuit can be drawn as below

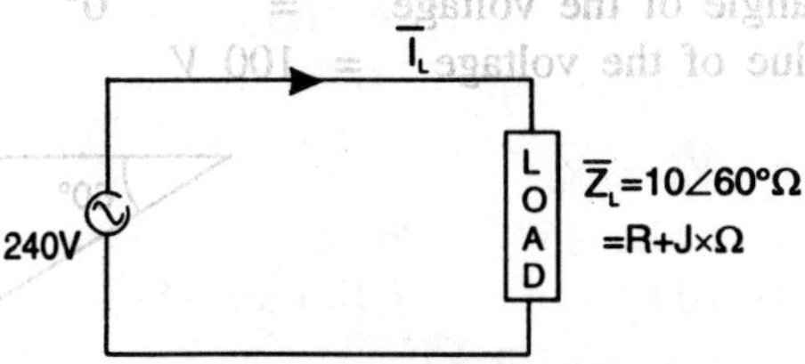

Fig. P. 1.2

Let us choose voltage as reference phasor. Hence

$$\overline{V} = 240\ \angle 0^\circ$$

Now, Current through the load $\overline{I} = \dfrac{\overline{I}}{\overline{Z}} = \dfrac{240\ \angle 0^\circ}{10\ \angle 60^\circ}$

$$= 24\ \angle{-60^\circ}\ \text{A}$$

To calculate R and X, we have $\overline{Z}_L = 10\ \angle 60^\circ\ \Omega$

$$= 10 \cos 60^\circ + j\ 10 \sin 60^\circ \Omega$$
$$= 5 + j\ 8.66\ \Omega$$
$$= R + j\ X\ \Omega$$

From this relation $R = 5\ \Omega$

$X = 8.66\ \Omega$ **Ans.**

Now, to calculate P and Q, Let us find complex power.

Complex power $= \overline{V}\ \overline{I}$

$$= 240\ \angle 0^\circ \times 24\ \angle{+\ 60^\circ}$$
$$= 5760\ \angle{+\ 60^\circ}$$
$$= 2880\ W + j\ 4988.31\ \text{VAr}$$
$$= P + j\ \theta$$

Again, from this relation, $P = 2880$ W

and $Q = 4988.3$ VAr **Ans.**

The positive sign of θ in the complex power expression shows that the circuit is inductive

Now, Power factor of the circuit (cos θ)

$$= \frac{P}{\sqrt{P^2 + \theta^2}}$$

$$= \frac{2880}{\sqrt{(2880)^2 + (-4988.3)^2}}$$

$$= 0.5000 \text{ (logging)}$$ **Ans.**

Problem 1.3. In a single-phase circuit $V_a = 120\ \angle 45^\circ$ V and $V_b = 100\ \angle{-15^\circ}$ V with respect to a reference node o. Find V_{ba} in polar form.

Solution: Given

$$\bar{V}_a = 120\ \angle 45°\ \text{V},\ \bar{V}_b = 100\underline{\angle -15°}\ \text{V}$$

with respect to a reference node.

Now, $\bar{V}_{ba} = \bar{V}_b - \bar{V}_a$

$= 100\angle -15° - 120\angle 45°$

$= 100 \cos 15° - j\ 100 \sin 15° - 120 \cos 45° - j\ 120 \sin 45°$

$= 11.74 - j\ 110.73$

$= 111.95\ \angle -83.95°\ \text{V}$

$= 111.35\ [\cos 83.95° + j \sin (-83.95°)]\ \text{V}$

This the rquired polar form. **Ans.**

Problem 1.4. A voltage source $E_{an} = -120\angle 210°\ V$ and the current through the source is given by $I_{na} = 10\angle 60°\ A$. Find the values of P and Q and state whether the source is delivering or receiving each.

Solution: Given

$$\bar{E}_{an} = -120\angle 210°\ \text{V and } \bar{I}_{na} = 10\angle 60°\ \text{A}$$

Now, Complex power $(\bar{S}) = \bar{E}_{an} \times \bar{I}_{an}$

$= -120\angle 210° \times 10\angle -60°$

$= -1200\angle 150°$

$= -1200 \cos 150° - j\ 1200 \sin 150°$

$= 1039.23\ \text{W} - j\ 600\ \text{VAr}$

$= P + j\ \theta$

From the complex power expression

$P = 1039.23\ \text{W}$ **Ans.**

and $\theta = -\ 600\ \text{VAr}$ **Ans.**

Since voltage source and current through the source have opposite sign, the source absorbs 1039.23 W and the –ve sign of the reactive power shows that the source supplies reactive power.

Problem 1.5. In a balanced three-phase system the Y-connected impedances are $10\angle 30°\ \Omega$. If $V_{bc} = 416\angle 90°$ V, specify I_{cn} in polar form.

Solution: Let us assume phase sequence as *abc*.

The phasor diagram of voltages in a balanced three phase circuit can be drawn as shown in fig. P. 1.5 (a) and (b).

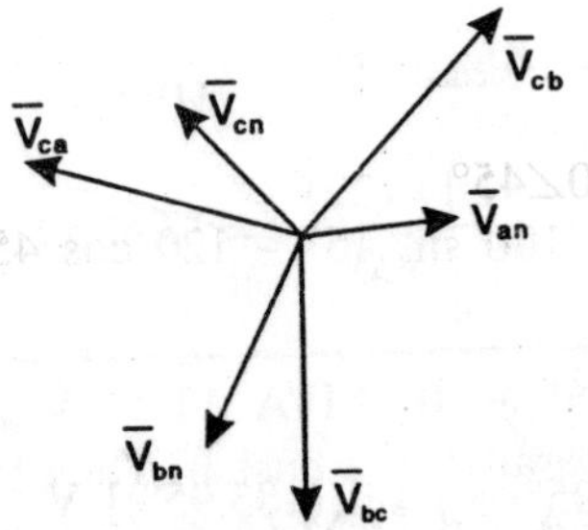

Fig. P. 1.5 (a)

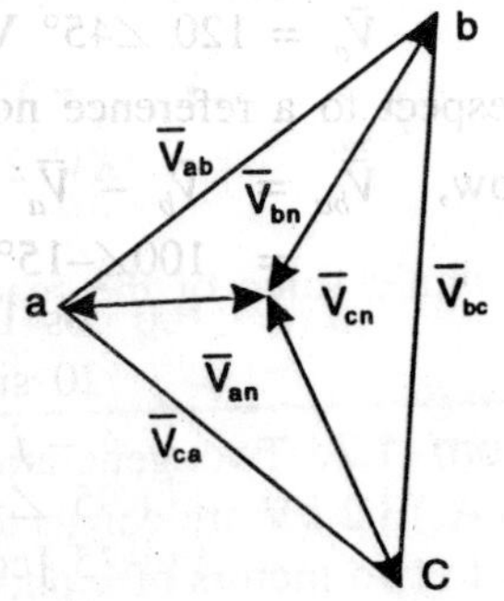

Fig. P. 1.5 (b)

Given, $\bar{Z}_L = 10\angle 30°$ and $\bar{V}_{bc} = 416\angle 90°$ V

For balanced three-phase circuit $\bar{V}_{ab} = 416\angle{-150°}$ V, $\bar{V}_{bc} = 416\angle 90°$ V and $\bar{V}_{ca} = 416\angle{-30°}$

For balanced three-phase circuit, the line voltages and phase voltages are related as

$$\bar{V}_{ab} = \sqrt{3}\, V_{an} \angle 30°$$

$$\therefore \bar{V}_{an} = \frac{416\angle -150°}{\sqrt{3}\angle 30°} = 240.18\angle{-150°} \text{ V}$$

$$\text{Similarly, } \bar{V}_{cn} = \frac{\bar{V}_{ca}}{\sqrt{3}\angle 30°} = \frac{416\angle -30°}{\sqrt{3}\angle 30°} = 240.18\angle{-60°}\ V$$

Now, Current through the phase C,

$$\bar{I}_{cn} = \frac{\bar{V}_{cn}}{\bar{Z}_L} = \frac{240.18\angle -60°}{10\angle 30°} = 24.02\angle{-90°}$$

$$= 24.02\,(\cos 90° - j \sin 90°) \text{ A, which is in polar form.}$$

> **Problem 1.6.** A generator (which may be represented by an emf in series with an inductive reactance) is rated 500 MVA, 22 kV. Its Y-connected windings have a reactance of 1.1 per unit. Find the ohmic value of the reactance of the windings.

Solution: Given,

Rated MVA of the generator = 500 MVA

Rated voltage of the generator = 22 kV

p.u. reactance of the windings = 1.1 p.u.

To calculate ohmic value of the reactance, apply the relation.

$$Z_{p.u.} = \frac{Z_{actual} \times MVA}{(KV)2} \quad \text{or} \quad Z_{actual} = \frac{Z_{p.u.} \times (KV)2}{MVA}$$

$\therefore$ Ohmic value of the reactance $= \dfrac{1.1 \times (22)^2}{500} = 1.065\ \Omega$ **Ans.**

Problem 1.7. Two generators rated at 10 MVA 13.2 kV and 15 MVA 13.2 kV are connected in parallel to a busbar. They feed supply to two motors of inputs 8 MVA and 12 MVA respectively. The operating voltage of motors is 12.5 kV. Assuming base quantities as 50 MVA and 13.8 kV draw the reactance diagram. The per cent reactance for generators is 15% and that for motors is 20%.

Solution: The diagram of the given situation can be drawn as

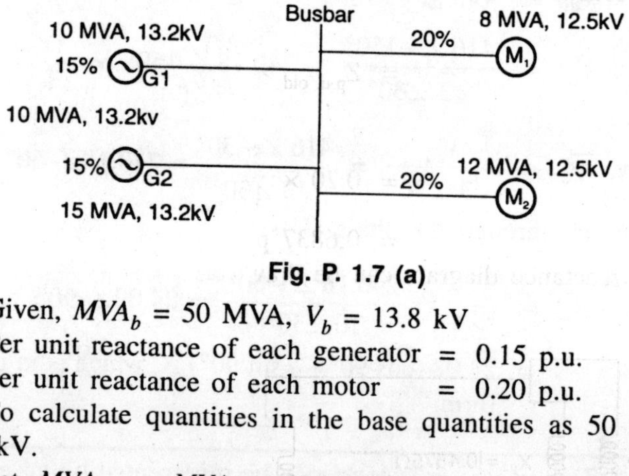

Fig. P. 1.7 (a)

Given, $MVA_b = 50$ MVA, $V_b = 13.8$ kV

Per unit reactance of each generator $= 0.15$ p.u.

Per unit reactance of each motor $= 0.20$ p.u.

To calculate quantities in the base quantities as 50 MVA and 13.8 kV.

Let, $MVA_{new} = MVA_b = 50$ MVA, $V_{new} = V_b = 13.8$ kV

For generator 1, $MVA_{old} = 10$ MVA, $KV_{old} = 13.2$ kV

$Z_{p.u.\ old} = 0.15$ p.u.

Applying the relation

$$Z_{p.u.\ new} = Z_{p.u.\ old} \times \frac{MVA_{new}}{MVA_{old}} \times \left(\frac{kV_{old}}{kV_{new}}\right)^2$$

$$= 0.15 \times \frac{50}{10} \times \left(\frac{13.2}{13.8}\right)^2 = 0.686 \text{ p.u.}$$

For generator 2, MVA_{old} = 15 MVA, kV_{old} = 13.2 kV,
$Z_{p.u.\ old} = 0.15$

$$\text{New p.u. reactance of generator 2} = 0.15 \times \frac{50}{10} \times \left(\frac{13.2}{13.8}\right)^2$$

$$= 0.4575 \text{ p.u.}$$

For motor M_1, MVA_{old} = 8 MVA, kV_{old} = 12.5 kV, $Z_{p.u.\ old} = 0.20$
New p.u. reactance of motor M_1

$$= Z_{p.u.\ old} \times \frac{MVA_{new}}{MVA_{old}} \times \left(\frac{kV_{old}}{kV_{new}}\right)^2$$

$$= 0.20 \times \frac{50}{8} \times \left(\frac{12.5}{13.8}\right)^2$$

$$= 1.0256 \text{ p.u.}$$

For motor M_2, MVA_{old} = 12 MVA, kV_{old} = 12.5 kV, $Z_{p.u.old} = 0.20$
New p.u. reactance of motor M_2

$$= Z_{p.u.\ old} \times \frac{MVA_{new}}{MVA_{old}} \times \left(\frac{kV_{old}}{kV_{new}}\right)^2$$

$$= 0.20 \times \frac{50}{12} \times \left(\frac{12.5}{13.8}\right)^2$$

$$= 0.6837 \text{ p.u.}$$

Now, the reactance diagram can be drawn as

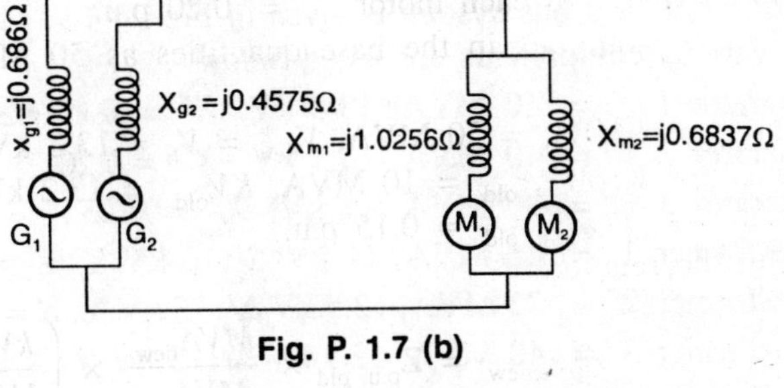

Fig. P. 1.7 (b)

Problem 1.8. Three generators are rated as follows: Generator 1–100 MVA, 33 kV, reactance 10%; Generator 2–150 MVA, 32 kV, reactance 8%; Generator 3–110 MVA, 30 kV, reactance

12%. Determine the reactance of the generator corresponding to base values of 200 MVA 35 kV.

Solution: Given, MVA_b = 200 MVA, V_b = 35 kV

Let, $MVA_{new} = MVA_b$ = 200 MVA and $V_{new} = V_b$ = 35 kV

Now, for generator 1, MVA_{old} = 100 MVA, V_{old} = 33 kV, $Z_{p.u.\ old} = 0.10$

To calculate reactance in given base quantities, apply the relation,

$$Z_{p.u.\ new} = Z_{p.u.\ old} \times \frac{MVA_{new}}{MVA_{old}} \times \left(\frac{V_{old}}{V_{new}}\right)^2$$

$$= 0.10 \times \frac{200}{100} \times \left(\frac{33}{35}\right)^2$$

$$= 0.1778 \text{ p.u.} \quad \textbf{Ans.}$$

For generator 2, MVA_{old} = 150 MVA, V_{old} = 32 kV, $Z_{p.u.\ old}$ = 0.08 p.u. reactance in the given base quantities

$$= 0.08 \times \frac{200}{150} \times \left(\frac{32}{35}\right)^2$$

$$= 0.0892 \text{ p.u.} \quad \textbf{Ans.}$$

For generator 3, MVA_{old} = 110 MVA, V_{old} = 30 kV, $Z_{p.u.\ old}$ = 0.12 p.u. reactance in the given base quantities

$$= 0.12 \times \frac{200}{110} \times \left(\frac{30}{35}\right)^2$$

$$= 0.1603 \text{ p.u.} \quad \textbf{Ans.}$$

Problem 1.9. A 3-bus system is given in Fig. P. 1.9. The ratings of the various components are listed below:

Generator 1 = 50 MVA, 13.8 kV, X'' = 0.15 p.u.

Generator 2 = 40 MVA, 13.2 kV, X'' = 0.20

Generator 3 = 30 MVA, 11 kV, X'' = 0.25

Transformer 1 = 45 MVA, 11 kV Δ/110 kV Y, X = 0.1 p.u.

Transformer 2 = 25 MVA, 12.5 kV Δ/115 kV Y, X = 0.15 p.u.

Transformer 3 = 40 MVA, 12.5 kV Δ/115 kV Y, X = 0.1 p.u.

The line impedances are shown in Fig. P. 1.9. Determine the reactance diagram based on 50 MVA and 13.8 kV as base quantities in Generator 1.

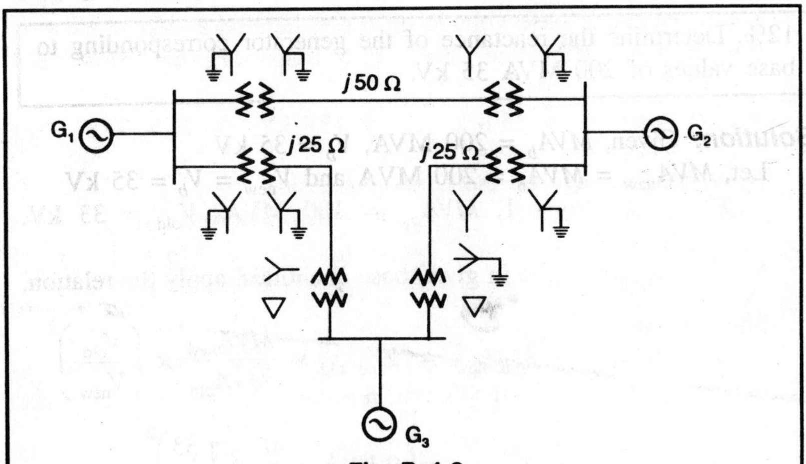

Fig. P. 1.9

Solution: The given power system can be redrawn as below:

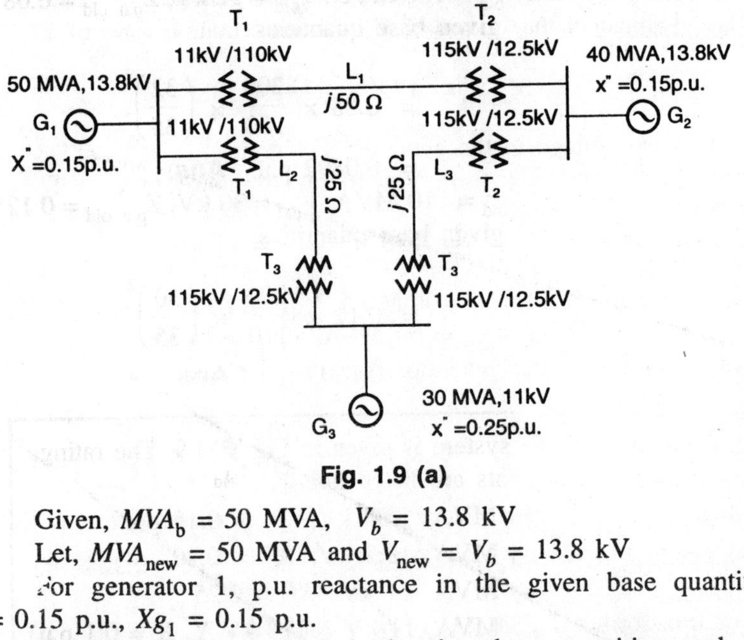

Fig. 1.9 (a)

Given, MVA_b = 50 MVA, V_b = 13.8 kV

Let, MVA_{new} = 50 MVA and $V_{new} = V_b$ = 13.8 kV

For generator 1, p.u. reactance in the given base quantities = 0.15 p.u., Xg_1 = 0.15 p.u.

To calculate the reactances in the given base quantities, apply the relation

$$Z_{\text{p.u. new}} = Z_{\text{p.u. old}} \times \frac{MVA_{\text{new}}}{MVA_{\text{old}}} \times \left(\frac{V_{\text{old}}}{V_{\text{new}}}\right)^2 \quad ...(1)$$

For generator 2, MVA_{old} = 40 MVA, V_{old} = 13.2 kV, $X_{\text{p.u. old}}$ = 0.20

Let us calcula.e base voltage on the generator 2 side.

Base voltage in the line L_1 = $V_b \times$ transformation ratio of T_1

$$V_{bL1} = V_b \times \frac{V_H}{V_L} = 13.8 \times \frac{110}{11} = 138 \text{ kV}$$

Now, base voltage on the generator 2 side

$$V_{bg2} = V_{bL} \times \text{transformation ratio of } T_2$$

$$= V_{bL} \times \frac{V_L}{V_H} = 138 \times \frac{12.5}{115} = 15 \text{ kV}$$

For generator 2,

$$MVA_{new} = MVA_n = 50 \text{ MVA}, \; V_b = V_{bg2} = 15 \text{ kV}$$

Now, Applying equation (1), p.u. reactance in the base quantities

$$X_{g2} = 0.20 \times \frac{50}{40} \times \left(\frac{13.2}{15}\right)^2$$

$$X_{g2} = 0.1936 \text{ p.u.}$$

Now, Let as calculate base voltage on the generator 3 side.

Base voltage in the line L_2 = $V_b \times$ transformation ratio of T_1

$$V_{bL2} = 13.8 \times \frac{110}{11} = 138 \text{ kV}$$

Now, base voltage in the generator 3 circuit

$$= V_{bL2} \times \text{transformation ratio of } T_3$$

$$= 138 \times \frac{12.5}{115} = 15 \text{ kV}$$

For generator 3, $MVA_{old} = 30$ MVA, $V_{old} = 11$ kV, $X_{p.u.\ old} = 0.25$,
$MVA_{new} = 50$ MVA, $V_{new} = 15$ kV

p.u. reactance of the generator 3 in the new quantities

$$= X_{p.u.\ old} \times \frac{MVA_{new}}{MVA_{old}} \times \left(\frac{V_{old}}{V_{new}}\right)^2$$

$$X_{g3} = 0.25 \times \frac{50}{30} \times \left(\frac{11}{15}\right)^2 = 0.2241 \text{ p.u.}$$

For transformer T_1, $MVA_{old} = 45$ MVA, $V_{old} = 11$ kV,
$X_{p.u.\ old} = 0.1$ p.u., $MVA_{new} = 50$ MVA,
$V_{new} = 13.8$ kV

p.u. reactance of transformer T_1 in the new quantities

$$= X_{p.u.\ old} \times \frac{MVA_{new}}{MVA_{old}} \times \left(\frac{V_{old}}{V_{new}}\right)^2$$

$$= 0.1 \times \frac{50}{45} \times \left(\frac{11}{13.8}\right)^2$$

$$X_1 = 0.0706 \text{ p.u.}$$

p.u. impedance of the line

$$L_1 = \frac{Z_{L_1} \times MVA_b}{(V_{b_1})^2} = \frac{50 \times 50}{(138)^2}$$

$$= 0.1313 \text{ p.u.}$$

For transformer T_2, $MVA_{\text{old}} = 25$ MVA, $V_{\text{old}} = 115$ kV, $X_{2 \text{ p.u. old}} = 0.15$ p.u., $MVA_{\text{new}} = 50$ MVA, $V_{\text{new}} = V_{bL1} = 138$ kV

p.u. reactance of the transformer T_2 in the new quantities

$$X_2 = X_{2 \text{ p.u. old}} \times \frac{MVA_{\text{new}}}{MVA_{\text{old}}} \times \left(\frac{V_{\text{old}}}{V_{\text{new}}}\right)^2$$

$$X_2 = 0.15 \times \frac{50}{25} \times \left(\frac{115}{138}\right)^2$$

$$X_2 = 0.2083 \text{ p.u.}$$

p.u. impedance of the line

$$L_2 = \frac{Z_{L_1} \times MVA_b}{(V_{bL2})^2} = \frac{25 \times 50}{(138)^2}$$

$$= 0.0656 \text{ p.u.}$$

For transformer T_3, $MVA_{\text{old}} = 40$ MVA, $V_{\text{old}} = 115$ kV, $X_{3 \text{ p.u. old}} = 0.1$ p.u., $MVA_{\text{new}} = 50$ MVA, $V_{\text{new}} = 138$ kV

p.u. reactance of the transformer T_3 in the new base quantities

$$= X_{3 \text{ p.u. old}} \times \frac{MVA_{\text{new}}}{MVA_{\text{old}}} \times \left(\frac{V_{\text{old}}}{V_{\text{new}}}\right)^2$$

$$= 0.1 \times \frac{50}{40} \times \left(\frac{115}{138}\right)^2 = 0.0868 \text{ p.u.}$$

p.u. impedance of the line

$$L_3 = \frac{Z_{L_3} \times MVA_b}{(V_{bL2})^2} = \frac{25 \times 50}{(138)^2}$$

$$= 0.0656 \text{ p.u.}$$

The reactance diagram of the power system can be drawn as below.

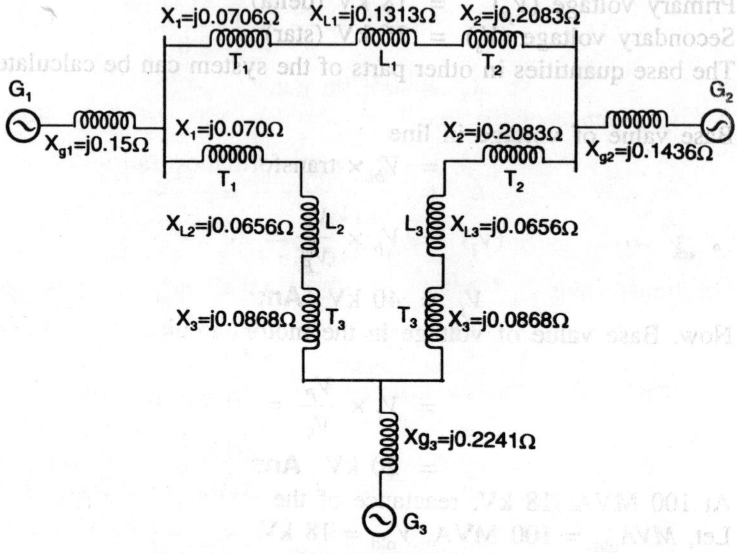

Fig. P. 1.9 (c)

Problem 1.10. A 80 MVA, 20 kV, 3–phase generator has a subtransient reactance of 10%. The generator is connected to the motor through a transmission line and transformers as shown in Fig. P. 1.10 (a).

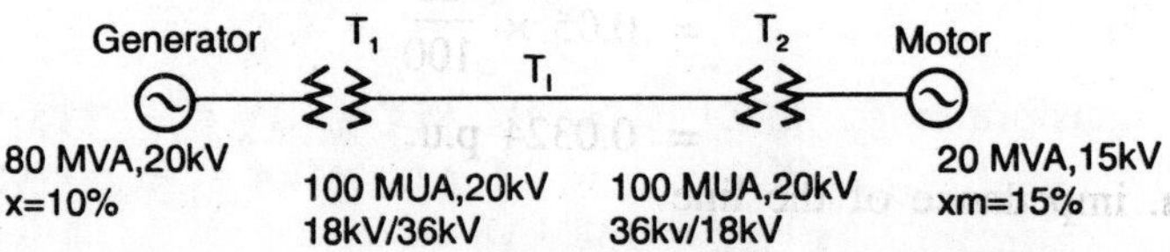

Fig. P. 1.10 (a)

The motor has rated input of 20 MVA at 15 kV with 15% subtransient reactance. The three-phase transformers are rated at 100 MVA, 18 kV Δ/36 kV with leakage reactance 5%. The line has a reactance of 10 Ω. Selecting the generating rating as base quantities in the generator circuit, determine the base quantities in other part of the system and calculate the corresponding p.u. values. Also draw the reactance diagram of the power system.

Solution: Since base quantities are generator ratings.

MVA_b = 80 MVA, V_b = 20 kV, p.u. reactance ($X_{g\ p.u.}$) = 0.10

Also, given, Line reactance (X_l) = 10 Ω, X_m = 0.15

Transformer rating = 100 MVA
Primary voltage (V_p) = 18 kV (delta)
Secondary voltage (V_s) = 36 kV (star)

The base quantities in other parts of the system can be calculated as:

Base value of voltage in line

$$= V_b \times \text{transformation ratio}$$

$$(V_l) = V_b \times \frac{V_S}{V_P} = 20 \text{ kV} \times \frac{36}{18}$$

$$V_l = 40 \text{ kV}$$ **Ans.**

Now, Base value of voltage in the motor circuit

$$= V_l \times \frac{V_P}{V_S} = 40 \text{ kV} \times \frac{18}{36}$$

$$= 20 \text{ kV}$$ **Ans.**

At 100 MVA, 18 kV, reactance of the transformer = 5%

Let, MVA_{old} = 100 MVA, V_{old} = 18 kV, X_{old} = 0.05

At base values, *i.e.* 80 MVA, 20 kV

P.u. reactance of the transformer

$$= X_{old} \times \frac{MVA_b}{MVA_{old}} \times \left(\frac{V_{old}}{V_b}\right)^2$$

$$= 0.05 \times \frac{80}{100} \times \left(\frac{18}{20}\right)^2$$

$$= 0.0324 \text{ p.u.}$$ **Ans.**

p.u. impedance of the line

$$= \frac{\text{Actual impedance of the line} \times \text{Base MVA}}{(\text{Base voltage of the line})^2}$$

$$= \frac{10 \times 80 \times 10^6}{(40 \times 10^3)^2} = 0.500 \text{ p.u.}$$ **Ans.**

Now, p.u. reactance of the motor

$$= \text{motor reactance} \times \frac{\text{Base rating}}{\text{motor MVA rating}} \times \left(\frac{\text{motor kV rating}}{\text{Base kV}}\right)^2$$

$$= 0.15 \times \frac{80}{20} \times \left(\frac{15}{20}\right)^2 = 0.3375 \text{ p.u.}$$ **Ans.**

The reactance diagram for the system can be drawn as below:

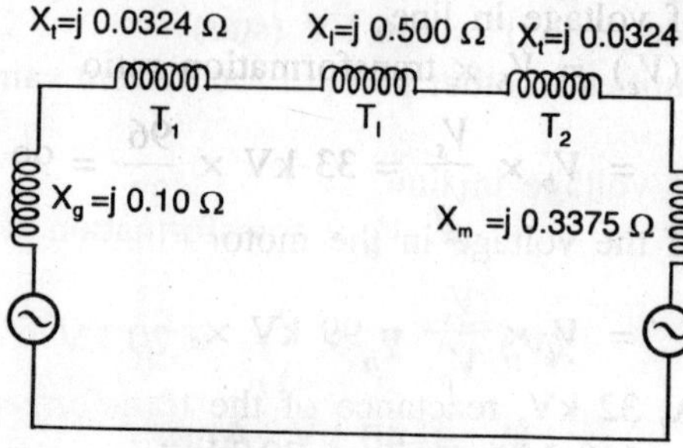

Fig. P. 1.10 (b)

Problem 1.11. A 100 MVA, 33 kV, 3-phase generator has a subtransient reactance of 10%. The generator is connected to two motors through a transmission line and transformers as shown in figure below:

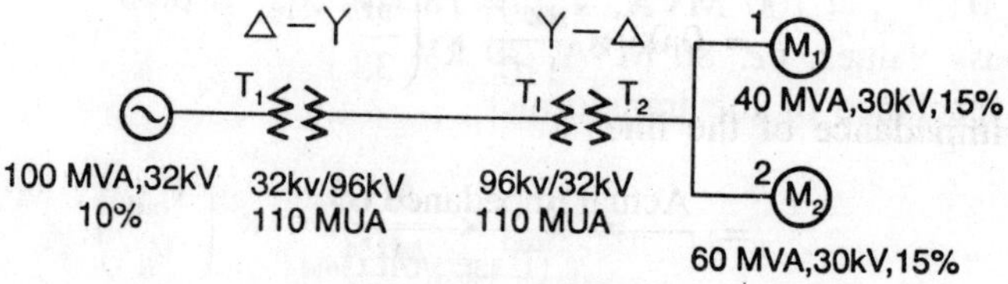

Fig. P. 1.11 (a)

The motor has rated input of 40 MVA and 60 MVA at 30 kV with 15% subtransient reactance. The three phase transformers are rated at 110 MVA, 32 kV Δ/96 kV(Y) with leakage reactance 5%. The line has a reactance of 200 hms. Selecting the generator rating as base quantities in generator circuit, determine the base quantities in other part of the system and calculate the corresponding p.u. values. Also draw the reactance diagram of the system.

Solution: Let us assume generator ratings as base values.

MVA_b = 100 MVA, V_b = 33 kV

p.u. reactance of generator (X_g) = 0.10

Given, Line reactance (X_l) = 20 Ω

Subtransient reactance of each motor = 15%

Transformer rating = 110 MVA

Primary voltage (V_p) of transformer (T_1) = 32 kV (Δ)

Secondary voltage (V_s) of transformer (T_2) = 96 kV (Y)

The base quantities in other parts of the system can be calculated as below:

Base value of voltage in line

$$(V_e) = V_b \times \text{transformation ratio}$$

$$= V_b \times \frac{V_s}{V_p} = 33 \text{ kV} \times \frac{96}{32} = 99 \text{ kV} \quad \textbf{Ans.}$$

Base value of the voltage in the motor circuit

$$= V_l \times \frac{V_s}{V_p} = 99 \text{ kV} \times \frac{32}{96} = 33 \text{ kV} \quad \textbf{Ans.}$$

At 110 MVA, 32 kV, reactance of the transformer = 5%

Let, MVA_{old} = 110 MVA, V_{old} = 32 kV, X_t = 0.05

Now, p.u. reactance of the transformer at base values, i.e. 100 MVA, 33 kV

$$= X_t \times \frac{MVA_b}{MVA_{\text{old}}} \times \left(\frac{V_{\text{old}}}{V_b}\right)^2$$

$$= 0.05 \times \frac{100}{110} \times \left(\frac{30}{33}\right)^2 = 0.0427 \text{ p.u.} \quad \textbf{Ans.}$$

p.u. impedance of the line

$$= \frac{\text{Actual impedance of the line} \times \text{Base MVA}}{(\text{Base voltage of the line})^2}$$

$$= \frac{20 \times 100}{(99)^2} = 0.2041 \text{ p.u.} \quad \textbf{Ans.}$$

To calculate p.u. reactance of each motor. Apply the relation.

$$Z_{\text{p.u. new}} = Z_{\text{p.u. old}} \times \frac{MVA_{\text{new}}}{MVA_{\text{old}}} \times \left(\frac{V_{\text{old}}}{V_{\text{new}}}\right)^2$$

For Motor 1, MVA_1 = 40 MVA, V_1 = 30 kV, $Z_{\text{p.u. 1}}$ = 0.15

Now, p.u. reactance of motor 1

$$= Z_{\text{p.u. 1}} \times \frac{MVA_b}{MVA_1} \times \left(\frac{V_1}{V_b}\right)^2$$

$$= 0.15 \times \frac{100}{40} \times \left(\frac{30}{33}\right)^2 = 0.3099 \text{ p.u.} \quad \textbf{Ans.}$$

p.u. reactance of motor 2

$$= Z_{\text{p.u. 2}} \times \frac{MVA_b}{MVA_2} \times \left(\frac{V_2}{V_b}\right)^2$$

$$= 0.15 \times \frac{100}{60} \times \left(\frac{30}{33}\right)^2 = 0.2066 \text{ p.u.} \quad \textbf{Ans.}$$

The reactance diagram for the system can be drawn as

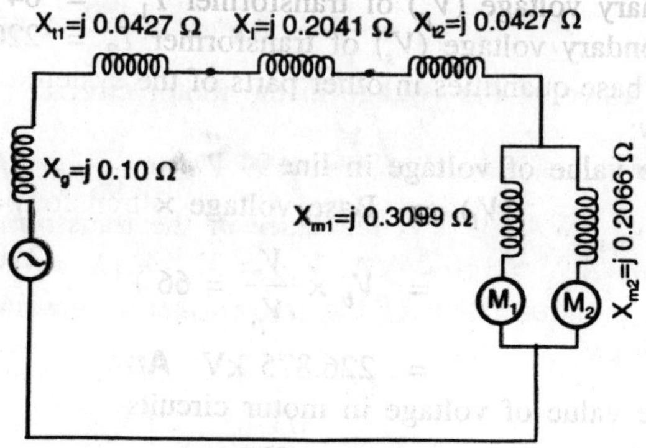

Fig. P. 1.11 (b)

Problem 1.12. A 200 MVA, 66 kV 3-phase generator has a subtransient reactance of 30%. The generator is connected to three motors through a transmission line and transformers as shown in Fig. P. 1.12 (a).

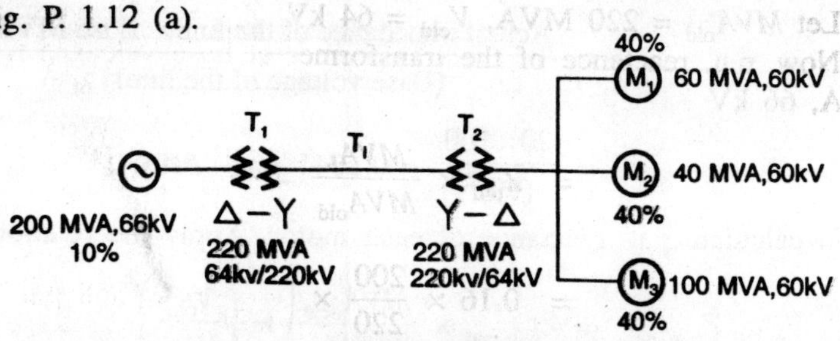

Fig. P. 1.12 (a)

The motors have rated inputs of 60 MVA, 40 MVA and 100 MVA at 60 kV with 40% subtransient reactance. The 3-phase transformers are rated at 220 MVA, 64 kV Δ/220 kV Y with leakage reactance 16%. The line has a reactance 100 ohms. Selecting the generator rating as the base quantities in the generator circuit, determine the base quantities in the other parts of the system and evaluate the corresponding p.u. values. Also draw the reactance diagram of the circuit.

Solution: Let us assume generator ratings as base values

$MVA_b = 200$ MVA, $V_b = 66$ kV

p.u. reactance of the generator (X_g) = 0.30
Given, line reactance (X_L) = 100 Ω
Subtransient reactance of each motor = 40%
Power rating of each transformer = 220 MVA
Primary voltage (V_p) of transformer T_1 = 64 kV (Δ)
Secondary voltage (V_s) of transformer T_2 = 220 kV (Y)

The base quantities in other parts of the system can be calculated as below:

Base value of voltage in line

$$(V_l) = \text{Base voltage} \times \text{transformation ratio}$$

$$= V_b \times \frac{V_s}{V_p} = 66 \text{ kV} \times \frac{220}{64}$$

$$= 226.875 \text{ kV}$$ **Ans.**

Base value of voltage in motor circuits

$$= V_l \times \frac{V_p}{V_s} = 226.875 \text{ kV} \times \frac{64}{220}$$

$$= 66 \text{ kV}$$ **Ans.**

At 220 MVA, 64 kV, reactance of each transformer

= 16%

Let MVA_{old} = 220 MVA, V_{old} = 64 kV, Z_{old} = 0.16 p.u.

Now, p.u. reactance of the transformer at base values, i.e. 200 MVA, 66 kV

$$= Z_{old} \times \frac{MVA_b}{MVA_{old}} \times \left(\frac{V_{old}}{V_b}\right)^2$$

$$= 0.16 \times \frac{200}{220} \times \left(\frac{64}{66}\right)^2 = 0.1368 \text{ p.u.}$$

p.u. impedance of the line

$$= \frac{\text{Actual impedance of the line} \times \text{Base MVA}}{(\text{Base voltage of the line})^2}$$

$$= \frac{100 \times 200}{(226.875)^2} = 0.3886 \text{ p.u.}$$

To calculate p.u. reactance of each motor, apply the relation

$$Z_{\text{p.u. new}} = Z_{\text{p.u. old}} \times \frac{MVA_{new}}{MVA_{old}} \times \left(\frac{V_{old}}{V_{new}}\right)^2$$

For motor 1, MVA_1 = 60 MVA, V_1 = 60 kV, $Z_{\text{p.u. 1}}$ = 0.40

New p.u. reactance of motor 1

$$= Z_{\text{p.u. }1} \times \frac{MVA_b}{MVA_1} \times \left(\frac{V_1}{V_b}\right)^2$$

$$= 0.40 \times \frac{200}{60} \times \left(\frac{60}{66}\right)^2 = 1.1019 \text{ p.u.} \quad \textbf{Ans.}$$

New p.u. reactance of motor 2

$$= 0.40 \times \frac{200}{40} \times \left(\frac{60}{66}\right)^2 = 1.65289 \text{ p.u.} \quad \textbf{Ans.}$$

New p.u. reactance of motor 3

$$= 0.40 \times \frac{200}{100} \times \left(\frac{60}{66}\right)^2 = 0.6612 \text{ p.u.} \quad \textbf{Ans.}$$

The reactance diagram of the system can be drawn as

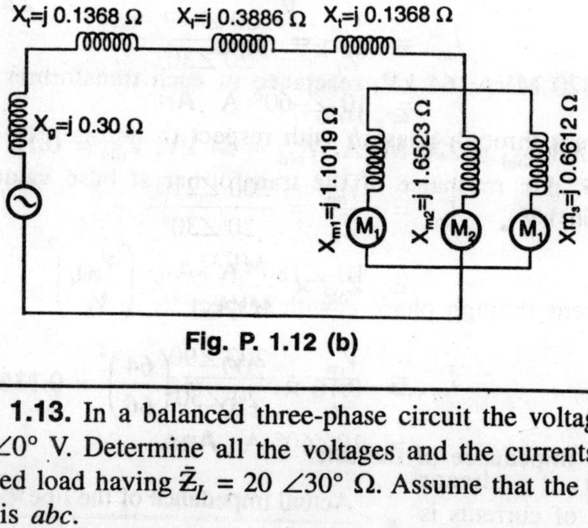

Fig. P. 1.12 (b)

> **Problem 1.13.** In a balanced three-phase circuit the voltage $\bar{V}_{ab}$ is 346.4 ∠0° V. Determine all the voltages and the currents in a Y-connected load having $\bar{Z}_L$ = 20 ∠30° Ω. Assume that the phase sequence is *abc*.

Solution: The phasor diagram of voltages is shown in the Fig. P. 1.13 (a).

Given, $\bar{V}_{ab}$ = 346.4 ∠0° V

Since phase sequence is *abc*

$\bar{V}_{bc}$ = 346.4 ∠240° V **Ans.**

and $\bar{V}_{ca}$ = 346.4 ∠120° V **Ans.**

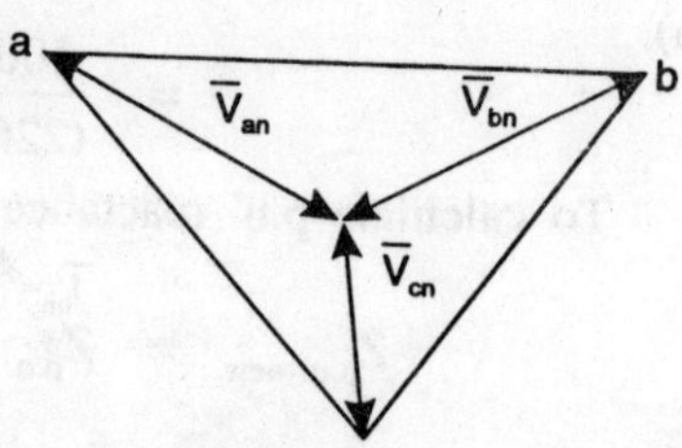

Fig. P. 1.13 (a)

For a balanced three-phase circuit,

$$\bar{V}_{ab} = \sqrt{3}\,\bar{V}_{an}\angle 30^\circ$$

$$\therefore \quad \bar{V}_{an} = \frac{346.4\angle 0^\circ}{\sqrt{3}\angle 30^\circ} = 200\angle{-30^\circ}\text{ V} \quad \textbf{Ans.}$$

Similarly,

$$\bar{V}_{bn} = \frac{\bar{V}_{bc}}{\sqrt{3}\angle 30^\circ} = \frac{346.4\angle 240^\circ}{\sqrt{3}\angle 30^\circ}$$

$$= 200\angle 210^\circ\text{ V} \quad \textbf{Ans.}$$

and

$$\bar{V}_{cn} = \frac{\bar{V}_{ca}}{\sqrt{3}\angle 30^\circ} = \frac{346.4\angle 120^\circ}{\sqrt{3}\angle 30^\circ}$$

$$= 200\angle 90^\circ\text{ V} \quad \textbf{Ans.}$$

Load impedance ($\bar{Z}_L$) = 20 ∠30° Ω.

Each current lags voltage across its load impedance by 30°.

Now, current through phase *a* with respect to neutral

$$\bar{I}_{an} = \frac{\bar{V}_{an}}{\bar{Z}_L} = \frac{200\angle -30^\circ}{20\angle 30^\circ}$$

$$= 10\angle{-60^\circ}\text{ A} \quad \textbf{Ans.}$$

Current through phase *b* with respect to the neutral

$$\bar{I}_{bn} = \frac{\bar{V}_{bn}}{\bar{Z}_L} = \frac{200\angle 210^\circ}{20\angle 30^\circ}$$

$$= 10\angle 180^\circ\text{ A} \quad \textbf{Ans.}$$

Current through phase *c* with respect to neutral

$$\bar{I}_{cn} = \frac{\bar{V}_{cn}}{\bar{Z}_L} = \frac{200\angle 90^\circ}{20\angle 30^\circ}$$

$$= 10\angle 60^\circ\text{ A} \quad \textbf{Ans.}$$

The phasor diagram of currents is shown in Fig. P. 1.13 (b).

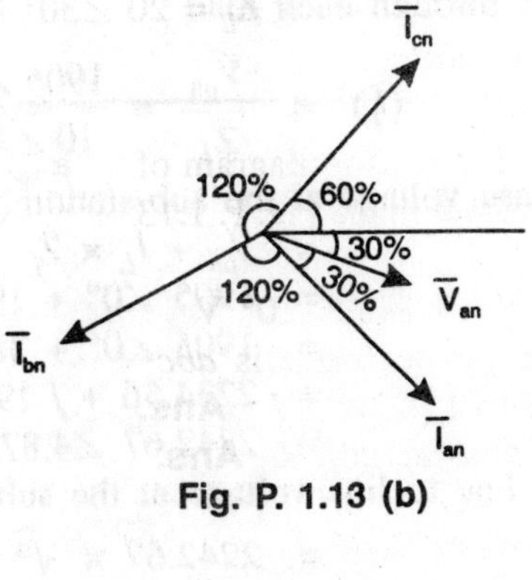

Fig. P. 1.13 (b)

Problem 1.14. The terminal voltage of a Y–connected load consisting of three equal impedances of 10 ∠30° Ω is 3.3 kV line to line. The impedances in each of the three lines connecting the load to a bus at a sub-station is $Z_l = 2\ \angle 60°\ \Omega$. Find the line to line voltage at the sub-station bus.

Solution:

The corresponding circuit diagram can be drawn as:

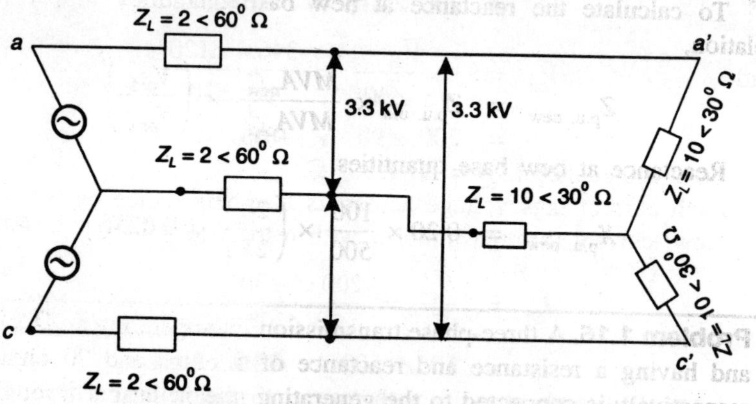

Fig. P. 1.14

Let us make calculation on per phase basic.

Given, line to line voltage across the load = 3.3 kV

$$\text{per phase voltage across each load} = \frac{3.3 \times 1000}{\sqrt{3}} = 1905 \text{ V}$$

Let us take per phase voltage as reference phasor

$$(\bar{V}_{ph}) = 1905\ \angle 0°\ \text{V}$$

Current through each load

$$(\bar{I}_L) = \frac{\bar{V}_{ph}}{\bar{Z}_L} = \frac{1905\ \angle 0°}{10\ \angle 30°} = 190.5\ \angle{-30°}\ \text{A}$$

Per phase voltage at the sub-station

$$\begin{aligned} &= \bar{V}_{ph} + \bar{I}_L \times \bar{Z}_L \\ &= 1905\ \angle 0° + 190.5\angle{-30°} \times 2\angle 60° \\ &= 1905\ \angle 0° + 381\ \angle 30° \\ &= 2234.56 + j\ 190.5 \text{ V} \\ &= 2242.67\ \angle 4.87° \text{ V} \end{aligned}$$

Hence, line to line voltage at the sub-station

$$= 2242.67 \times \sqrt{3} \text{ V} = 3.9 \text{ kV} \quad \textbf{Ans.}$$

Problem 1.15. The reactance of a generator designated X is given as 0.20 p.u. based on the generator's name plate of 20 kV, 500 MVA. Calculate the reactance of the generator at base quantities as 25 kV, 100 MVA.

Solution:

Let, MVA_{new} = 100 MVA, V_{new} = 25 kV

Given, $X_{p.u.\ old}$ = 0.20, MVA_{old} = 500 MVA, V_{old} = 20 kV

To calculate the reactance at new base quantities. Apply the relation,

$$Z_{p.u.\ new} = Z_{p.u.\ old} \times \frac{MVA_{new}}{MVA_{old}} \times \left(\frac{V_{old}}{V_{new}}\right)^2$$

Reactance at new base quantities,

$$X_{p.u.\ new} = 0.20 \times \frac{100}{500} \times \left(\frac{20}{25}\right)^2 = 0.0256 \text{ p.u.} \quad \textbf{Ans.}$$

Problem 1.16. A three-phase transmission line operating at 33 kV and having a resistance and reactance of 5 ohms and 20 ohms respectively is connected to the generating station busbar through a 5000 kVA step-up transformer which has a reactance 6%. Connected to the busbars are two alternators A and B, one 10 MVA having 10% reactance and another 5 MVA having 7.5% reactance. Calculate the single line-diagram of the power system. Selecting base quantity as 10 MVA.

Solution:

The diagram of the given power system can be drawn as

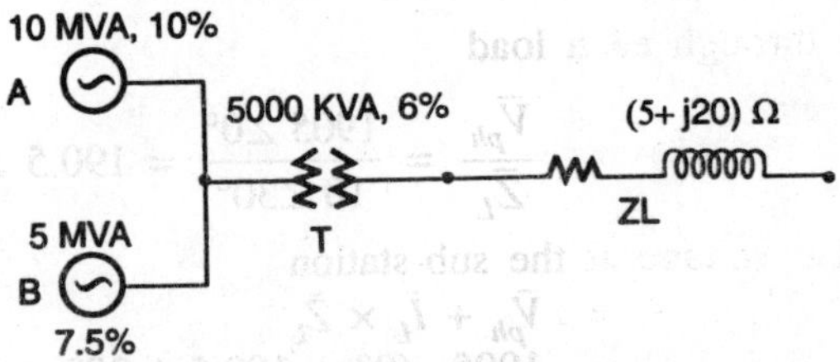

Fig. P. 1.16 (a)

Given, MVA_b = 10 MVA,

p.u. reactance of the generator 10 MVA = 0.10 p.u.

for 5 MVA, generator,

p.u. reactance in the base quantity

$$= \frac{X_{g_2} \times MVA_b}{MVA_2} = \frac{0.075 \times 10}{5} = 0.150 \text{ p.u.}$$

for transformer,

p.u. reactance of the transformer in the base quantity

$$= \frac{X_{old} \times MVA_b}{MVA_{old}} = \frac{0.06 \times 10}{5} = 0.12 \text{ p.u.}$$

Now, Impedance of the line = 5 + j 20

Operating voltage of the line = 33 kV

Now, Applying the relation

$$Z_{p.u.\ new} = \frac{Z_L \times (KVA)}{10 \times (KV)^2} = \frac{(5 + j\,20) \times 1000}{10 \times (33)^2}$$

$$= (0.04592 + j\,0.18368) \text{ p.u.}$$

The reactance diagram of the power system is drawn as below

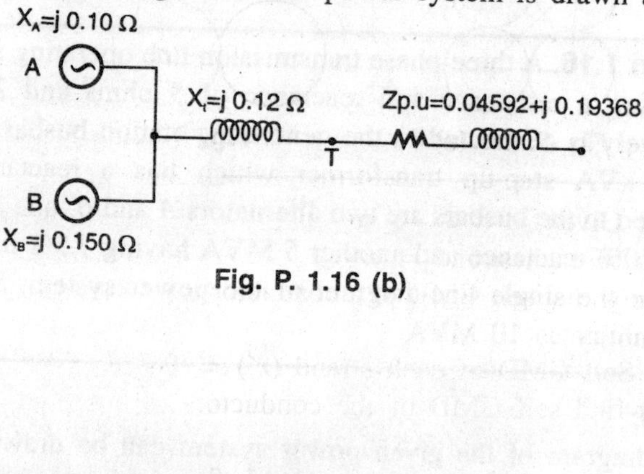

Fig. P. 1.16 (b)

Chapter 2

Line Constant Calculations

Problem 2.1. Determine the self GMD of the following types of conductors in terms of the radius r of an individual strand

Fig. P. 2.1 (a)

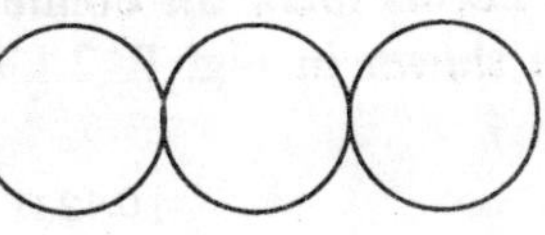

Fig. P. 2.1 (b)

Solution: Given,

Radius of each strand = r

We know that self GMD of a conductor is 0.7788 times the radius of the conductor.

$\therefore$ Self GMD of each strand $(r') = 0.7788\ r$

(*i*) To find self GMD of the conductors in this strand.

Let us mark the conductors as 1, 2, 3 and 4 as shown in Fig. P. 2.1 (c).

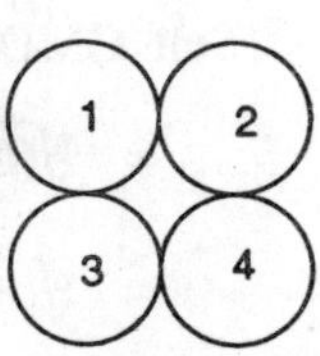

Fig. P. 2.1 (c)

From Fig. P. 2.1 (d), distance between any two adjacent strands

$$= r + r = 2r$$

Since centres of any three strands form equilatoral tringles distance between diagonal strands

$$= \sqrt{(2r)^2 + (2r)^2}$$

$$= 2\sqrt{2}\ r$$

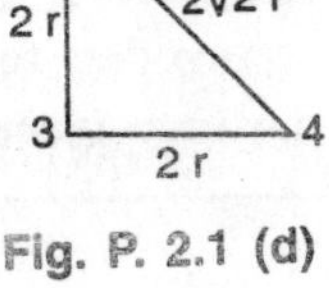

Fig. P. 2.1 (d)

From Fig. P. 2.1 (d), $D_{11} = D_{22} = D_{33} = D_{44} = r' = 0.7788\ r$

$D_{12} = D_{13} = D_{31} = D_{21} = D_{24} = D_{42} = D_{43} = D_{34} = 2r$

and $D_{14} = D_{41} = D_{23} = D_{32} = 2\sqrt{2}\ r$

Here, number of conductors (n) = 4

Now, self GMD of the conductors.

$$= \sqrt[16]{(D_{11}\ D_{12}\ D_{13}\ D_{14})\ (D_{21}\ D_{22}\ D_{23}\ D_{24})\ (D_{31}\ D_{32}\ D_{33}\ D_{34})\ (D_{41}\ D_{42}\ D_{43}\ D_{44})}$$

$$= \sqrt[16]{(r' \times 2r \times 2r \times 2\sqrt{2}r)\ (r' \times 2r \times 2r \times 2\sqrt{2}r)\ (r' \times 2r \times 2r \times 2\sqrt{2}r)\ (r' \times 2r \times 2r \times 2\sqrt{2}r)}$$

$$= \sqrt[4]{(r' \times 2r \times 2r \times 2\sqrt{2}r)}$$

Since $r' = 0.7788\ r$

$$= \sqrt[4]{0.7788r \times 2r \times 2r \times 2\sqrt{2}r} = \sqrt[4]{0.7788 \times 8\sqrt{2}} \times r$$

$$= 1.7229\ r \quad \textbf{Ans.}$$

(*ii*) To find the self GMD of this conductor. Let us mark the conductors as 1, 2 and 3 as shown in Fig. P. 2.1 (e).

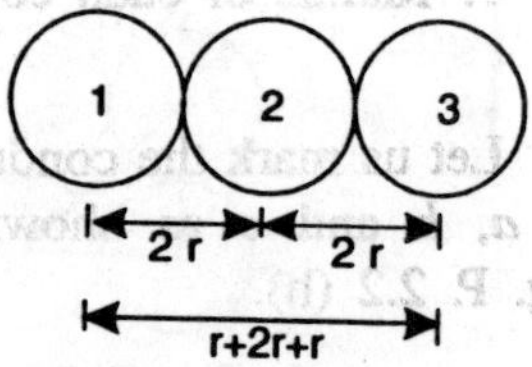

Fig. P. 2.1 (e)

From Figure P. 2.1 (e),

Distance between any two adjacent strands = 2r

Distance between the strands at extreme ends = r + 2r + r = 4r

Now, $D_{11} = D_{22} = D_{33} = r' = 0.7788\ r$

$D_{13} = D_{31} = 4r$

$D_{21} = D_{23} = D_{12} = D_{32} = 2r$

Self GMD of

$$= \sqrt[9]{(D_{11}\ D_{12}\ D_{13})\ (D_{21}\ D_{22}\ D_{23})\ (D_{31}\ D_{32}\ D_{33})}$$

$$= \sqrt[9]{(r' \times 2r \times 4r)\ (2r \times r' \times 2r)\ (4r \times 2r \times r')}$$

$$= \sqrt[9]{(0.7788r \times 2r \times 4r)(2r \times 0.7788r \times 2r)(4r \times 2r \times 0.7788r)}$$

$$= \sqrt[9]{(0.7788r \times 64 \times 4r)} = 1.704\ r \quad \textbf{Ans.}$$

Problem 2.2. A single circuit 3-phase line operated at 50 Hz is arranged as follows. The conductor diameter is 0.6 cm. Determine the inductance and inductive reactance per km.

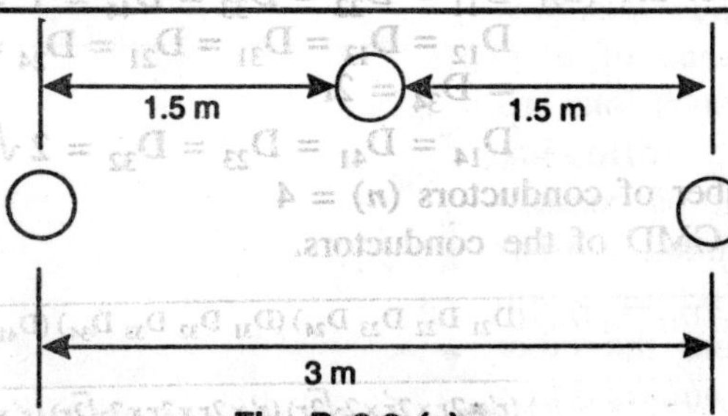

Fig. P. 2.2 (a)

Solution: Given,

Supply frequency (f) = 50 Hz

Diameter of each conductor = 0.6 cm

$\therefore$ Radius of each conductor (r) $= \dfrac{0.6}{2} = 0.3$ cm

$= 0.3 \times 10^{-2}$ m

Let us mark the condutors as *a*, *b* and *c* as shown in Fig. P. 2.2 (b).

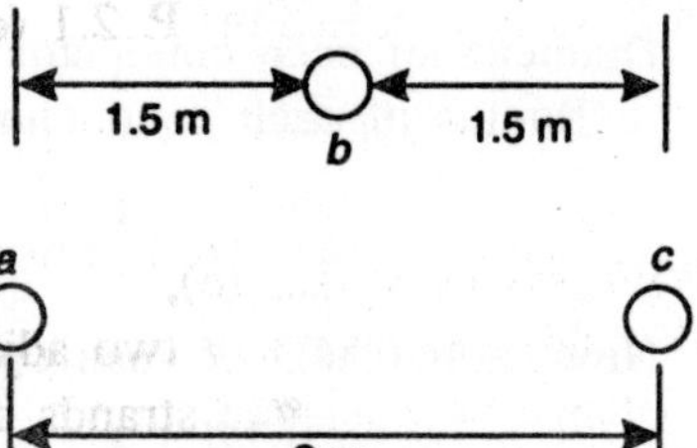

Fig. P. 2.2 (b)

Self GMD of each conductor

(r') $= 0.7788 \times r$

$D_s = 0.7788 \times 0.3 \times 10^{-2}$ m $= 2.336 \times 10^{-3}$ m

Mutual GMD of three conductors

$$(D_m) = \sqrt[3]{D_{ab}\, D_{bc}\, D_{ca}} = \sqrt[3]{1.5 \times 3 \times 1.5} = 1.890 \text{ m.}$$

To calculate inductance of this circuit. Apply the relation

Inductance per km

$$= 2 \times 10^{-4} \ln \frac{D_m}{D_s} \text{ H/km}$$

$$= 2 \times 10^{-4} \ln \frac{1.890}{2.336 \times 200^{-2}} \text{ H/km} = 1.339 \text{ m H/km}$$

$\therefore$ Inductance of the line (L) = 1.339 mH/km **Ans.**

We know that reactance of the line

$= \omega L = 2\pi f L$

$= 2 \times \pi \times 50 \times 1.339 \times 10^{-3}$ Ω/km

$= 0.421$ Ω/km **Ans.**

Problem 2.3. Determine the inductance of a 1-phase transmission line having the following arrangement of conductors. One circuit consists of three wires of 2 mm dia each and the other circuit two wires of 4 mm dia each.

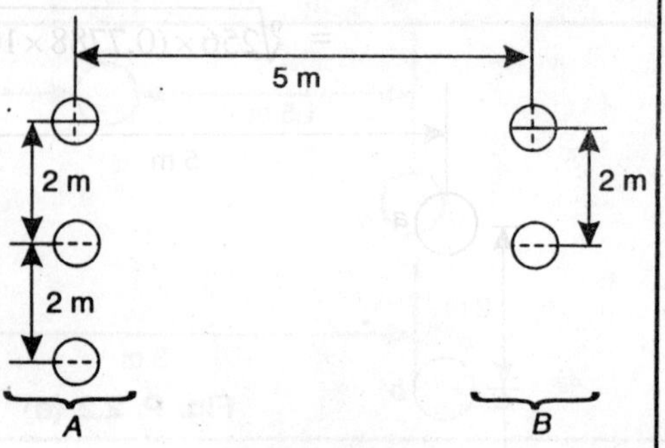

Fig. P. 2.3 (a)

Solution: Given,

Diameter of each conductor in the circuit A = 2 mm

∴ Radius of each conductor in the circuit A

$$= \frac{2}{2} = 1 \text{ mm} = 1 \times 10^{-3} \text{ m}$$

Diameter of each conductor in the circuit B = 4 mm

∴ Radius of each conductor in the circuit B

$$= \frac{4}{2} = 2 \text{ mm} = 2 \times 10^{-3} \text{ m}$$

Now, Self GMD of each conductor in the circuit A

$$\begin{aligned}(r'_A) &= 0.7788 \times r_A \\ &= 0.7788 \times \times 10^{-3} \text{ m} \\ &= 0.7788 \times 10^{-3} \text{ m}\end{aligned}$$

Self GMD of each conductor in the circuit B

$$\begin{aligned}(r'_B) &= 0.7788 \times r_B \\ &= 0.7788 \times 2 \times 10^{-3} \text{ m} \\ &= 1.558 \times 10^{-3} \text{ m}\end{aligned}$$

Let us mark the conductors in both the circuits as *a*, *b*, *c*, *d* and *e* as shown in Fig. P. 2.3 (b).

From figure, $D_{ab} = D_{bc} = D_{ba} = D_{cb} = 2$ m

$D_{ac} = D_{ca} = 4$ m

$D_{aa} = D_{bb} = D_{cc} = r'_A$

Now, self-GMD of the conductors in Group A (D_{SA})

$$= \sqrt[9]{(D_{aa}\, D_{ab}\, D_{ac})\,(D_{ba}\, D_{bb}\, D_{bc})\,(D_{ca}\, D_{cb}\, D_{cc})}$$

$$= \sqrt[9]{(r'_A \times 2 \times 4)\,(2 \times 2 \times r'_A)\,(4 \times 2 \times r'_A)}$$

$$= \sqrt[9]{64 \times 4 \times r'^3_A}$$

$$= \sqrt[9]{256 \times (0.7788 \times 10^{-3})^3} = 0.1704 \text{ m}$$

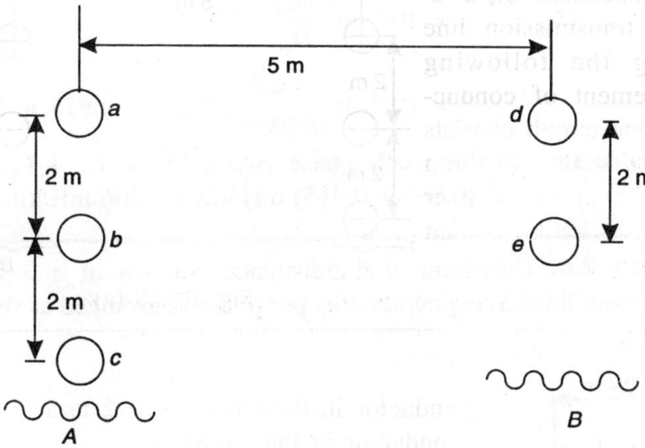

Fig. P. 2.3 (b)

For Group B, $D_{dd} = D_{ee} = r'_B = 1.558 \times 10^{-3}$ m

$D_{de} = D_{ed} = 2$ m

Self GMD of the conductors in Group B

$$(D_{SB}) = \sqrt[4]{(D_{dd} \times D_{de})(D_{ee} \times D_{ed})}$$

$$= \sqrt[4]{1.558 \times 10^{-3} \times 2} = 0.0558 \text{ m}$$

For mutual GMD between Groups A and B,

$$D_{ae} = D_{ce} = D_{bd} = D_{db} = D_{ec} = D_{ea}$$

$$= \sqrt{(5)^2 + (2)^2} = \sqrt{29} \text{ m} = 5.385 \text{ m}$$

$$D_{ad} = D_{be} = D_{eb} = D_{da} = 5 \text{ m}$$

$$D_{cd} = D_{dc} = \sqrt{(5)^2 + (4)^2} = \sqrt{41} \text{ m} = 6.403 \text{ m}$$

Mutual GMD between the Groups A and B

$$D_{mA} = D_{mB} = \sqrt[6]{(D_{ad} \times D_{ae}) \times (D_{bd} \times D_{be}) \times (D_{cd} \times D_{ce})}$$

$$= \sqrt[6]{(5 \times 5.385)(5.385 \times 5)(6.403 \times 5.385)}$$

$$= 5.407 \text{ m}$$

Now, Inductance of Group A conductors

$$(L_A) = 2 \times 10^{-4} \ln \frac{D_{mA}}{S_{bA}} \text{ H/km}$$

$$= 2 \times 10^{-4} \ln \frac{5.407}{0.1704} \text{ H/km} = 0.691 \text{ mH/km}$$

Inductance of Group B conductors

$$(L_B) = 2 \times 10^{-4} \ln \frac{D_{mB}}{D_{sB}} \text{ H/km}$$

$$= 2 \times 10^{-4} \ln \frac{5.407}{0.0558} \text{ H/km} = 0.915 \text{ mH/km}$$

$\therefore$ Inductance of this single-phase circuit line = $L_A + L_B$

$= (0.691 + 0.915)$ mH/km = 1.606 mH/km **Ans.**

Problem 2.4. Determine the inductance per km of a 3-phase transmission line having conductors per phase as arranged as shown in figure.

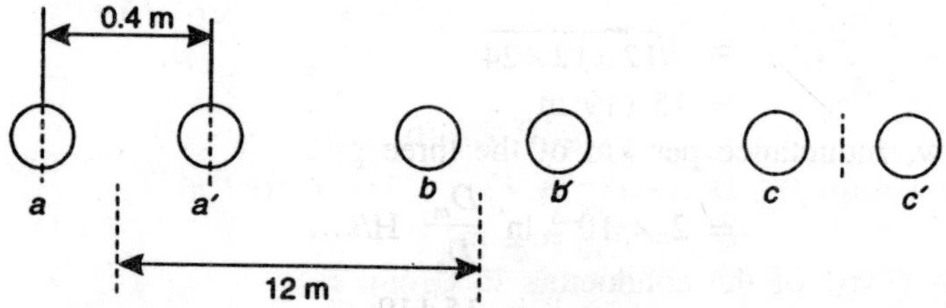

Fig. P. 2.4 (a)

The dia of each conductor is 25 mm and carries 50% of the phase current.

Solution: Given,

Diameter of each conductor = 25 mm

$\therefore$ Radius of each conductor $= \frac{25}{2} \times 10^{-3}$ m

$= 12.5 \times 10^{-3}$ m

Self GMD of each conductor $(r'_a = r'_{a'} = r'_b = r'_{b'} = r'_c = r'_{c'})$

$= 0.7788 \times 12.5 \times 10^{-3}$ m

$= 9.735 \times 10^{-3}$ m

Self GMD of each phase $(r'_A = r'_B = r'_C = D_S)$

$$= \sqrt[4]{(D_{aa}\, D_{aa'})\,(D_{a'a'}\, D_{aa})}$$

$$= \sqrt[4]{(r'_a \times 0.4)\,(r'_{a'} \times 0.4)}$$

$$= \sqrt{0.4 \times 9.735 \times 10^{-3}} \text{ m} = 0.0624 \text{ m}$$

From the circuit, $D_{AB} = D_{BC} = 12$ m, $D_{AC} = D_{CA} = 24$ m

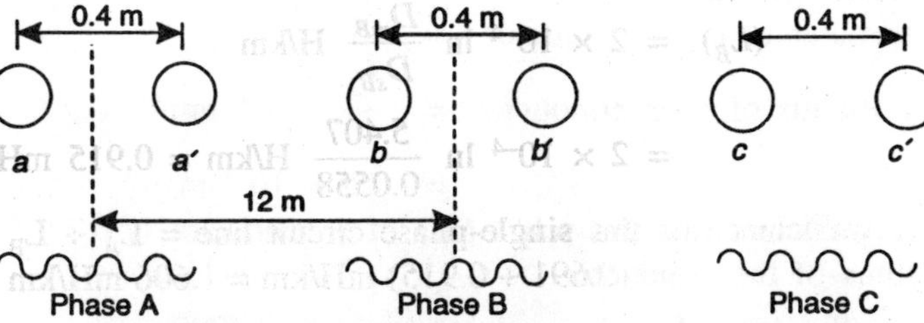

Fig. P. 2.4 (b)

Mutual GMD between the three phases (D_m)

$$= \sqrt[3]{D_{AB} \times D_{BC} \times D_{CA}}$$

$$= \sqrt[3]{12 \times 12 \times 24}$$

$$= 15.119 \text{ m}$$

Now, Inductance per km of the three phase transmission line

$$= 2 \times 10^{-4} \ln \frac{D_m}{D_s} \text{ H/km}$$

$$= 2 \times 10^{-4} \ln \frac{15.119}{0.0624} \text{ H/km}$$

$$= 1.098 \text{ mH/km}$$ **Ans.**

Problem 2.5. Determine the inductance per km of a double circuit 3-phase line as shown in figure below. The transmission line is transposed within each circuit and each circuit remains on its own side. The dia of each condutor is 15 mm. Explain why the given arrangement is better as compared to when conductors of the same phase are placed in the same horizontal plane.

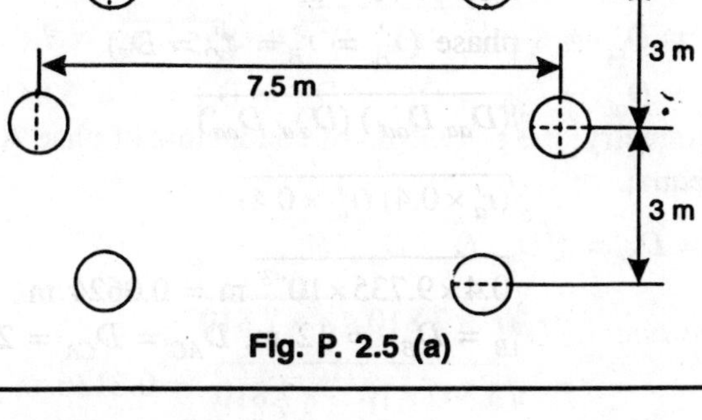

Fig. P. 2.5 (a)

Solution: Given,

Diameter of each conductor = 15 mm

∴ Radius of each conductor = $\frac{15}{2}$ = 7.5 mm

= 7.5×10^{-3} m

As we know that the self GMD of a conductor is 0.7788 times the radius of the conductor.

∴ self GMD of each conductor (r') = $0.7788 \times r$

= $0.7788 \times 7.5 \times 10^{-3}$ m

= 5.841×10^{-3} m

Let us mark the phase conductors as *aa*′, *bb*′ and *cc*′ as shown in Fig. P. 2.5 (b).

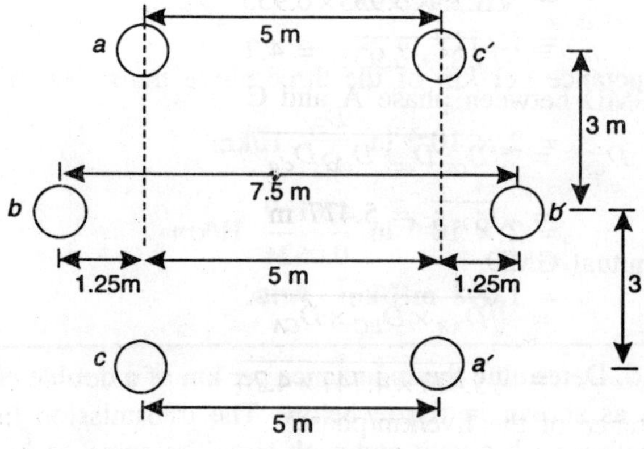

Fig. P. 2.5 (b)

From the given circuit,

$$D_{ab} = D_{b'c'} = D_{bc} = D_{a'b'} = \sqrt{3^2 + (1.25)^2} = 3.25 \text{ m}$$

$$D_{ab'} = D_{bc'} = D_{b'c} = D_{a'b} = \sqrt{(5+1.25)^2 + (3)^2} = 6.933 \text{ m}$$

$$D_{ca} = D_{a'c'} = 3 + 3 = 6 \text{ m}$$

$$D_{aa'} = D_{cc'} = \sqrt{(5)^2 + (3+3)^2} = \sqrt{25+36} = 7.810 \text{ m}$$

$$D_{aa} = D_{a'a'} = D_{bb} = D_{b'b'} = D_{cc} = D_{c'c'} = r' = 5.841 \times 10^{-3} \text{ m}$$

From symmetry, the self GMD of conductors of phae A and phase C must be equal.

$$\therefore D_{SA} = D_{SC} = \sqrt[4]{D_{aa}\, D_{aa'}\, D_{a'a'}\, D_{a'a}}$$

$$= \sqrt[4]{r' \times 7.810 \times r' \times 7.810}$$

$$= \sqrt{5.841 \times 10^{-3} \times 7.810} = 0.2136 \text{ m}$$

Self GMD of conductors of phase B

$$D_{SB} = \sqrt[4]{D_{bb}\, D_{bb'}\, D_{b'b'}\, D_{b'b}} = \sqrt[4]{r' \times 7.5 \times 7.5 \times r'}$$

$$= \sqrt{5.841 \times 10^{-3} \times 7.5} = 0.2093 \text{ m}$$

Now, net self GMD of the circuit

$$D_S = \sqrt[3]{D_{SA} \times D_{SB} \times D_{SC}}$$

$$= \sqrt[3]{0.2093 \times 0.2136 \times 0.2136} = 0.2122 \text{ m}$$

Again from symmetry, the mutual GMD between phase A and B must be equal to that between phase B and C.

$$\therefore\ D_{AB} = D_{BC} = \sqrt[4]{D_{ab}\, D_{ab'}\, D_{a'b}\, D_{a'b'}}$$

$$= \sqrt[4]{3.25 \times 6.933 \times 6.933 \times 3.25}$$

$$= \sqrt{3.25 \times 6.933} = 4.747 \text{ m}$$

Mutual GMD between phase A and C

$$D_{CA} = \sqrt[4]{D_{ca}\, D_{ca'}\, D_{c'a}\, D_{c'a'}} = \sqrt[4]{6 \times 5 \times 5 \times 6}$$

$$= \sqrt{6 \times 5} = 5.477 \text{ m}$$

$\therefore$ Net mutual GMD

$$= \sqrt[3]{D_{AB} \times D_{BC} \times D_{CA}}$$

$$= \sqrt[3]{5.477 \times 4.747 \times 4.747} = 4.979 \text{ m}$$

$\therefore$ Inductance of the live/km/phase

$$= 2 \times 10^{-4} \ln \frac{D_m}{D_s} \text{ H/km/phase}$$

$$= 2 \times 10^{-4} \ln \frac{4.979}{0.2122} \text{ H/km/phase}$$

$$= 0.6311 \text{ mH/km/phase} \quad \textbf{Ans.}$$

Problem 2.6. Determine the inductance per km/phase of a double circuit 3–ϕ line. The radius of each conductor is 15 mm.

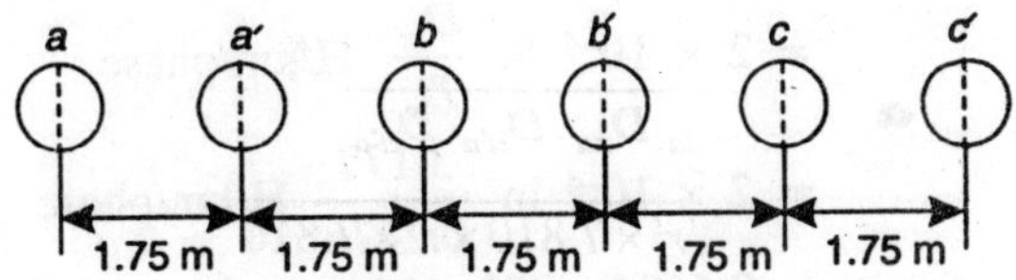

Fig. P. 2.6

Solution: Given,

Radius of each conductor (r) = 15 mm

= 15×10^{-3} m

∴ Self GMD of each conductor (r') = $0.7788 \times 15 \times 10^{-3}$ m

From given circuit, $D_{aa'} = D_{bb'} = D_{cc'} = 3 \times 1.75$ m

From symmetrical arrangement, the self GMD of each phase must be identical.

$$\therefore D_{SA} = D_{SB} = D_{SC} = \sqrt[4]{D_{aa} D_{aa'} D_{a'a'} D_{a'a}}$$

$$= \sqrt[4]{r' \times 3 \times 1.75 \times r' \times 3 \times 1.75}$$

$$= \sqrt{r' \times 3 \times 1.75}$$

$$= \sqrt{0.7788 \times 3 \times 1.75 \times 15 \times 10^{-3}}$$

$$= 0.2476 \text{ m}$$

∴ Net self GMD of this circuit

$$= \sqrt[3]{D_{SA} \times D_{SB} \times D_{SC}}$$

$$= \sqrt[3]{0.2476 \times 0.2476 \times 0.2476} = 0.2476 \text{ m}$$

From given circuit, the mutual GMD between phase A and B must be equal to that between phase B and C.

$$\therefore D_{AB} = D_{BC} = \sqrt[4]{D_{ab} D_{ab'} D_{a'b} D_{a'b'}}$$

$$= \sqrt[4]{1.75 \times 4 \times 1.75 \times 2 \times 1.75 \times 1.75}$$

$$= \sqrt[4]{8} \times 1.75 = 2.943 \text{ m}$$

Mutual GMD between phase C and A,

$$D_{CA} = \sqrt[4]{D_{ca} D_{ca'} D_{c'a} D_{c'a'}}$$

$$= \sqrt[4]{2 \times 1.75 \times 1.75 \times 5 \times 1.75 \times 2 \times 1.75}$$

$$= \sqrt[4]{20} \times 1.75 = 3.701 \text{ m}$$

Net mutual GMD of the given line

$$D_m = \sqrt[3]{D_{AB} \times D_{BC} \times D_{CA}} = \sqrt[3]{2.943 \times 2.943 \times 3.701}$$

$$= \sqrt[3]{(2.943)^2 \times 3.701} = 3.177 \text{ m}$$

We know that, Inductance per km per phase

$$= 2 \times 10^{-4} \ln \frac{D_m}{D_s} \text{ H/km/phase}$$

$$= 2 \times 10^{-4} \ln \frac{3.177}{0.2476} \text{ H/km/phase}$$

$$= 0.5104 \text{ mH/km/phase}$$ **Ans.**

Problem 2.7. An all-aluminum conductor is composed of 37 strands each having a diameter of 0.333 cm. Compute the dc resistance in other per kilometer at 75° C.

Solution: Given,

Diameter of each strand $= 0.333$ cm

$= 0.333 \times 10^{-2}$ m

No. of strands in the conductor $=$ 37

Temprature (t) $= 75°$ C

Area of each strand $= \frac{\pi}{4} \times (0.333 \times 10^{-2})^2$

Area of the conductor $= 37 \times \frac{\pi}{4} \times 10^{-4} \times (0.333)^2$ sq. meter

$(A) = 3.222 \times 10^{-4} \times$ sq. meter

Resistinity of aluminium at 20° C.

$(\rho) = 2.83 \times 10^{-3}$ Ω/m

Now, Resistance of the conductor per km at 20° C

$$= \rho \frac{l}{A} \times 1.02$$

The factor 1.02 due to spiraling

$$= \frac{2.83 \times 10^{-6} \times 1000}{3.222 \times 10^{-4}} \text{ Ω/km} \times 1.02$$

$$= 89.59 \text{ Ω/km}$$

To calculate dc resistance at 75° C, apply the relation

$$\frac{R_{75}}{R_{20}} = \frac{T+75}{T+20}$$

where T is constant of the aluminium conductor and equal to 228° C.

$$\therefore \text{ Resistance at } 75° \text{ C} = \left(\frac{228+75}{228+20}\right) \times 89.59 \text{ Ω/km}$$

$$= 109.46 \text{ Ω/km} \quad \text{Ans.}$$

Problem 2.8. A three-phase 60-Hz transmission line has its conductors arranged in a triangular formation so that two of the distances between conductors are 25 ft and the third distance is 42 ft. The conductors are ACSR Osprey. Determine the inductance and inductive reactance per mile.

Solution: For ASCR Osprey,

Self GMD of each conductor (D_s) = 0.0284 ft.

Given, supply frequency (f) = 60 Hz

According to question, the conductors can be arranged as shown in Fig. P. 2.8.

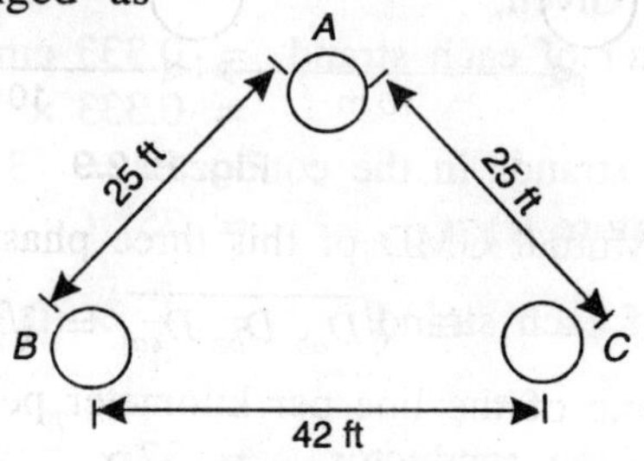

Fig. P. 2.8

The mutual GMD of this arrangement

$$D_m = \sqrt[3]{D_{AB}\, D_{BC}\, D_{CA}} = \sqrt[3]{25 \times 25 \times 42}$$

$$= \sqrt[3]{(25)^2 \times 42} = 29.72 \text{ ft.}$$

Now, Inductance of the line per km per phase

$$= 2 \times 10^{-4} \ln \frac{D_m}{D_s} \text{ H/km/phase}$$

$$= 2 \times 10^{-4} \ln \frac{29.72}{0.0284} \text{ H/km/phase}$$

$$= 1.391 \text{ mH/km/phase}$$

Inductance of this line/mile = 1.391 × 1.609 mH/mile/phase

= 2.238 mH/mile/phase **Ans.**

Inductance reactance of the line per mile per phase

$= 2\pi fL$

$= 2 \times \pi \times 60 \times 2.238 \times 10^{-3}$ Ω/mile/phase

= 0.844 Ω/mile/phase **Ans.**

Problem 2.9. A three-phae 60-Hz line has flat horizontal spacing. The conductors have a GMR of 0.0133 m with 10 m between adjacent conductors. Determine the inductive reactance per phase in ohms per kilometer. What is the name of this conductor?

Solution: Given,

Supply frequency (f) = 60 Hz

Self GMR of the conductors = 0.0133 m

According to the question, the arrangements of conductors can be drawn as below.

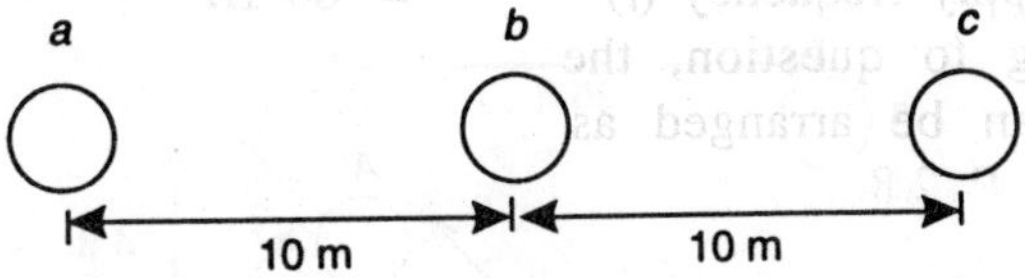

Fig. P. 2.9

Now, Mutual GMD of this three phase circuits

$$= \sqrt[3]{D_{ab}\, D_{bc}\, D_{ca}} = \sqrt[3]{10 \times 10 \times 20} = 12.599 \text{ m}$$

Inductane of the line per kilometer per phase

$$= 2 \times 10^{-4} \ln \frac{D_m}{D_s} \text{ H/km/phase}$$

$$= 2 \times 10^{-4} \ln \frac{12.599}{0.0133} \text{ H/km/phase}$$

$$= 1.37 \text{ mH/km/phase}$$

Now, to calculate inductive reactance, apply the relation

$$X_L = 2\, \pi f L$$

∴ Inductance reactance of the line per kilometer per phase

$$= 2 \times \pi \times 60 \times 1.37 \times 10^{-3}\ \Omega\text{/km/phase}$$

$$= 0.5165\ \Omega\text{/km/phase} \quad \textbf{Ans.}$$

Since the self GMD of the conductor is 0.0133 m, i.e. $\frac{0.0133}{0.3048}$ ft or 0.0436 ft.

From GMR table, the conductors having GMR, 0.0436 ft are *ASCR Fench* **Ans.**

> **Problem 2.10.** Six conductors of ASCR Drake constitute a 60-Hz double-circuit three-phase line arranged as shown in Fig. P. 2.10. The vertical spacing, however, is 14 ft; the longer horizontal distance is 32 ft; and the shorter horizontal distances are 25 ft. Find the inductance per phase per mile and the inductive reactance in ohms per mile.

Solution: Given,

Operating frequency (f) = 60 Hz
Vertical spacing = 14 ft
Longer horizontal distance = 32 ft
Shorter horizontal distance = 25 ft

Since the conductors are ASCR drake, from self GMD table
Self GMD of each conductor (r') = 0.0373 ft

$\therefore\ D_{aa} = D_{a'a'} = D_{bb} = D_{b'b'} = D_{cc} = D_{c'c'} = 0.0373$ ft

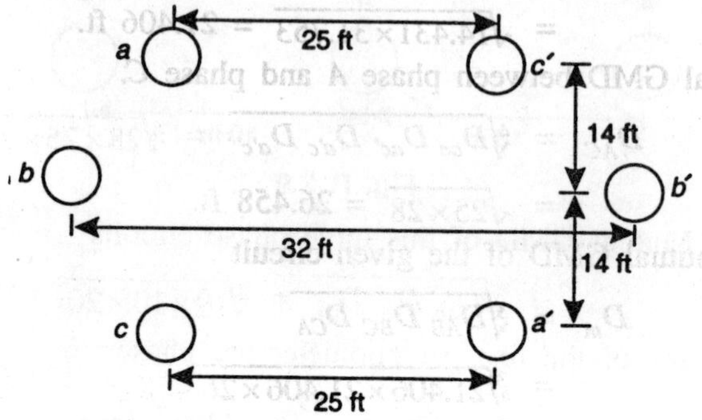

Fig. P. 2.10

$$D_{aa'} = D_{cc'} = \sqrt{(25)^2 \times (14+14)^2} = \sqrt{(25)^2 + (28)^2} = 37.537 \text{ ft.}$$

The self GMD of conductors of phase A and phase C are equal

$$\therefore\ D_{SA} = D_{SC} = \sqrt[4]{D_{aa}\, D_{aa'}\, D_{a'a'}\, D_{a'a}}$$

$$= \sqrt[4]{0.0373 \times 37.537 \times 0.0373 \times 37.537}$$

$$= \sqrt{0.0373 \times 37.537} = 1.183 \text{ ft}$$

Self GMD of the conductors of phase B

$$D_{SB} = \sqrt[4]{D_{bb}\, D_{bb'}\, D_{b'b'}\, D_{b'b}}$$

$$= \sqrt[4]{0.0373 \times 32 \times 0.0373 \times 32} = \sqrt{0.0373 \times 32}$$

$$= 1.093 \text{ ft.}$$

Now, Net self GMD of this circuit,

$$D_s = \sqrt[3]{D_{SA}\, D_{SB}\, D_{SC}}$$

$$= \sqrt[3]{1.183 \times 1.093 \times 1.183} = 1.152 \text{ ft.}$$

From the given circuit, $D_{ab} = D_{ba} = D_{cb'} = D_{b'c'} = D_{bc} = D_{b'a'}$

$$= \sqrt{(14)^2 + (3.5)^2} = 14.431 \text{ ft.}$$

$$D_{ab'} = D_{bc'} = D_{ba'} = D_{b'c}$$

$$= \sqrt{(14)^2 + (25+3.5)^2} = \sqrt{(14)^2 + (28.5)^2}$$

$$= 31.753 \text{ ft.}$$

From symmetry of the circuit, the mutual GMD of between phase *A* and phase *B* must be equal to that between phase *B* and phase *C*.

$$\therefore D_{AB} = D_{BC} = \sqrt[4]{D_{ab}\,D_{ab'}\,D_{a'b'}\,D_{a'b'}}$$

$$= \sqrt[4]{14.431\times31.753\times31.753\times14.431}$$

$$= \sqrt{14.431\times31.753} = 21.406 \text{ ft.}$$

Natural GMD between phase A and phase C.

$$D_{AC} = \sqrt[4]{D_{ac}\,D_{ac'}\,D_{a'c}\,D_{a'c'}} = \sqrt[4]{28\times25\times25\times28}$$

$$= \sqrt{25\times28} = 26.458 \text{ ft.}$$

Net mutual GMD of the given circuit

$$D_m = \sqrt[3]{D_{AB}\,D_{BC}\,D_{CA}}$$

$$= \sqrt[3]{21.406\times21.406\times26.458} = 22.973 \text{ ft.}$$

Now, Inductance of the line per phase per meter (L)

$$= 2\times10^{-7}\,ln\,\frac{D_m}{D_s} \text{ H/km/phase}$$

$$= 2\times10^{-7}\,ln\,\frac{22.973}{1.152} \text{ H/km/phase}$$

$$= 5.986\times10^{-7} \text{ H/m/phase}$$

$$= 5.986\times1609\times10^{-7} \text{ H/mile/phase}$$

$$= 0.9631 \text{ mH/mile/phase}$$

Inductive reactance of the line in ohm per mile per phase

$$= 2\,\pi fL$$

$$= 2\times\pi\times60\times0.9631\times10^{-3}\ \Omega\text{/mile/phase}$$

$$= 0.3631\ \Omega\text{/mile/phase}$$

Inductive reactance of the line in ohm per mile

$$= 3\times X_L \text{ per phase} = 3\times0.3631\ \Omega\text{/mile}$$

$$= 1.089\ \Omega\text{/mile}$$ **Ans.**

Problem 2.11. A single-circuit three phase line operated at 60 Hz is arranged as shown in figure. The conductors are ASCR drake. Find the inductive reactance per mile per phase.

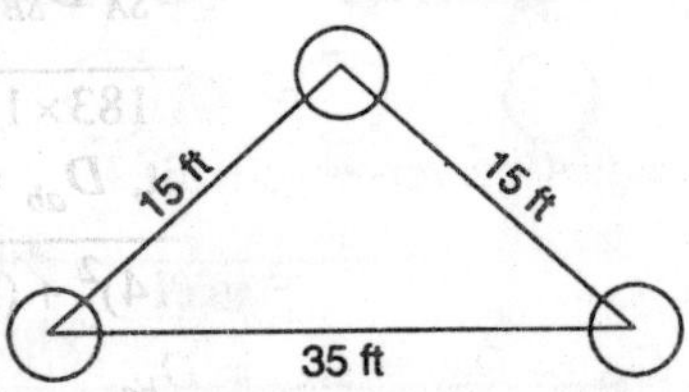

Fig. P. 2.11 (a)

Solution: Given,

Supply frequency (f) = 60 Hz

From GMR table, self GMR of ASCR Drake conductors,

$$D_S = 0.0373 \text{ ft}$$

Let mark conductors as a, b and c respectively.

From this circuit,

$$D_{ab} = 15 \text{ ft},$$
$$D_{bc} = 15 \text{ ft},$$
$$D_{ca} = 35 \text{ ft}$$

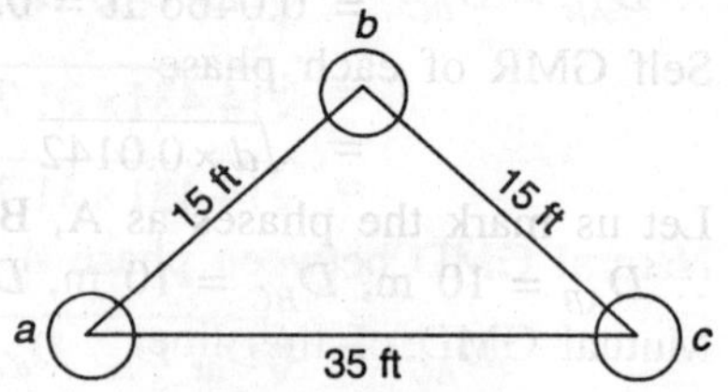

Fig. P. 2.11 (b)

Mutual GMD of this circuit

$$D_m = \sqrt[3]{D_{ab} \times D_{bc} \times D_{ca}} = \sqrt[3]{15 \times 15 \times 35}$$
$$= 19.895 \text{ ft.}$$

Inductance per m of this line per phase

$$L = 2 \times 10^{-7} \ln \frac{D_m}{D_s} \text{ H/m/phase}$$
$$= 2 \times 10^{-7} \ln \frac{19.895}{0.0373} \text{ H/m/phase}$$
$$= 12.558 \times 10^{-7} \text{ H/m/phase}$$

Now, Reactance of the line per phase

$$X_L = \omega L, X_L = 2\,\pi f L$$
$$X_L = 2 \times p \times 60 \times 1609 \times 12.558 \times 10^{-7}\ \Omega\text{/mile/phase}$$
$$= 0.7617\ \Omega\text{/mile/phase}$$

Problem 2.12. Each conductor of the bundled-conductor line shown in figure below is ASCR, 1,272,000-c mile phasant. Find the inductive reactance in ohms/km per phase for d = 30 cm. Also find the per unit of series reactance of the line if the length is 100 km and base is 100 MVA 330 kV. Take supply frequency 60 Hz.

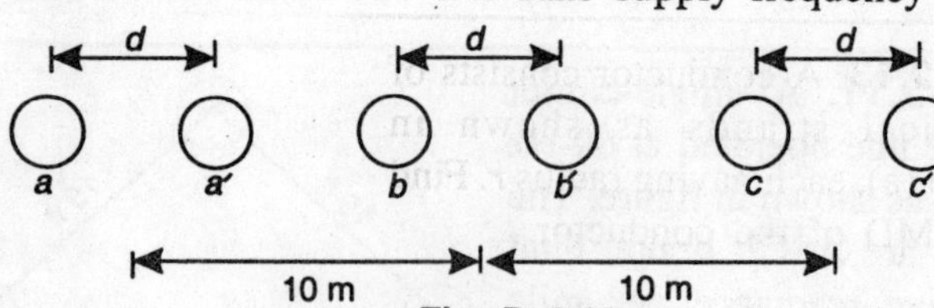

Fig. P. 2.12

Solution: Given,

Supply frequency	=	60 Hz
Length of the line	=	100 km
Base MVA	=	100 MVA
Base kV	=	330 kV

From the GMR Chart, Self GMR of given conductors

$$= 0.0466 \text{ ft} = 0.0466 \times 0.3048 \text{ m} = 0.0142 \text{ m}$$

Self GMR of each phase

$$= \sqrt{d \times 0.0142} = \sqrt{0.30 \times 0.0142} = 0.0653 \text{ m}$$

Let us mark the phases as A, B and C,

$\therefore D_{AB} = 10$ m, $D_{BC} = 10$ m, $D_{CA} = 20$ m

Mutual GMD of this line

$$D_m = \sqrt[3]{D_{AB} \times D_{BC} \times D_{CA}} = \sqrt[3]{10 \times 10 \times 20}$$

$$= 12.599 \text{ m}$$

Inductance of the line per km

$$= 2 \times 10^{-4} \ln \frac{D_m}{D_s} \text{ H/km/phase}$$

$$= 2 \times 10^{-4} \ln \frac{12.599}{0.0653} \text{ H/km/phase}$$

$$= 1.0524 \text{ mH/km/phase}$$

Reactance of the line per phase per km = 2 πfL

$$= 2 \times \pi \times 60 \times 1.0524 \times 10^{-3} \ \Omega\text{/km/phase}$$

$$= 0.397 \ \Omega\text{/km/phase} \quad \text{Ans.}$$

Now, Reactance of 100 km line

$$X = X_L \times l = 100 \times 0.397 \ \Omega\text{/phase}$$

$$= 39.7 \ \Omega\text{/phase}$$

$$\text{Now, Base impedance (Z)} = \frac{(\text{Base kV})^2}{\text{Base MVA}} = \frac{(330)^2}{100} = 1089 \ \Omega$$

$\therefore$ p.u. series reactance of the line

$$= \frac{X}{\text{Base Z}} = \frac{39.7}{1089} \text{ p.u.} = 0.0365 \text{ p.u.} \quad \textbf{Ans.}$$

Problem 2.13. A conductor consists of six identical strands as shown in Fig. P. 2.13 (a), each having radius *r*. Find the self GMD of the conductor.

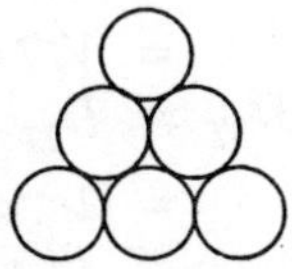

Fig. P. 2.13 (a)

Solution: Given,

Radius of each strand = *r*

We know that self GMD of a conductor is 0.7788 times the radius of the conductor.

$\therefore$ self GMD of each strand $(r') = 0.7788$ r

To find the self GMD of the conductor. Let us mark the strands as 1, 2, 3, 4, 5 and 6 as shown in Fig. P. 2.13 (b).

Fig. P. 2.13 (b)

Here, $D_{11} = D_{22} = D_{33} = D_{44}$
$= D_{55} = D_{66} = r' = 0.7788\ r$

From figure,

$$D_{12} = D_{13} = D_{24} = D_{36} = D_{25} = D_{35} = D_{45} = D_{56} = 2r$$
$$D_{14} = D_{16} = D_{46} = r + r + 2r = 4r$$

D_{15}, D_{34} and D_{26} can be calculated by using cosine formula.

From Fig. P. 2.13 (c) , Distance between 1 and 5

$$= \sqrt{D_{25}^2 + D_{12}^2 - 2D_{12}D_{25}\cos\angle 120^\circ}$$

$$= \sqrt{(2r)^2 + (2r)^2 - 2\times 2r\times 2r\cos 120^\circ}$$

$$= \sqrt{4r^2 + 4r^2 - 8\sqrt{2}\times\left(-\frac{1}{2}\right)}$$

$$= \sqrt{12r^2} = 2\sqrt{3}\ \mathrm{r}$$

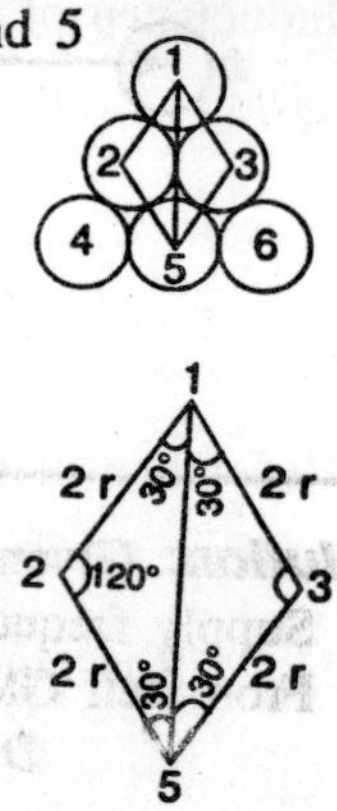

Fig. P. 2.13 (c)

Hence, $D_{15} = D_{34} = D_{26} = 2\sqrt{3}\ r$

Now, self GMD of the conductor

$$= \sqrt[36]{\begin{matrix}(D_{11}\ D_{12}\ D_{13}\ D_{14}\ D_{15}\ D_{16})\\(D_{21}\ D_{22}\ D_{23}\ D_{24}\ D_{25}\ D_{26})\\ \ldots\ldots.(D_{61}\ D_{62}\ D_{63}\ D_{64}\ D_{65}\ D_{66})\end{matrix}}$$

Now, Let us find

$$D_{s1} = D_{11}\ D_{12}\ D_{13}\ D_{14}\ D_{15}\ D_{16}$$
$$= 0.7788r \times 2r \times 2r \times 4r \times 2\sqrt{3}\ r \times 4r$$
$$= 172.662\ r^6$$
$$D_{S2} = D_{21}\ D_{22}\ D_{23}\ D_{24}\ D_{25}\ D_{26}$$
$$= 2r \times 0.7788\ r \times 2r \times 2r \times 2r \times 2\sqrt{3}\ r$$
$$= 43.165\ r^6$$

Similarly,

$$D_{S3} = 43.165\ r^6,\ D_{S4} = 172.662\ r^6,\ D_{S6} = 172.662\ r^6,$$
$$D_{S5} = D_{54}\ D_{56}\ D_{53}\ D_{52}\ D_{51}\ D_{66}$$
$$= 2r \times 2r \times 2r \times 2r \times 2\sqrt{3}\ r \times 0.7788 \times r$$
$$= 43.165\ r^6$$

$\therefore$ Self GMD of the conductor

$$= \sqrt[36]{D_{S1} \times D_{S2} \times D_{S3} \times D_{S4} \times D_{S5} \times D_{S6}}$$

$$= \sqrt[36]{(172.662)^3 \times r^{18} \times (43.165)^3 \times r^{18}}$$

$$= \sqrt[36]{172.662 \times 43.165} \times r = 181.492 \text{ r} \quad \textbf{Ans.}$$

Problem 2.14. A three-phase double circuit line is composed of 300,000-cmil 26/7 ASCR ostrich conductors arranged as shown. Find 60-Hz inductive reactance in ohms per km/phase.

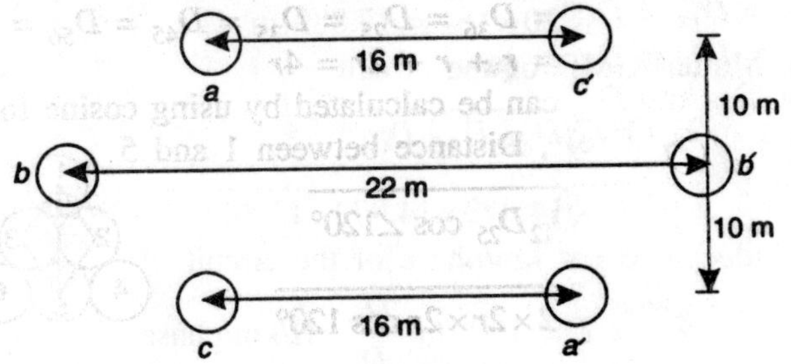

Fig. P. 2.14 (a)

Solution: Given,

Supply frequency = 60 Hz

From self GMR table, for ostrich

$$D_s = 0.0229 \text{ ft.}$$

$$= 0.0229 \times 0.3048 \text{ m} = 6.98 \times 10^{-3} \text{ m}$$

The circuit can be drawn as

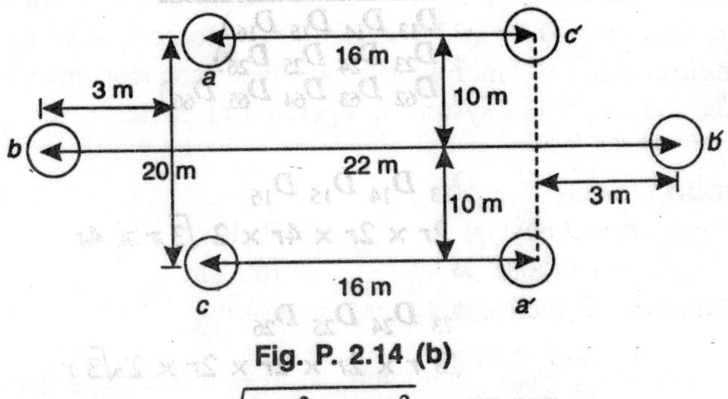

Fig. P. 2.14 (b)

Now, $D_{aa'} = D_{cc'} = \sqrt{(16)^2 + (20)^2} = 25.612 \text{ m}$

$$D_{ab'} = D_{bc'} = D_{b'c} = D_{a'b} = \sqrt{(16+3)^2 + (10)^2} = 21.471 \text{ m}$$

$$D_{cb'} = D_{ab} = D_{bc} = D_{a'b'} = \sqrt{(10)^2 + (3)^2} = 10.440 \text{ m}$$

Now, Mutual GMD of this circuit

$$D_m = \sqrt[3]{D_{AB}\, D_{BC}\, D_{CA}}$$

Now, $D_{AB} = \sqrt[4]{D_{ab}\, D_{ab'}\, D_{a'b}\, D_{a'b'}}$

$= \sqrt[4]{10.440 \times 21.447 \times 21.471 \times 10.440}$

$= \sqrt[4]{10.440 \times 21.471} = 14.972$ m. $= D_{BC}$

Now, $D_{AC} = \sqrt[4]{D_{ac}\, D_{ac'}\, D_{a'c}\, D_{a'c'}} = \sqrt[4]{20 \times 16 \times 16 \times 20}$

$= \sqrt[4]{20 \times 16} = 17.889$ m

Again, Mutual GMD of the circuit

$$D_m = \sqrt[3]{D_{AB}\, D_{BC}\, D_{CA}}$$

$$= \sqrt[3]{14.972 \times 14.972 \times 17.889} = 15.760 \text{ m}$$

Now, Inductance per km/phase of the circuit

$$L = 2 \times 10^{-4} \ln \frac{D_m}{D_s} \text{ H/km/phase}$$

$$= 2 \times 10^{-4} \ln \frac{15.760}{6.98 \times 10^{-3}} \text{ H/km/phase}$$

$$= 1.544 \text{ mH/km/phase}$$

Reactance of the line in ohms per km/phase $= 2\,\pi f L$

$= 2 \times \pi \times 60 \times 1.544 \times 10^{-3}$ Ω/km/phase

$= 0.582$ Ω/km/phase **Ans.**

Problem 2.15. A three-phase transmission line 100 km long has its conductor of 0.4 cm dia spaced at the corners of an equilateral triangle of side 110 cm. Find the inductance and reactance per phase of the system if the system is operated at 50Hz.

Solution: Given,

Supply frequency (f) = 50 Hz

Length of the line (l) = 100 km

Diameter of each conductor = 0.4 cm

$\therefore$ Radius of each conductor $= \frac{0.4}{2} = 0.2 \text{ cm} = 0.2 \times 10^{-2}$ m

We know that self GMD of a conductor is 0.7788 times the radius of the conductor.

$\therefore$ Self GMD of each conductor (r)

$= 0.7788 \times r$

$= 0.7788 \times 0.2 \times 10^{-2}$ m

$= 1.558 \times 10^{-3}$ m

The system diagram can be drawn as

Given, $D_{AB} = D_{BC} = D_{CA}$

$= 110 \text{ cm} = 1.1 \text{ m}$

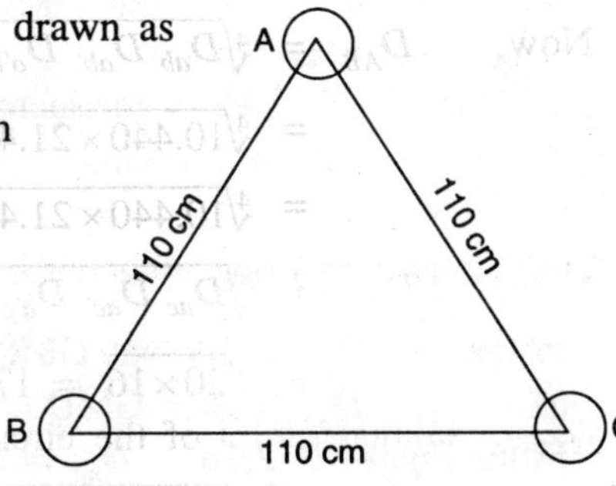

Mutual GMD of this system

Fig. P. 2.15

$$D_m = \sqrt[3]{D_{AB}\, D_{BC}\, D_{CA}} = \sqrt[3]{1.1 \times 1.1 \times 1.1} = 1.1 \text{ m.}$$

Now, Inductance per km per phase of the system

$$= 2 \times 10^{-4} \ln \frac{D_m}{D_s} \text{ H/km/phase}$$

$$= 2 \times 10^{-4} \ln \frac{1.1}{1.558 \times 10^{-3}} \text{ H/km/phase}$$

$$= 1.312 \text{ mH/km/phase}$$

Inductance of 100 km line per phase $L_{100} = L \times 1$

$$L_{100} = 100 \times 1.312 \text{ mH/phase}$$

$= 131.2$ mH/phase **Ans.**

Reactance of 100 km line per phase $= 2\,\pi f \times L_{100}$

$$= 2 \times \pi \times 50 \times 131.2 \times 10^{-3}\ \Omega\text{/phase}$$

$= 41.218\ \Omega$/phase **Ans.**

Problem 2.16. In a single phase line as shown in Fig. P. 2.16 conductors a and a' in parallel form one conductor while conductors b and b' in parallel form return path, calculate the total inductance of the line per km assuming that current is equally shared by the two paralled conductors, conductor dia is 2.0 cm.

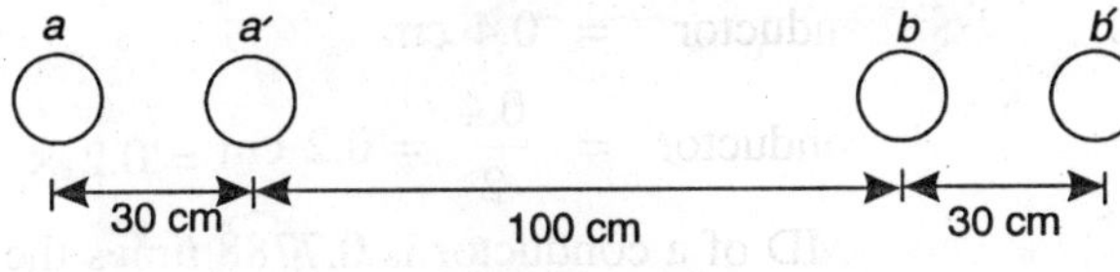

Fig. P. 2.16

Solution: Given,

Diameter of each conductor $= 2.0$ cm

$\therefore$ Radius of each conductor $= \frac{2}{2} = 1$ cm

$$r = 1 \times 10^{-2} \text{ m}$$

As we know that the self GMD of a conductor is 0.7788 times the radius of the conductor.

$\therefore$ Self GMD of each conductor = 0.7788 × r

$$r' = 0.7788 \times 1 \times 10^{-2} \text{ m} = 7.788 \times 10^{-3} \text{ m}$$

We know that Loop inductance per km = $4 \times 10^{-4} \ln \dfrac{D_m}{D_s}$ H/km

Where D_m = Mutual GMD

and D_S = Self GMD

From Figure P. 2.16, $D_{aa'} = D_{bb'} = 30$ cm = 0.30 m,

$D_{ab} = D_{ab'} = 130$ cm = 1.30 m

Here, self GMD (D_s)

$$= \sqrt[4]{D_{aa}\, D_{aa'}\, D_{a'a'}\, D_{a'a}} = \sqrt[4]{r' \times 0.30 \times r' \times 0.30}$$

$$= \sqrt[4]{.788 \times 10^{-3} \times 0.30} = 0.0483 \text{ m}$$

And, Mutual GMD of the line (D_m)

$$= \sqrt[4]{D_{ab}\, D_{ab'}\, D_{a'b}\, D_{a'b'}}$$

$$= \sqrt[4]{1.30 \times 1.60 \times 1.00 \times 1.30} = 1.282 \text{ m}$$

Now, Total inductance of the line per km

$$= 4 \times 10^{-4} \ln \frac{D_m}{D_s} \text{ H/km}$$

$$= 4 \times 10^{-4} \ln \frac{1.282}{0.0463} \text{ H/km}$$

$$= 1.3114 \text{ mH/km} \quad \textbf{Ans.}$$

Problem 2.17. Determine the inductance of the double circuit line shown in figure. The diameter of each conductor is 4 cm.

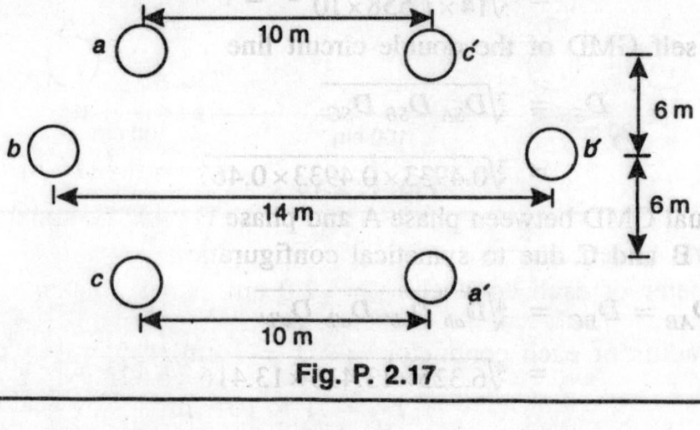

Fig. P. 2.17

Solution: Given,

Diameter of each conductor $= 4$ cm

$\therefore$ Radius of each conductor $(r) = \frac{4}{2} = 2$ cm

$r = 1 \times 10^{-2}$ m

As we know that the self GMD of a conductor is 0.7788 times the radius of the conductor.

$\therefore$ Self GMD of each conductor $(r') = 0.7788 \times r$

$= 0.7788 \times 2 \times 10^{-2}$ m $= 1.558 \times 10^{-2}$ m

From Fig. P. 2.17 $D_{ab} = D_{bc} = D_{b'c'} = D_{b'a'} = \sqrt{(6)^2 + (2)^2}$

$= 6.325$ m

$$D_{ab'} = D_{bc'} = D_{b'c} = D_{ba'}$$

$$= \sqrt{(10+2)^2 + 6^2} = \sqrt{144+36}$$

$$= \sqrt{180} = 13.416 \text{ m}$$

$$D_{aa'} = D_{cc'} = \sqrt{(6+6)^2 + (10)^2} = \sqrt{12^2 + 10^2}$$

$$= 15.620 \text{ m}$$

Since self GMD of phase a is equal to that of phase C

$$\therefore \quad D_{sa} = D_{sc} = \sqrt[4]{D_{aa}\, D_{aa'}\, D_{a'a'}\, D_{aa'}}$$

$$= \sqrt[4]{r' \times 15.620 \times r' \times 15.620}$$

$$= \sqrt{15.620 \times 1.558 \times 10^{-2}} = 0.4933 \text{ m}$$

Self GMD of phase B

$$D_{SB} = \sqrt[4]{D_{bb}\, D_{bb'}\, D_{b'b'}\, D_{b'b}} = \sqrt[4]{r' \times 14 \times r' \times 14}$$

$$= \sqrt{14 \times 1.558 \times 10^{-2}} = 0.467 \text{ m}$$

Net self GMD of the double circuit line

$$D_S = \sqrt[3]{D_{SA}\, D_{SB}\, D_{SC}}$$

$$= \sqrt[3]{0.4933 \times 0.4933 \times 0.467} = 0.4844 \text{ m.}$$

Mutual GMD between phase A and phase B must be equal to that between B and C due to symetical configuration

$$\therefore D_{AB} = D_{BC} = \sqrt[4]{D_{ab}\, D_{ab'}\, D_{a'b}\, D_{a'b'}}$$

$$= \sqrt[4]{6.325 \times 13.416 \times 13.416 \times 6.325}$$

$$= \sqrt{6.325 \times 13.416} = 9.212 \text{ m}$$

$$D_{CA} = \sqrt[4]{D_{ac}\, D_{ac'}\, D_{a'c}\, D_{a'c'}} = \sqrt[4]{12 \times 10 \times 10 \times 12}$$

$$= \sqrt{120} = 10.954 \text{ m.}$$

Net mutual GMD of the double circuit $= \sqrt[3]{D_{AB}\, D_{BC}\, D_{CA}}$

$$= \sqrt[3]{4.212 \times 9.212 \times 10.954} = 9.760 \text{ m.}$$

Now, Inductance of the double circuit line per phase

$$= 2 \times 10^{-4} \ln \frac{D_m}{D_s} \text{ H/km}$$

$$= 2 \times 10^{-4} \ln \frac{9.760}{0.4844} \text{ H/km}$$

$= 0.601$ mH/km **Ans.**

Chapter 3

Capacitance of Transmission Lines

Problem 3.1. Determine the capacitance and charging current per unit length of the line when the arrangement of the conductors is as shown in Fig. P. 3.1 (a)

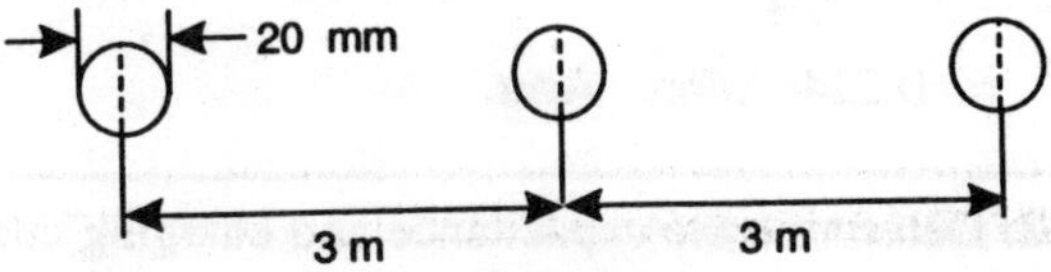

Fig. P. 3.1 (a)

The operating voltage is 132 kV.

Solution: Given,

Operating voltage (v) = 132 kV

Diameter of each conductor = 20 mm

Radius of each conductor $= \frac{20}{2} = 10$ mm

$= 10 \times 10^{-3}$ m $= 0.01$ m

Distance between adjacent conductors = 3 m

Distance between two extreme ends conductors = (3 + 3) m = 6 m

Let us mark conductors as a, b and c as shown below

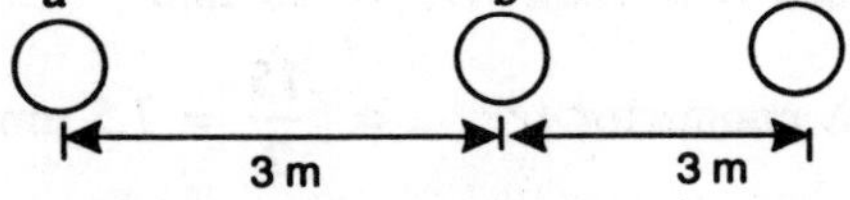

Fig. P. 3.1 (b)

Mutual GMD of these conductors

$$D_m = \sqrt[3]{D_{ab}\,D_{bc}\,D_{ca}} = \sqrt[3]{3\times3\times6} = 3.78\text{m}$$

To calculate capacitance per unit length of the line. Apply the relation.

Capacitance of the line (*c*)

$$= \frac{2\pi\varepsilon_0}{\ln\frac{GMD}{r}}\text{ F/m} = \frac{10^{-9}}{18\ln\frac{3.78}{0.01}} = \frac{1}{18\ln\frac{3.78}{0.01}}$$

$$= 9.361\times10^{-3}\ \mu\text{F/km} = 9.361\ \eta\text{F/km} \quad \textbf{Ans.}$$

Now, Charging current (I_e)

$$= \frac{V/\sqrt{3}}{X_C} = \frac{V}{\sqrt{3}}\,2\pi fc$$

$$= \frac{132\times10^3}{\sqrt{3}}\times 2\pi50\times9.361\times10^{-9}\ \text{A/km}$$

$$= 0.224\ \text{A/km} \quad \textbf{Ans.}$$

Problem 3.2. Determine the capacitance and charging current per unit length of the line when the arrangment of the conductor is shown in Fig. P. 3.2. The line is completely transposed.

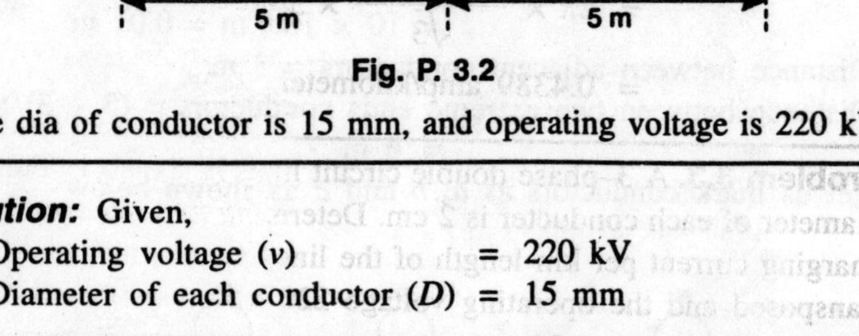

Fig. P. 3.2

The dia of conductor is 15 mm, and operating voltage is 220 kV.

Solution: Given,

Operating voltage (v) = 220 kV

Diameter of each conductor (D) = 15 mm

Radius of each conductor (r) $= \frac{15}{2} = 7.5$ mm

$= 7.5\times10^{-3}$mm $= 0.0075$ m.

From symmetry, the modified GMD of each phase is same.

$$D_{SA} = D_{SB} = D_{SC} = \sqrt[4]{D_{aa}\, D_{aa'}\, D_{a'a'}\, D_{aa'}} = \sqrt[4]{r \times 0.4 \times r \times 0.4}$$

$$= \sqrt{r \times 0.4} = \sqrt{0.0075 \times 0.4} = 0.0548 \text{ m}$$

Net modified GMD $(D_S) = \sqrt[3]{(0.0548)^3} = 0.0548$ m

The mutual GMD between phase A and phase B must be equal to that between phase B and phase C due to symmetry

$$D_{AB} = D_{BC} = \sqrt[4]{D_{ab}\, D_{ab'}\, D_{a'b}\, D_{a'b'}} = \sqrt[4]{5 \times 5.4 \times 4.6 \times 5}$$

$$= 4.992 \text{ m}$$

Mutual GMD between phase A and phase C

$$D_{CA} = \sqrt[4]{D_{ca}\, D_{ca'}\, D_{c'a}\, D_{c'a'}} = \sqrt[4]{10 \times 9.6 \times 10.4 \times 10}$$

$$= 9.996 \text{ m}$$

Net mutual GMD $(D_m) = \sqrt[3]{D_{AB}\, D_{BC}\, D_{CA}}$

$$= \sqrt[3]{9.996 \times 9.946 \times 4.992} = 7.931 \text{ m}$$

Capacitance of the line per km

$$= \frac{10^{-6}}{18 \ln \dfrac{D_m}{D_S}} \text{ F/km} = \frac{10^{-6}}{18 \ln \dfrac{7.931}{0.0548}} \text{ F/km}$$

$$= 0.011 \ \mu\text{F/km}$$

Charging current per unit length $(I_c) = \dfrac{V}{\sqrt{3} \times X_C} = \dfrac{2\pi f v c}{\sqrt{3}}$

$$= 2\pi \times \frac{220 \times 1000}{\sqrt{3}} \times 50 \times 0.011 \times 10^{-6} \text{ amp/km}$$

$$= 0.4389 \text{ amp/kilometer} \quad \textbf{Ans.}$$

Problem 3.3. A 3–phase double circuit lines is shown in Fig. The diameter of each conducter is 2 cm. Determine the capacitance and charging current per km length of the line; assume that the line is transposed and the operating voltage 220 kV.

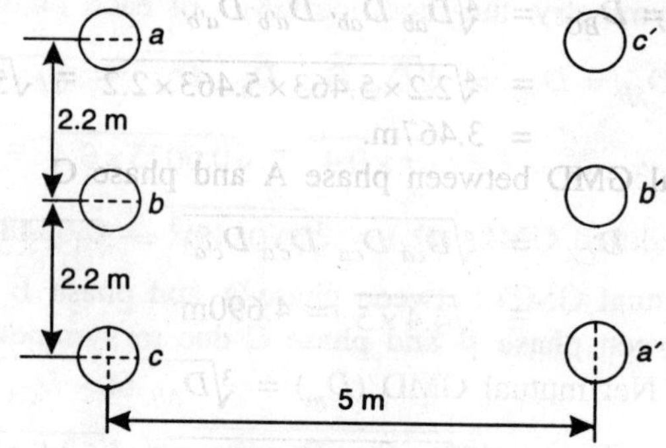

Fig. P. 3.3

Solution: Given,

Operating voltage (V) = 220 kV
Diameter of each conductor = 2 cm

$$\text{Radius of each conductor} = \frac{2}{2} = 1 \text{ cm} = 1 \times 10^{-2} \text{ m} = 0.01 \text{ m}$$

The modified GMDS of phase A and phase C should be equal due to symmetry.

$$D_{aa'} = D_{cc'} = \sqrt[4]{(5)^2 + (2.2+2.2)^2} = \sqrt[4]{(5)^2 + (4.4)^2}$$
$$= 6.6603 \text{ m}$$

$$D_{SA} = D_{SC} = \sqrt[4]{D_{aa}\, D_{aa'}\, D_{a'a'}\, D_{a'a}}$$
$$= \sqrt[4]{r \times 6.6603 \times r \times 6.6603} = \sqrt{r \times 6.6603}$$
$$= \sqrt{0.01 \times 6.6603} = 0.2581 \text{ m}$$

Modified GMD of phase B,

$$D_{SB} = \sqrt[4]{D_{bb}\, D_{bb'}\, D_{b'b'}\, D_{b'b}} = \sqrt[4]{r \times 5 \times 5 \times r}$$
$$= \sqrt{5r} = \sqrt{5 \times 0.01} = 0.2236 \text{ m}$$

Net modified GMD of the circuit (D_S)

$$= \sqrt[3]{D_{SA} \times D_{SB} \times D_{SC}}$$
$$= \sqrt[3]{0.2581 \times 0.2236 \times 0.2581} = 0.2460 \text{ m}$$

Again from symmetry, the mutual GMD between phase A and phase B must be equal to that between phase B and phase C

$$D_{ab'} = D_{bc'} = D_{ba'} = D_{cb'} = \sqrt{(5)^2 + (2.2)^2} = 5.463 \text{ m}$$

$$D_{AB} = D_{BC} = \sqrt[4]{D_{ab}\, D_{ab'}\, D_{a'b}\, D_{a'b'}}$$

$$= \sqrt[4]{2.2 \times 5.463 \times 5.463 \times 2.2} = \sqrt{5.463 \times 2.2}$$

$$= 3.467\text{m}.$$

Mutual GMD between phase A and phase C,

$$D_{CA} = \sqrt[4]{D_{ca}\, D_{ca'}\, D_{c'a}\, D_{c'a'}} = \sqrt[4]{4.4 \times 5 \times 5 \times 4.4}$$

$$= \sqrt{4.4 \times 5} = 4.690\text{m}$$

Now, Net mutual GMD (D_m) = $\sqrt[3]{D_{AB}\, D_{BC}\, D_{CA}}$

$$= \sqrt[3]{3.467 \times 3.467 \times 4.690} = 3.834 \text{ m}$$

Capacitance of the line per unit length

$$= \frac{10^{-6}}{18 \ln \frac{D_m}{D_S}} \text{ F/km} = \frac{1}{18 \ln \frac{3.834}{0.2460}} \text{ F/km}$$

$$= 0.02023 \ \mu\text{F/km} \quad \textbf{Ans.}$$

Charging current per unit length of the line (I_c) = $\dfrac{V}{\sqrt{3} \times X_C}$

$$= \frac{V}{\sqrt{3}} \times 2\pi fc$$

$$= \frac{220 \times 10^{-3}}{\sqrt{3}} \times 2\pi 50 \times 0.02023 \times 10^{-6} \text{ amp/km}$$

$$= 0.8072 \text{ amp/km} \quad \textbf{Ans.}$$

Problem 3.4. Determine the capacitance and charging current per km length of the doubts circuit 3–phase line as shown in Fig P. 3.4. The transmission line is transposed within each circuit and each circuit remains on its own side. The dia of such conductor is 15 mm and operating voltage 220 kV.

Solution: Given,

Operating voltage (v) = 220 kV

Diameter of each conductor (D) = 15 mm

radius of each conductor (r) $= \dfrac{D}{2} = \dfrac{15}{2} = 7.5$ mm

$= 7.5 \times 10^{-3}$m = 0.0075 m

Let us mark the phase conductors aa', bb' and cc' as shown below:

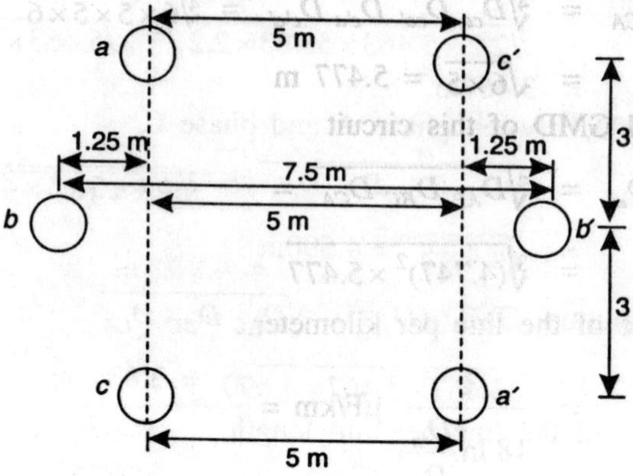

Fig. P. 3.4

$$D_{aa'} = D_{cc'} = \sqrt{(5)^2 + (3+3)^2} = \sqrt{(5)^2 + (6)^2} = 7.810 \text{ m}$$

From the given circuit, the modified GMD of conductors of phase *A* and phase *B* must be equal.

$$D_{SA} = D_{SC} = \sqrt[4]{D_{aa}\, D_{aa'}\, D_{a'a'}\, D_{a'a}} = \sqrt[4]{r \times 7.810 \times r \times 7.810}$$

$$= \sqrt{r \times 7.810} = \sqrt{0.0075 \times 7.810} = 0.2420 \text{ m}$$

Modified GMD of conductors of phase B.

$$D_{SB} = \sqrt[4]{D_{bb}\, D_{bb'}\, D_{b'b'}\, D_{b'b}} = \sqrt[4]{r \times 7.5 \times r \times 7.5}$$

$$= \sqrt{r \times 7.5} = \sqrt{7.5 \times 0.0075} = 0.2372 \text{ m}$$

Net modified GMD of the given circuit,

$$D_S = \sqrt[3]{D_{SA}\, D_{SB}\, D_{SC}} = \sqrt[3]{0.2420 \times 0.2372 \times 0.2420}$$

$$= 0.2404 \text{ m}$$

Again from the given circuit,

$$D_{ab} = D_{bc} = D_{b'c'} = D_{b'a'} = \sqrt{(3)^2 + (1.25)^2} = 3.25 \text{ m}$$

$$D_{ab'} = D_{bc'} = D_{ba'} = D_{cb'} = \sqrt{(3)^2 + (5+1.25)^2}$$

$$= \sqrt{(3)^2 + (6.25)^2} = 6.933 \text{ m}$$

Due to symmetry in arrangement, the mutual GMD between phase *A* and phase *B* must be equal to that between *B* and phase *C*.

$$D_{AB} = D_{BC} = \sqrt[4]{D_{ab}\, D_{ab'}\, D_{a'b}\, D_{a'b'}}$$

$$= \sqrt[4]{3.25 \times 6.933 \times 6.933 \times 3.25}$$

$$= \sqrt{3.25 \times 6.933} = 4.747 \text{ m}$$

Mutual GMD between phase *A* and phase *C*.

$$D_{CA} = \sqrt[4]{D_{ca}\,D_{ca'}\,D_{c'a}\,D_{c'a'}} = \sqrt[4]{6\times5\times5\times6}$$

$$= \sqrt{6\times5} = 5.477 \text{ m}$$

Net mutual GMD of this circuit

$$D_m = \sqrt[3]{D_{AB}\,D_{BC}\,D_{CA}} = \sqrt[3]{4.747\times4.747\times5.477}$$

$$= \sqrt[3]{(4.747)^2\times5.477} = 4.979 \text{ m}$$

Capacitance of the line per kilometer

$$= \frac{1}{18\ln\dfrac{D_m}{D_S}}\ \mu\text{F/km} = \frac{1}{18\ln\dfrac{4.979}{0.2404}}$$

$$= 0.01833\ \mu\text{F/km}$$

Now, Charging current of the line per km

$$= \frac{V}{\sqrt{3}\times X_C} = \frac{V\times2\pi fc}{\sqrt{3}}$$

$$= \frac{220\times10^3}{\sqrt{3}}\times2\pi50\times0.01833\times10^{-6}\ \text{amp/km}$$

$$= 0.7314 \text{ amp/km}$$ **Ans.**

Problem 3.5. Calculate the capacitance to neutral in farads perameter of a single-phase line composed of two-single Strand conductors each having a diameter of 0.229 in the conductors are 10 ft apart 25 ft above the ground.

Solution: Given

Diameter of each conductor = 0.229 inch

$$= \frac{0.229}{12}\text{ ft} = \frac{0.229\times0.3048}{12}\text{ m}$$

$$\text{Radius of each conductor } (r) = \frac{0.229\times0.3048}{12\times2} = 2.91\times10^{-3}\text{ m.}$$

When the effect of earth is neglected:

Butal GMD of this circuit (D_m)

= distance between the conductors

= 10 ft = 10 × .3048 m = 3.048 m

In this case, Capacitance of the line to neutral

$$= \frac{10^{-9}}{18 \ln \frac{D}{r}} \text{ F/m} = \frac{10^{-9}}{18 \ln \frac{3.048 \times 10^3}{2.91}} \text{ F/m}$$

$$= 7.989 \times 10^{-12} \text{ F/m} \quad \textbf{Ans.}$$

When the effect of earth is considered.

Height of the conductor above the ground (h)

$$= 25 \times 0.3048 \text{ m} = 7.62 \text{ m}$$

Capacitance of the live to neutral is given by

$$C_h = \frac{2\pi\varepsilon}{\ln\left(\dfrac{D}{r\sqrt{1+\dfrac{D^2}{4h^2}}}\right)} \text{ F/m}$$

$$= \frac{10^{-9}}{18 \ln\left(\dfrac{3.048 \times 10^3}{2.91 \times \sqrt{1+\dfrac{(3.048)^2}{4 \times (7.62)^2}}}\right)} \text{ F/m}$$

$$= 0.0124 \times 10^{-9} \text{ F/m} = 12.4 \times 10^{-12} \text{ F/m} \quad \textbf{Ans.}$$

Problem 3.6. A three-phase 60-Hz transmission line has its conductors arranged in a triangular formation so that two of the distances between conductors are 25 ft and the third is 42 ft. The conductors are ACSR Osprey. Determine the capacitance to neutral in microfarads per mole and the capacitive reactance to neutral in ohm-miles. If the line is 150 mile long, find the capacitance to neutral and capacitive reactance of the line.

Solution: Given,

Operating frequency (f) = 60 Hz,

Length of the line (l) = 150 mile

Since the conductors are ASCR osprey, From GMR table

Diameter of each conductor = 0.879 inch

$$\text{Radius of each conductor } (r) = \frac{0.892}{2 \times 12} = 0.0372 \text{ ft}$$

Acording to the question the conductors can be arranged as below.

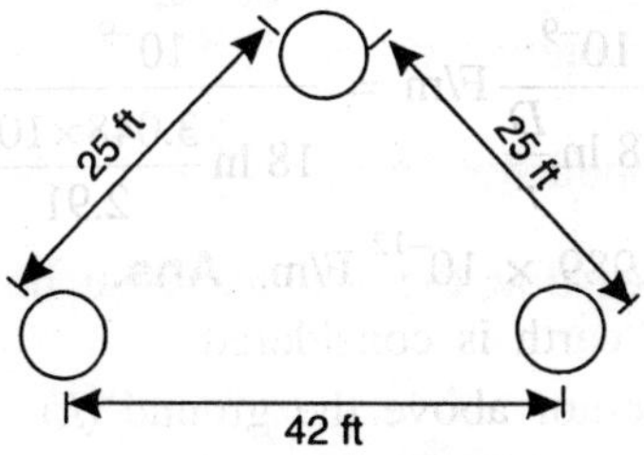

Fig. P. 3.6

Mutual GMD of this circuit (D_m) = $\sqrt[3]{25 \times 25 \times 42}$

$$= 29.72 \text{ ft}$$

Now, Capacitance of the line to neutral per phase

$$= \frac{10^{-9}}{18 \ln \frac{D_m}{r}} \text{F/m} = \frac{10^{-9}}{18 \ln \frac{29.72}{0.0.372}} \text{F/m}$$

$$= 8.313 \times 10^{-12} \text{ F/m} = 8.313 \times 1609 \times 10^{-12} \text{ F/m}$$

$$= 0.01338 \ \mu\text{F/mile} \quad \textbf{Ans.}$$

Capacitive reactance of the line to neutral (X_c)

$$= \frac{1}{2\pi f C_n} = \frac{1}{2 \times \pi \times 60 \times 0.01338 \times 10^{-6}} \Omega \text{ mile}$$

$$= 0.198 \ \Omega \text{ mile per phase to neutral} \quad \textbf{Ans.}$$

Now, Given, length of the line (l) = 150 mile

Capacitance of 150 mile long line to neutral (C)

= capacitance per unit length × length of the line

= 0.01338 × 150 μF = 2.007 μF **Ans.**

Capacitive reactance of 150 mile long line

$$= \frac{1}{2\pi fc} = \frac{1}{2 \times \pi \times 60 \times 2.007} 10^6 \ \Omega$$

$$= 1.322 \text{ k}\Omega \quad \textbf{Ans.}$$

Problem 3.7: A three-phase 60-Hz line has flat horizontal spacing. The conductors have an outside diameter of 3.28 cm with 12 m between conductors. Determine the capacitive reactance to neutral in ohmmeters and the capacitive reactance of the line in ohms if its length is 125 mile.

Solution: Given,

Operating frequency (f) = 60 Hz

Diameter of each conductor = 3.28 cm

Radius of each conductor (r) = $\frac{3.28}{2}$ = 1.64 cm = 0.0164 m

According to the question, the conductors can be arranged as below

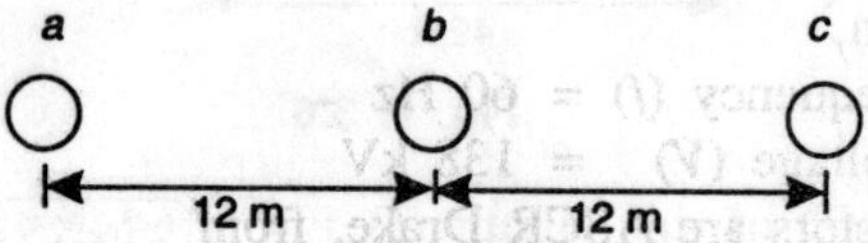

Fig. P. 3.7

Mutual GMD of this circuit (D_m)

$$= \sqrt[3]{D_{ab}\, D_{bc}\, D_{ca}} = \sqrt[3]{12\times12\times24} = 15.119\text{m}$$

Now, Capacitance of the line to neutral

$$C_n = \frac{10^{-9}}{18\ln\frac{D_m}{r}}\ \text{F/m} = \frac{10^{-9}}{18\ln\frac{15.119}{0.0164}}\ \text{F/m}$$

$$= 8.138\times10^{-12}\ \text{F/m}$$

Capacitive reactance of the line to neutral (X_e)

$$= \frac{1}{2\pi fc}$$

$$= \frac{1}{2\times\pi\times60\times8.138\times10^{-12}}\ \Omega\text{m/phase/neutral}$$

$$= 3.26\times10^{8}\ \Omega\text{m}$$ **Ans.**

Now, Capacitance of 125 mi long line

= Capacitance per unit length × length of the line

$= 8.138 \times 1609 \times 10^{-12}$ F/mi × 125 mi

= 1.037 μF

Capacitive reactance of 125 mi long line

$$= \frac{1}{2\pi fc} = \frac{1}{2\times\pi\times60\times1.637\times10^{-6}}\ \Omega$$

= 1.6204 kΩ **Ans.**

Problem 3.8. Six conductors of ACSR Drake constitute a 60-Hz double-circuit three-phase line arranged as shown in Fig. The vertical spacing, however, is 14 ft; the longer horizontal distance is 32 ft; and the shorter horizontal distances are 25 ft. Find the capacitive reactance to neutral in ohm-miles and the charging current per mile per phase and per conductor at 138 kV.

Solution: Given,

Operating frequency (f) = 60 Hz

Operating voltage (V) = 138 kV

Since conductors are ASCR Drake, from GMR table,

Diameter of this type of conductors = 1.108 inch

$$\text{Radius of each conductor } (r) = \frac{1.108}{2} = \frac{1.108}{2\times 12}\text{ ft}$$

$$= 0.0462 \text{ ft}$$

According to the question, the conductors can be arranged as given below.

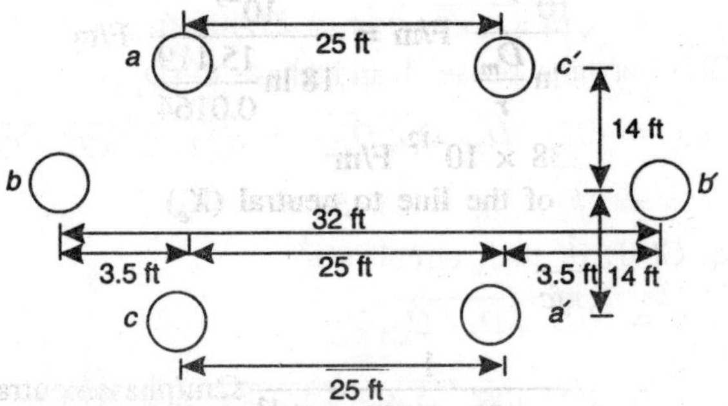

Fig. P. 3.8

Due to symmetry in arrangements, the modified GMRs of conductors of phases A and B are equal.

$$D_{ad'} = D_{cc'} = \sqrt{(25)^2 + (14+14)^2} = \sqrt{(25)^2 + (28)^2}$$

$$= 37.537 \text{ ft}$$

$$D_{sa} = D_{sc} = \sqrt[4]{D_{aa}\, D_{aa'}\, D_{a'a'}\, D_{a'a}}$$

$$= \sqrt[4]{r\times 37.537\times r\times 37.537} = \sqrt{r\times 37.537}$$

$$= \sqrt{0.0462\times 37.537} = 1.3169 \text{ ft}$$

Modified GMR of phase B conductors,

$$D_{SB} = \sqrt[4]{D_{bb}\, D_{bb'}\, D_{b'b'}\, D_{b'b}} = \sqrt[4]{r\times 32\times r\times 32}$$

$$= \sqrt{r\times 32} = \sqrt{0.0462\times 32} = 1.2159 \text{ ft}$$

Net modified GMR of the conductors

$$D_S = \sqrt[3]{D_{SA}\,D_{SB}\,D_{SC}}$$

$$= \sqrt[3]{1.3169\times1.2159\times1.3169} = 1.2823 \text{ ft.}$$

The Mutual GMD between phase *A* and phase *B* must be equal to that between phase *B* and phase *C*.

$$D_{ab} = D_{b'c'} = D_{bc} = D_{a'b'} = \sqrt{(14)^2+(3.5)^2} = 14.481 \text{ ft}$$

$$D_{aa'} = D_{cc'} = \sqrt{(14+14)^2+(25)^2} = \sqrt{(28)^2+(25)^2}$$

$$= 37.537 \text{ ft}$$

$$D_{ab'} = D_{bc'} = D_{b'c} = D_{ba'} = \sqrt{(25+3.5)^2+(14)^2}$$

$$= \sqrt{(28.5)^2+(14)^2} = 31.753 \text{ ft}$$

$$D_{AB} = D_{BC} = \sqrt[4]{D_{ab}\,D_{ab'}\,D_{a'b}\,D_{a'b'}}$$

$$= \sqrt[4]{14.431\times31.753\times31.753\times14.431}$$

$$= \sqrt{14.431\times31.753} = 21.406 \text{ ft.}$$

Mutual GBD between phase *A* and phase *C* (D_{CA})

$$= \sqrt[4]{D_{ca}\,D_{ca'}\,D_{c'a}\,D_{c'a'}} = \sqrt[4]{28\times25\times28\times25}$$

$$= \sqrt{25\times28} = 26.458 \text{ ft.}$$

Net mutual GMD of their circuit (D_m)

$$= \sqrt[3]{D_{AB}\,D_{BC}\,D_{CA}}$$

$$= \sqrt[3]{21.406\times21.406\times26.458} = 22.973 \text{ ft}$$

To calculate capacitive reactance to neutral in ohm-miles, apply the relation

$$X_c = 2.965\times10^4 \ln\frac{D_m}{D_s}\ \Omega.\text{mile to neutural}$$

$$= 2.965\times10^4 \ln\frac{22.973}{1.2823}\ \Omega.\text{mile to neutral}$$

$$= 8.556\times10^4\ \Omega - \text{mile to neutural}$$ **Ans.**

$$\text{Now, charging current } (I_c) = \frac{V}{\sqrt{3}\times X_C} \text{ amp/mi/phase}$$

$$= \frac{138\times10^3}{\sqrt{3}\times8.556\times10^4} \text{ amp/mi/phase}$$

$$= 0.9312 \text{ amp/mi/phase}$$

Again, charging current per mile per conductor

$$= \frac{0.9312}{2} \text{ amp/mi/conductor}$$

$$= 0.4606 \text{ amp/mi/conductor} \quad \textbf{Ans.}$$

Problem 3.9. Find the capacitive susceptance per kilometer of a single-phase line operating at 60 Hz. The conductor is patridge and spacing is 25 m between the centres

Solution: Given,

Operating frequency (f) = 60 Hz

Mutual GMD i.e. spacing between the conductors (D_m) = 25 m

For diameter of patridge conductors, using GMR Tables,

Diameter of such type of conductors = 0.642 inch

$$= \frac{0.642}{12} \text{ ft} = \frac{0.642}{12} \times 0.3048 \text{ m}$$

Radius of such type of conductor (r)

$$= \frac{0.642}{12 \times 2} \times 0.3048 \text{ m} = 0.00815\text{m}$$

To calculate capacitive reactance, apply the formula.

$$X_e = \frac{2.862}{f} \times 10^9 \ln \frac{D_m}{r} \ \Omega.\text{m to neutral}$$

$$= \frac{2.862}{60} \times 10^6 \ln \frac{25}{0.00815} \ \Omega.\text{km to neutral}$$

$$= 0.3830 \times 10^6 \ \Omega.\text{km to neutral}$$

Capacitive susceptance per km of their line

$$B_c = \frac{1}{X_C} = \frac{1}{0.3830} \times 10^{-6} \ \mho/\text{km to neutral}$$

$$= 2.611 \times 10^{-6} \ \mho/\text{km to neutral}$$

Problem 3.10. A single-circuit three-phase line operated at 50 Hz is arranged as shown in figure. The diameter of each conductor is 20 mm. If the length of the line is 100 km and normal operating voltage is 220 kV. Find the capacitance, capacitive reactance and charging current in the line.

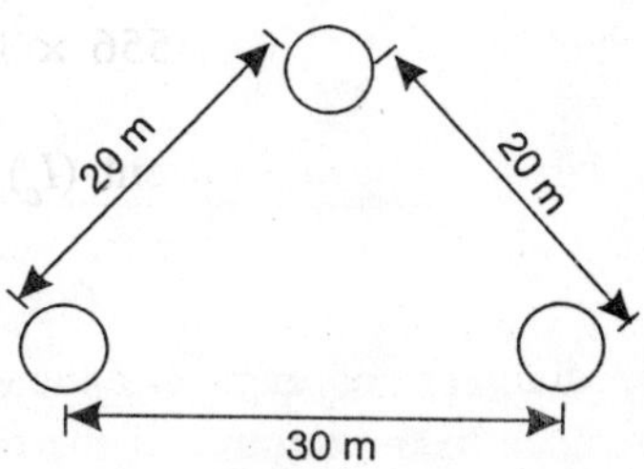

Fig. P. 3.10

Sulution: Given,

Operating voltage (V)	= 220 kV
Operating frequency (f)	= 50 Hz
Diameter of each conductor	= 20 mm

$$\text{Radius of each conductor } (r) = \frac{20}{2} = 10 \text{ mm}$$

$$= 10\times10^{-3} \text{ m} = 0.01\text{m}$$

Mutual GMD of this circuit (D_m)

$$= \sqrt[3]{20\times20\times30} = \sqrt[3]{400\times30} = 22.894 \text{ m}$$

Capacitance of the line per km

$$C = \frac{1}{18 \ln \frac{D_m}{r}} \mu\text{F/km.} = \frac{1}{18 \ln \frac{22.894}{0.01}} \mu\text{F/km}$$

$$= 7.18\times10^{-9} \text{ F/km}$$

Capacitance of 100 km long line

= capacitance per unit length × length of the line

$= 7.18 \times 10^{-9} \times 100$ F $= 0.718$ μF **Ans.**

Capacitive reactance of the line (X_c)

$$= \frac{1}{2\pi fc} = \frac{1}{2\times0.718\times50\times\pi} 10^6 = 4.433 \text{ k}\Omega \textbf{ Ans.}$$

Charging current of the line (I_c)

$$= \frac{V}{\sqrt{3}\times X_C} = \frac{138\times10^3}{\sqrt{3}\times4.443\times10^3} \text{ amp}$$

$= 28.653$ amp **Ans.**

Problem 3.11. A three-phase double-circuit line is arranged as shown in figure given below.

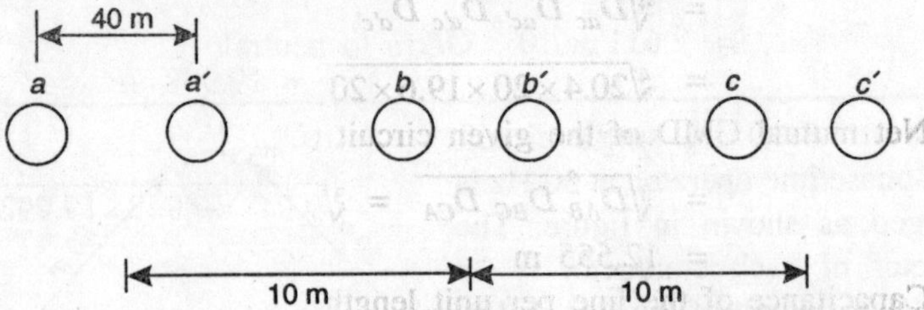

Fig. P. 3.11

The diameter of each conductor is 20 mm. Find the capacitive reactance to the neutral of the line of operating voltage is 220 kV.

Also find the per unit capacitive reactance of the line of the length is 100 km and base is 1 kVA, 330 kV. The operating frequency is 50 Hz.

Solution: Given,

Operating frequency (f) = 50 Hz
Length of the line (l) = 100 km
Operating voltage (V) = 220 kV
Diameter of each conductor = 20 mm

Radius of each conductor (r) = $\frac{20}{2}$ = 10 mm

$= 10 \times 10^{-3}$ m = 0.01m

Due to symmetry in arrangments, the modified, GMR of conductors of each phase must be identical

$$D_{SA} = D_{SB} = D_{SC} = \sqrt[4]{D_{aa} D_{aa'} D_{a'a'} D_{a'a}}$$

$$= \sqrt[4]{r \times 0.4 \times r \times 0.4} = \sqrt{r \times 0.4}$$

$$= \sqrt{0.4 \times 0.01} = 0.0632 \text{ m}$$

Net modified GMR of the given circuit

$$= \sqrt[3]{D_{SA} \times D_{SB} \times D_{SC}}$$

$$= \sqrt[3]{0.0632 \times 0.0632 \times 0.0632}$$

$$= 0.0632 \text{ m}$$

Now, Mutual GMD between phase A and B (D_{AB})

$$= \sqrt[4]{D_{ab} D_{ab'} D_{a'b} D_{a'b'}}$$

$$= \sqrt[4]{10 \times 10.2 \times 9.6 \times 10}$$

Mutual GMD between phase C and phase A (D_{CA})

$$= \sqrt[4]{D_{ac} D_{ac'} D_{a'c} D_{a'c'}}$$

$$= \sqrt[4]{20.4 \times 20 \times 19.6 \times 20} = 19.998 \text{ m}$$

Net mutual GMD of the given circuit (D_m)

$$= \sqrt[3]{D_{AB} D_{BC} D_{CA}} = \sqrt[3]{9.948 \times 9.948 \times 19.998}$$

$$= 12.555 \text{ m}$$

Capacitance of the line per unit length

$$= \frac{1}{18 \ln \frac{D_m}{D_s}} \mu\text{F/km} = \frac{1}{18 \ln \frac{12.555}{0.0632}} \mu\text{F/km}$$

$= 0.0105$ μF/km **Ans.**

Capacitive Reactance of the line per km (X_c)

$$= \frac{1}{2\pi fc} = \frac{1}{2\times\pi\times 50\times 0.0105} 10^6 \ \Omega/\text{km}$$

$= 303.152$ kΩ/km **Ans.**

Capacitive reactance of 100 km line

$= X_c \times$ length of the line

$= 303.152 \times 10^3 \ \Omega \times 100$

$= 303.152 \times 10^5 \ \Omega$ **Ans.**

To calculate per unit reactance, apply the formulas

$$Z_{p.u.} = \frac{\text{Zactual} \times \text{Base MVA}}{(\text{Base kV})^2}$$

$$= \frac{303.152\times 10^5 \times 1.00\times 10^3}{(330)^2 \times 10^6}$$

$= 0.2784$ p.u. **Ans.**

Problem 3.12. Find the capacitance of a single phase line 50 km long consisting of two paralel wires each 6 mm in dia and 2.0 m apart. Also find the charging current if the operating voltage is 220 kV and frequency is 50 Hz.

Solution: Given,

Diameter of each conductor = 6 mm

Radius of each conductor (r) $= \frac{6}{2} = 3$ mm $= 3 \times 10^{-3}$ m

Distance between the line (D) = 2 m

Mutual GMD of the line (D_m) = D = 2 m

Capacitance of the line per kilometer

$$= \frac{1}{18 \ln \frac{D}{r}} \mu\text{F/km} = \frac{1}{18 \ln \frac{2}{3\times 10^{-3}}} \mu\text{F/km}$$

$= 8.544\times 10^{-9}$ μF/Km

Capacitance of the 50 km line

= capacitance/km × length of the line

$= 8.544 \times 10^{-9} \times 50$ F $= 427.2 \times 10^{-9}$F

= 0.4272 μF

Now, Charging current of the line (I_e)

$$= \frac{V}{\sqrt{3} \times X_C} = \frac{V \times 2\pi fc}{\sqrt{3}}$$

$$= \frac{220 \times 10^3 \times 2 \times \pi \times 50 \times 0.4272 \times 10^{-6}}{\sqrt{3}} \text{ amp}$$

$= 17.047$ amp **Ans.**

Problem 3.13. Find the capacitance of a single phase line 100 km long consisting of 2 parallel wires each 6 mm in dia and 3 m apart. Take effect of the ground if the height of the conductors above the ground is 7 m.

Solution: Given,

Diameter of each conductor = 6 mm

Radius of each conductor $= \frac{6}{2} = 3 \times 10^{-3}$ m

Distance between the wires (D) = 3 m

Height of the wires above the ground (h) = 7 m

The circuit diagram are to presence of earth can be drawn as shown in Fig. P. 3.13.

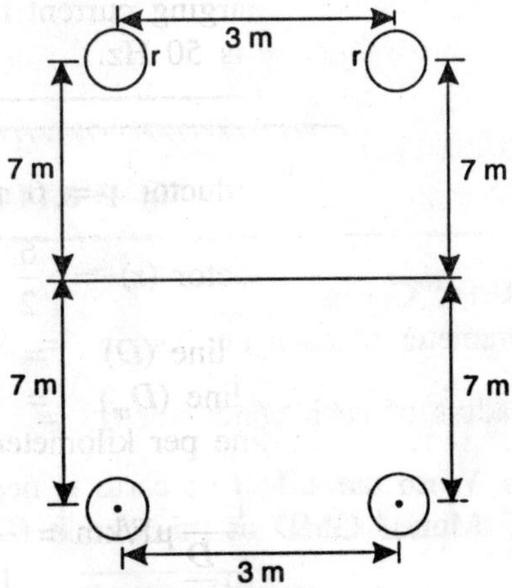

Fig. P. 3.13

Capacitance of the line in this case is given by

$$C_n = \frac{2\pi\varepsilon}{\ln\left(\dfrac{D}{r\sqrt{1+\dfrac{D^2}{4h^2}}}\right)} \text{ F/m} = \frac{10^{-6}}{18 \ln\left(\dfrac{3 \times 10^3}{3\sqrt{1+\dfrac{(3)^2}{4 \times (7)^2}}}\right)}$$

$$= \frac{1}{18 \ln\left(\dfrac{1000}{\sqrt{1+\dfrac{9}{4\times 49}}}\right)} = 8.069 \times 10^{-9} \text{ F/km}$$

∴ Capacitance of 100 km line

$$= \text{capacitance perunit length} \times \text{length of the line}$$

$$= 8.069 \times 100 \text{ F}' = 0.8069 \ \mu\text{F} \quad \textbf{Ans.}$$

Problem 3.14. Find the capacitance per km per phase to neutral of a three phase line as shown in Fig. P. 3.14 (a). (i) neglecting earth's effect and (ii) taking effect of the earths. The line is completely transposed. The diameter of each conductor is 3 cm.

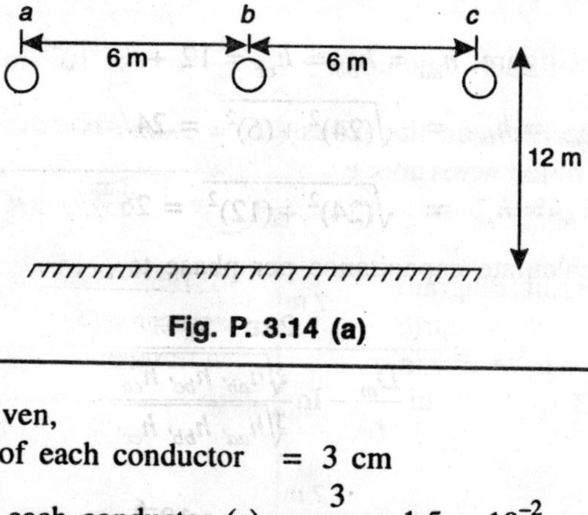

Fig. P. 3.14 (a)

Solution: Given,

Diameter of each conductor $= 3$ cm

Radius of each conductor (r) $= \dfrac{3}{2} = 1.5 \times 10^{-2}$ m $= 0.015$ m

(i) When the effect of earth is neglected.

Mutual GMD of this circuit (D_m)

$$= \sqrt[3]{D_{ab}\, D_{bc}\, D_{ca}} = \sqrt[3]{6\times 6\times 12}$$

$$= \sqrt[3]{(6)^2 \times 12} = 7.560 \text{ m}$$

Capacitance of the line per km per phase to neutral

$$= \frac{1}{18 \ln \dfrac{D_m}{r}} = \frac{1}{18 \ln \dfrac{7.560}{0.015}} \ \mu\text{F/km}$$

$$= 8.928 \times 10^{-9} \ \mu\text{F/km} \quad \textbf{Ans.}$$

(ii) When the effect of earth is neglected, the effect of the earth on the line can drawn as

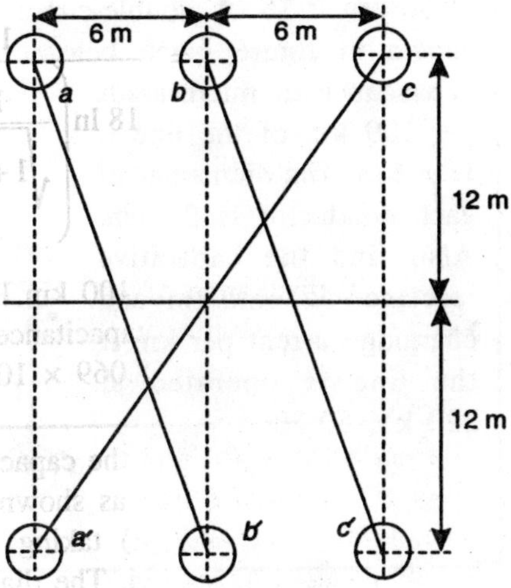

Fig. P. 3.14 (b)

From figure, $h_{aa'} = h_{bb'} = h_{cc'} = 12 + 12 = 24$ m

$$h_{ab'} = h_{bc'} = \sqrt{(24)^2 + (6)^2} = 24.739 \text{ m}$$

$$h_{ca'} = h_{ac'} = \sqrt{(24)^2 + (12)^2} = 26.833 \text{ m}$$

To calculate capacitance per phase to neutral, apply the relation

$$C_n = \frac{2\pi\varepsilon}{\ln\frac{D_m}{r} - \ln\frac{\sqrt[3]{h_{ab'}\, h_{bc'}\, h_{ca'}}}{\sqrt[3]{h_{aa'}\, h_{bb'}\, h_{cc'}}}} \text{ F/m}$$

$$= \frac{10^{-6}}{18\, ln\left(\frac{7.560}{0.015}\right) - ln\left(\frac{\sqrt[3]{(24.739)^2 \times 26.833}}{\sqrt[3]{24 \times 24 \times 24}}\right)} \text{ F/km}$$

$$= \frac{10^{-6}}{18\left[ln\left(\frac{7.560}{0.015}\right) - ln\left(\frac{\sqrt[3]{(24.739)^2 \times 26.833}}{24}\right)\right]} \text{ F/km}$$

$= 9.0112 \times 10^{-9}$ F/km

Problem 3.15. A double-circuit three-phase line is arranged as shown in figure given below. Calculate the phase to neutral capacitance in microfarads per 100 km of the line if D = 2 m. The diamenter of each conductor is 20 cm. Also find the capacitive reactance in ohm-km and charging current per km if the line is operated at 220 kV, 50 Hz.

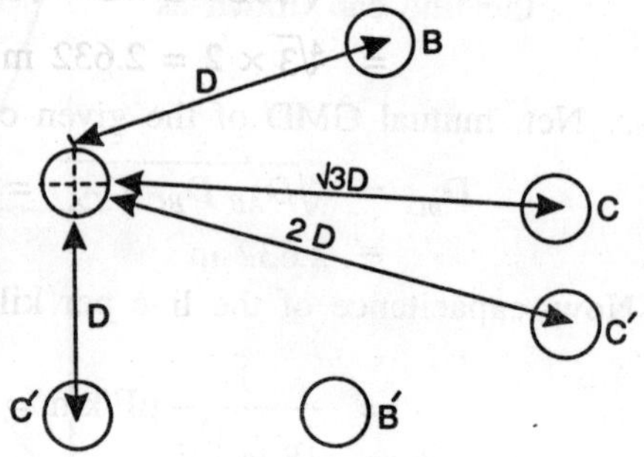

Fig. P. 3.15

Solution: Given

Diameter of each conductor = 20 cm

$$\text{Radius of each conductor } (r) = \frac{20}{2} = 10 \text{ cm}$$

$$= 10 \times 10^{-2} \text{ m} = 0.1 \text{ m}$$

Due to symmetry in arrangement of conductors, the modified GMR of conductors of phase A phase B and phase C are equal.

$$D_{SA} = D_{SB} = D_{SC} = \sqrt[4]{D_{AA}\, D_{AA'}\, D_{A'A'}\, D_{A'A}}$$

$$= \sqrt[4]{r \times 2D \times r \times 2D} = \sqrt{r \times 2D}$$

$$= \sqrt{0.1 \times 2 \times 2} = 0.632 \text{ m}$$

Net modified GMR of the conductors

$$D_S = \sqrt[3]{D_{SA}\, D_{SB}\, D_{SC}}$$

$$= \sqrt[3]{0.632 \times 0.632 \times 0.632}$$

$$= \sqrt[3]{(0.632)^3} = 0.632 \text{ m}$$

The mutual GMD between phase A and phase B must be equal to that between phase B and phase C.

$$D_{AB} = D_{BC} = \sqrt[4]{D_{ab}\, D_{ab'}\, D_{a'b}\, D_{a'b'}}$$

$$= \sqrt[4]{D \times \sqrt{3}D \times \sqrt{3}D \times D} = \sqrt[4]{3} \times \text{D}$$

$$= \sqrt[4]{3}\text{ D} = \sqrt[4]{3} \times 2 = 2.632 \text{ m}$$

Mutual GMD between phase A and phase C

$$D_{CA} = \sqrt[4]{D_{ca}\, D_{ca'}\, D_{c'a}\, D_{c'a'}}$$

$$= \sqrt[4]{\sqrt{3}D \times D \times D \times \sqrt{3}D} = \sqrt[4]{3}\ \text{D}$$

$$= \sqrt[4]{3} \times 2 = 2.632 \text{ m}$$

$\therefore$ Net, mutual GMD of the given circuit

$$D_m = \sqrt[3]{D_{AB}\, D_{BC}\, D_{CA}} = \sqrt[3]{2.632 \times 2.632 \times 2.632}$$

$$= 2.632 \text{ m}$$

Now, capacitance of the line per kilometer per phase (C_n)

$$= \frac{1}{18 \ln \frac{D_m}{D_s}} \mu\text{F/km} = \frac{1}{18 \ln \frac{2.632}{0.632}} \mu\text{F/km}$$

$$= 0.0389 \ \mu\text{F/km}$$

Capacitance of the line having length 100 km per phase

$$C = 0.0389 \times 100 \ \mu\text{f} = 3.89 \ \mu\text{F}$$ **Ans.**

Capacitive reactance per km (X_c)

$$= \frac{1}{2\pi f c_n} = \frac{1}{2 \times \pi \times 50 \times .0389} \times 10^6 \ \Omega\text{.km}$$

$$= 8.183 \times 104 \ \Omega\text{.km}$$ **Ans.**

Charging current per kilometer (I_c)

$$= \frac{V}{\sqrt{3} \times X_C} = \frac{220 \times 10^3}{\sqrt{3} \times 8.183 \times 10^4} \text{ amp/km}$$

$$= 1.552 \text{ amp/km}$$ **Ans.**

Chapter 4

Performance of Lines

Problem 4.1. Determine the sending end voltage, current, power factor of a 1-phase 50 Hz, 76.2 kV transmission delivering a load of 12 MW at 0.8 p.f. The line constants are R = 25 ohm, inductance 200 mH and capacitance between lines 2.5 μF. Also determine the regulation and η of transmission. Use nominal-π method. Draw phasor diagram.

Solution: Given,

Operating frequency (f) = 50 Hz, V_r = 76.2 kV, P_r = 12 MW, Cos ϕ_r = 0.8, R = 25 Ω, L = 200 mH, C = 2.5 MF

Now, Series impedance of the line (Z) = $R + j\omega l$

$$= 25 + j\, 2 \times \pi \times 50 \times 200 \times 10^{-3}\ \Omega$$

$$= 25 + j\ 62.83\ \Omega = 67.621\angle 68.30\ \Omega$$

Shunt admittance of the line(Y) = $j\omega c$

$$= j\ 2 \times \pi \times 50 \times 2.5 \times 10^{-6}\ \mho$$

$$= 7.854 \times 10^{-4} \angle 90°\ \mho$$

$$\text{Receiving end current } (I_r) = \frac{P_r}{V_r \, Cos\, \phi_r} = \frac{12 \times 10^6}{76.2 \times 10^3 \times 0.8}$$

$$= 196.85\text{A}$$

Let us take receiving voltage as reference phasar

$$\bar{V}_r = 76.2 \times 10^3 \angle 0°\ \text{V} = 76200\ \angle 0°\ \text{V}$$

$$\bar{I}_r = 196.85\ \angle{-\text{Cos}}\ 0.8 = 196.85\ \angle{-36.8°}\ \text{A}$$

The nominal π-circuit of the transmission line can be drawn as shown in Fig. P. 4.1.

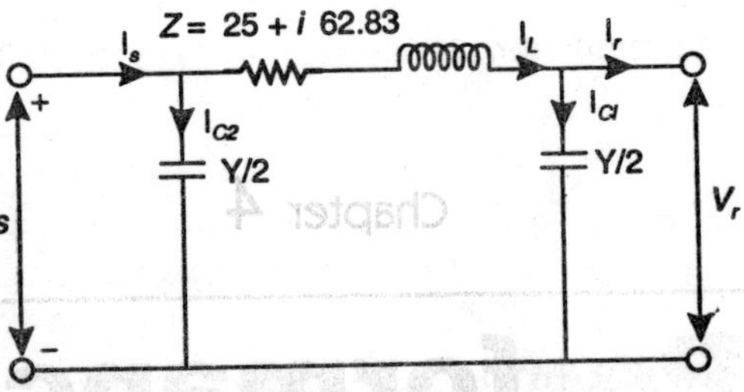

Fig. P. 4.1

Current through receiving end shunt admittance

$$\bar{I}_{c1} = \frac{\bar{V}_r Y}{2} = \frac{76200 \angle 0° \times 7.854 \times 10^{-4} \angle 90°}{2}$$

$$= 29.91 \angle 90° \text{A}$$

Current through the series impedance ($\bar{I}_L$)

$$= \bar{I}_{c1} + \bar{I}_r = 29.91 \angle 90° + 196.85 \angle -36.8°$$

$$= 157.62 - j\ 88.01 \text{ A} = 180.527 \angle -29.18° \text{ A}$$

Now, Voltage across receiving end admittance ($\bar{V}_{C_2}$)

$$= \bar{Z} \times \bar{I}_L + \bar{V}_{C_1}$$

$$= 67.621 \angle 68.30 \times 180.527 \angle -29.18 + 76200 \angle 0° \text{V}$$

since $\bar{V}_{C_1} = \bar{V}_r$

$$= 12207.416 \angle 39.12° + 76200 \angle 0° \text{ V}$$

$$= 85670.83 + j\ 7702.23 \text{ V} = 86015.541 \angle 5.14° \text{ V}$$

The sending end voltage is same as the voltate $\bar{V}_{C_2}$

Sending end voltage (V_s) = 86015.54 V = 86.016 kV **Ans.**

Now, Current through the sending end admittance ($\bar{I}_{c_2}$)

$$= \bar{V}_{C_2} \times \frac{Y}{2}$$

$$= \frac{86015.514 \angle 5.14° \times 7.854 \angle 90° \times 10^{-4}}{2} \text{A}$$

$$= 33.778 \angle 95.14° \text{ A}$$

Sending end current ($\bar{I}_s$) = $\bar{I}_L + \bar{I}_{c_2}$

$$= 33.778 \angle 95.14° + 180.527 \angle -29.18°$$

$$= 154.591 - j\ 54.375 = 163.8.8 \angle -19.38° \text{A}$$

∴ Sending end current ($\bar{I}_s$) = 163.88 A **Ans.**

p.f. = Cos (–19.38°) = 0.9433 **Ans.**

To calculate regulation, let us find no-load receiving end voltage $(\bar{V}_r')$

$$= \frac{|V_s|\left(-\frac{2j}{\omega C}\right)}{R + j\times\frac{-2j}{\omega C}} = \frac{86015.54\left(-\frac{2j}{2\pi\times 50\times 2.5\times 10^{-6}}\right)}{25 + j62.83 - \frac{2j}{2\pi\times 50\times 2.5\times 10^{-6}}}$$

$$= \frac{219.037\times 10^6\angle -90^\circ}{25 + j62.83 - j2543.48} = \frac{219.037\times 10^6\angle -90^\circ}{25 - j2483.65}$$

$$= \frac{219.037\angle -90^\circ \times 10^6}{2483.776\angle -89.42} = 88187.1 \angle -0.58^\circ \text{ V}$$

No-load receiving end voltage $(V_r') = 88187.1$ V

$\therefore$ % regulating of the line $= \dfrac{V_r' - V_r}{V_r} \times 100$

$$= \frac{88187 - 76200}{76200} \times 100$$

$= 15.731\%$ **Ans.**

Total ohmic losses in the line $= I_L^2 \times R = 25 \times (180.527)^2$

$= 0.8147$ MW

Hence, Efficiency of the line

$$= \frac{\text{Power output}}{\text{Power output + Total losses}} \times 100\%$$

$= \dfrac{12}{12.8147} \times 100\% = 93.642\%$ **Ans.**

Problem 4.2. Determine the efficiency and regulation of a 3-phase, 50 Hz, 150 kms long transmission line having three conductors spaced 3.5 meters delta formation when the receiving end delivers 25 MVA at 120 kV and p.f. 0.9 lagging. The resistance of the conductor is 0.25 ohm per km and the effective dia is 0.75 cm. Neglect leakance and use (i) nominal-T, (ii) nominal-π

Solution: Given,

Length of the line (l) = 150 kms

Operating frequency (f) = 50 Hz

$S_r = 25$ MVA, $V_{rL} = 120$ kV, $\cos\phi_r = 0.9$ (logging),

$r = 0.25\ \Omega$ per km

Diameter of each conductor = 0.75 cm

radius of each conductor (r) = 0.375 cm = 0.375×10^{-2} m

Butual GMD of the circuit (D_m) = $\sqrt[3]{3.5\times3.5\times3.5}$

$$= \sqrt[3]{(3.5)^3} = 3.5 \text{ m}$$

Self GMD of each conductor (D_s)

$$= 0.7788 \times 0.375 \times 10^{-2} \text{ m} = 0.2921 \times 10^{-2} \text{ m}$$

Inductance of the line per km per phase

$$= 2 \times 10^{-4} \ln \frac{D_m}{D_s} \text{ H/km} = 2 \times 10^{-4} \ln \frac{3.5\times10^2}{0.2921} \text{ H/km}$$

$$= 1.418 \text{ mH/km}$$

Inductance of 150 km long line (L)

= Inductance/km × lenght of the line

= 1.418 mH/km × 150 km = 0.213 H

Capacitance of the line per phase per km

$$= \frac{10^{-6}}{18 \ln \frac{D_m}{r}} \text{ F/km} = \frac{10^{-6}}{18 \ln \frac{2.632\times10^2}{0.375}} \text{ F/km}$$

$$= 8.124 \times 10^{-9} \text{ F/km}$$

Capacitance of 150 km long line (C)

= Capacitance per km × length of the line

$= 8.124 \times 10^{-9} \times 150$ F = 1.219 μF

Total resistance of the line (R) = $r \times l = 0.25 \times 150 = 37.5\ \Omega$

Series impedance of the line (τ) = $R + j\omega l$

$$= 37.5 + j\,2\pi fL = 37.5 + j\,2 \times \pi \times 50 \times 0.213\ \Omega$$

$$= 37.5 + j\,66.916\ \Omega = 76.707\ \angle 60.73°\ \Omega$$

Shunt admittance of the line (Y) = $j\omega c = j\,2\pi \times f \times c$

$$= j\,2 \times \pi \times 50 \times 1.219 \times 10^{-6}\ \mho$$

$$= 3.83 \times 10^{-4}\ \angle 90°\ \mho$$

Receiving end current ($\bar{I}_r$) = $\dfrac{S_r}{\sqrt{3}\,V_r} = \dfrac{25\times10^6}{\sqrt{3}\times120\times10^3}$ A

$$= 120.28 \text{ A}$$

p.f angle (ϕ_r) = $\cos^{-1}(0.9) = 25.84°$

In phasor form, $\bar{I}_r = 120.28\ \angle{-25.84}$ A

(1) Nominal-T-Method:

The circuit diagram for nominal-T method can be drawn as given Fig. P. 4.2 (a).

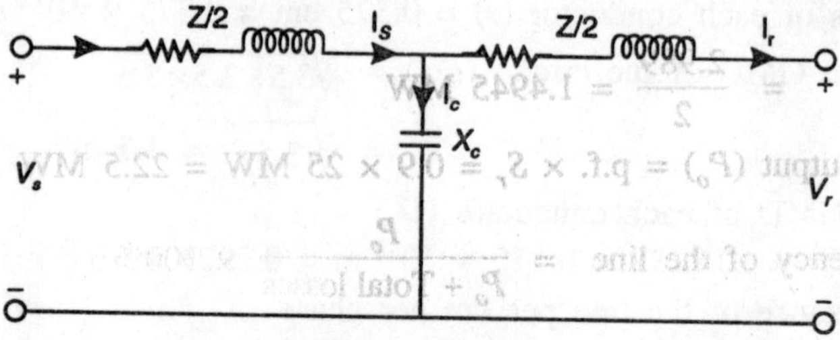

Fig. P. 4.2 (a)

Receiving end voltage per phase $(\bar{V}_r) = \dfrac{120 \times 10^3}{\sqrt{3}}$ V = 69282 V

Let us take receiving end voltage as reference phasor.

$\therefore \quad \bar{V}_r = 69282 \angle 0°$ V

Voltage across shunt admittance $(V_c) = V_r + \dfrac{\bar{I}\,\bar{Z}}{2}$

$= 69282 \angle 0° + 120.28 \angle -25.84° \times \dfrac{76.707}{2} \angle 60.73°$

$= 69282 \angle 0° + 4613.159 \angle 34.89°$

$= 73065.95 + j\ 2638.74\ \Omega$

$= 73.113.853 \angle 2.07°$ V

Current through shunt admittance $(I_c) = V_c \times$ Y

$= 73113.853 \angle 2.07 \times 3.83 \times 10^{-4} \angle 90°$

$= 28.009 \angle 92.07°$ A

Sending end current $(\bar{I}_s) = \bar{I}_c + \bar{I}_r$

$= 28.003 \angle 92.07° + 120.28 \angle -25.84°$

$= 107.24 - j\ 24.44$ A $= 109.99 \angle -12.84°$ A

Now, Sending end voltage $(\bar{V}_s) = \bar{I}_s \dfrac{\bar{Z}}{2} + \bar{V}_c$

$= \dfrac{109.99}{2} \angle -12.84° \times 76.707 \angle 60.73° + 73113.853$

$\angle 2.07°$

$= 4218.50 \angle 47.89° + 73113.853 \angle 2.07°$

$= 75894.883 + j\ 5770.437$ V

$= 76113.936 \angle 4.35°$ V

Total losses in the line $= 3 \times \left(I_s^2 \dfrac{R}{2} + I_r^2 \dfrac{R}{2} \right)$

$= 3 \times [(120.28)^2 + (109.99)^2] \times \dfrac{37.5}{2}$ W

$$= \frac{2.989}{2} = 1.4945 \text{ MW}$$

Power output (P_o) = p.f. × S_r = 0.9 × 25 MW = 22.5 MW

$$\% \text{ efficiency of the line } = \frac{P_o}{P_o + \text{Total losses}} \times 100\%$$

$$= \left(\frac{22.5}{22.5 + 1.495}\right) = 93.77\% \quad \textbf{Ans.}$$

Now, No-load receiving end voltage can be calculated by using the relation

$$V_r' = \frac{|V_s|\left(-\frac{j}{\omega c}\right)}{\frac{R}{2} + j\frac{x}{2} - \frac{j}{\omega c}} = \frac{76113.936\left(-\frac{j}{3.83 \times 10^{-4}}\right)}{\frac{37.5}{2} + j\frac{66.916}{2} - \frac{j}{3.83 \times 10^{-4}}}$$

$$= \frac{19873.09\angle -90° \times 10^4}{18.75 + j\,33.458 - j\,2610.966}$$

$$= \frac{19873.09\angle -90° \times 10^4}{18.75 - j\,2577.51}$$

$$= \frac{19873.09\angle -90° \times 10^4}{2577.578\angle -89.58°} = 77099.859 \angle -0.42° \text{ V}$$

$$\% \text{ regulation } = \frac{V_r' - V_r}{V_r} 100 = \left(\frac{77099.859 - 69282}{69282}\right) 100$$

$$= 11.28\% \quad \textbf{Ans.}$$

(2) Nominal π-Method :

The circuit diagram for nominal π-Method can be drawn as given in Fig. P. 4.2 (b).

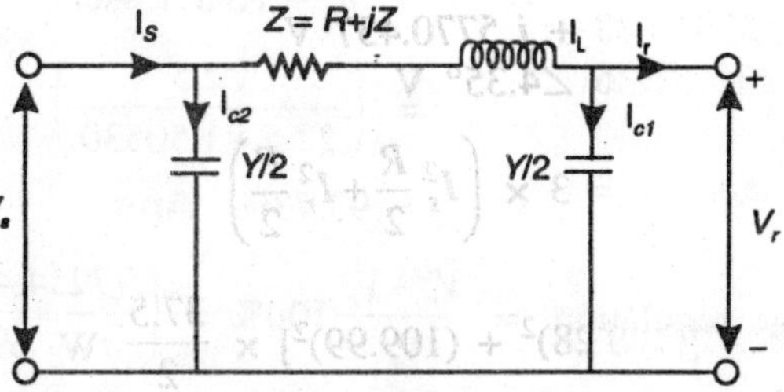

Fig. P. 4.2 (b)

Let us take receiving end voltage as reference phasor

$\therefore \quad \bar{V}_r = 69282 \angle 0° \text{ V}, \bar{I}_r = 120.84 \angle -25.84° \text{ A}$

Now, Current through the shunt admittance at receiving end

$$\bar{I}_{c1} = \bar{V}_r \frac{\bar{Y}}{2} = \frac{69282 \angle 0° \times 3.83 \times 10^{-4} \angle 90°}{2}$$

$$= 13.268 \angle 90° \text{ A}$$

Current through series impedance $(\bar{I}_L) = \bar{I}_{c1} + \bar{I}_r$

$$= 13.268 \angle 90° + 120.84 \angle -25.84°$$

$$= 108.758 - j\,39.40 \text{ A} = 115.674 \angle -19.91° \text{ A}$$

Voltage across receiving end admittance $(\bar{V}_{c2}) = \bar{I}_L \bar{Z} + \bar{V}_{c1}$

$$= 115.674 \angle -19.91° \times 76.707 \angle 60.63° + 69282 \angle 0°$$

$[\therefore \ \bar{V}_{c1} = \bar{V}_r]$

$$= 8873.01 \angle 40.72° + 69282 \angle 0°$$

$$= 76006.91 + j\,5788.42$$

$$= 76227 \angle 4.36° \text{ V}$$

Sending end voltage $(V_s) = V_{c2} = 76227$ V

For nominal π-method No-load receiving end voltage

$$V_r = \frac{|V_s|\left(-\frac{2j}{\omega c}\right)}{R + j\omega L - \frac{2j}{\omega c}} = \frac{76227\left(-\frac{2j}{3.837 \times 10^{-4}}\right)}{37.5 + j66.916 - \frac{2j}{3.83 \times 10^{-4}}}$$

$$= \frac{39805.22 \times 10^{-4} \angle -90°}{37.5 - j5155.016}$$

$$= \frac{39805.22 \times 10^{-4} \angle -90°}{5155.152 \angle -89.58°} = 77214.44 \angle -0.42° \text{ V}$$

$\therefore$ Power output $(P_o) = 25 \times 0.9$ MW $= 22.5$ MW

$\therefore$ Total transmission losses $= I_L^2 \times R \times 3$

$$= (115.674)^2 \times 37.5 \times 3$$

$$= 1.5053 \text{ MW}$$

$$\text{Efficiency of the line} = \frac{P_o}{P_o + \text{Total losses}} \times 100\%$$

$$= \left(\frac{22.5}{22.5 + 1.50530}\right) \times 100\%$$

$= 93.73\%$ **Ans.**

$$\% \text{ voltage regulation} = \frac{V_r' - V_r}{V_r} 100\% = \left(\frac{77214.44 - 69282}{69282}\right)$$

$$= 11.45\%$$

Problem 4.3. A 3-phase 50 Hz transmission line has resistance, inductance and capacitance per phase of 10 ohm, 0.1H and 0.9 μF respectively and delivers a load of 35 MW at 332 kV and 0.8 p.f. lag. Determine the efficiency and regulation of the line using (i) nominal-T, (ii) nominal-π and

Solution: Given,

R = 10 Ω/phase, L = 0.1 H/phase, C = 0.9 μF/phase

P_r = 35 MW, V_{Lr} = 132 kV, p.f. ($\cos \phi_r$) = 0.8 (lagging)

$$\therefore \text{ Receiving end current } (I_r) = \frac{P_r}{\sqrt{3}V_{Lr} \times Cos\, \phi_r}$$

$$= \frac{35 \times 10^6}{\sqrt{3} \times 132 \times 10^3 \times 0.8}$$

$$= 191.36 \text{ A}$$

$$\text{Receiving end voltage per phase } (V_r) = \frac{132 \times 1000}{\sqrt{3}} = 76210 \text{ V}$$

(2) Nominal-T-Method:

The circuit diagram for nominal-T method can be drawn as given below.

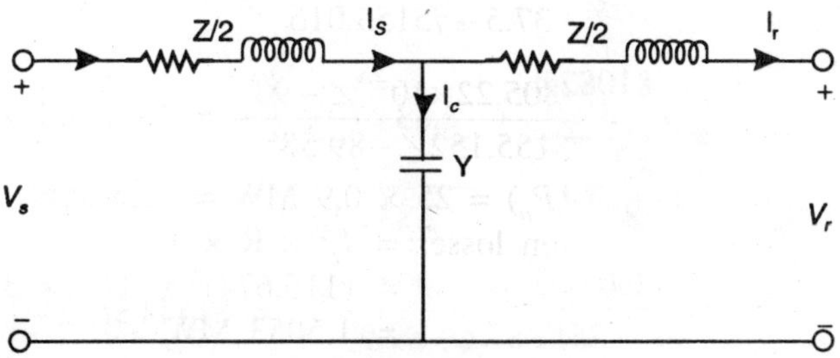

Fig. P. 4.3 (a)

Series impedance of the line (Z)

$= \text{R} + j\omega l = 10 + \text{j } 2 \times \pi \times 50 \times 0.1$

$= 10 + \text{j}31.4 \ \Omega = 32.95 \angle 72.33° \ \Omega$

Shunt admittance of the line (Y) = $j\omega c$

$= \text{j}2\pi \times 50 \times 0.9 \times 10^{-6} = 2.83 \times 10^{-4} \angle 90° \ \mho$

Let us take receiving end voltage as reference phasor.

$\bar{V_r} = 76210 \angle 0° \text{ V}$

$\bar{I_r} = 191.36 \angle -36.9$

Voltage across shunt admittance (V_c) = $\dfrac{I_r \bar{Z}}{2} + \bar{V}_r$

$$= \frac{32.95\angle 72.33°}{2} \times 191.36\ \angle -36.9° + 76210\ \angle 0°$$

$$= 3152.66\ \angle 35.43 + 76210\ \angle 0°$$

$$= 78778.86 + j\ 1827.62 = 78800\ \angle 1.33°\ \text{V}$$

Current through shunt admittance (I_c)

$$= V_c \times Y = 78800\ \angle 1.33° \times 2.83 \times 10^{-4}\ \angle 90°$$

$$= 22.30\ \angle 91.33°\ \text{A}$$

Sending end current (I_s) = $\bar{I}_c + \bar{I}_r$

$$= 22.30\ \angle 91.93 + 191.36\ \angle -36.9°$$

$$= 152.51 - j\ 92.60 = 178.42\ \angle -31.27°\ \text{A}$$

Sending end voltage ($\bar{V}_s$) = $\bar{V}_c + \bar{I}_s \dfrac{\bar{Z}}{2}$

$$= \frac{32.95\angle 72.33°}{2} \times 178.42\ \angle -31.27° + 78800\ \angle 1.33°$$

$$= 2939.47\ \angle 41.06 + 78800\ \angle 1.33°$$

$$= 80995.20 + j\ 3759.80 = 81082.42\ \angle 2.66°\ \text{V}$$

No-load receiving end voltage (V'_r) = $\dfrac{|V_s|\left(-\dfrac{j}{\omega c}\right)}{\dfrac{R}{2} + \dfrac{j\omega L}{2} - \dfrac{j}{\omega c}}$

$$= \frac{81082.42\left(-\dfrac{j}{2.83 \times 10^{-4}}\right)}{\dfrac{10}{2} + j\dfrac{31.4}{2} - \dfrac{j}{2.83 \times 10^{-4}}}$$

$$= \frac{28651.032\angle -90° \times 10^4}{3517.87\angle -89.9°} = 81444.26\ \angle -0.1\ \text{V}$$

$$= 81444.26\ \angle -0.08°\ \text{V}$$

Power output (P_o) = 35 MW

Total transmission losses = $3\ (I_r^2 + I_s^2)\dfrac{R}{2}$

$$= 3\ [(191.36)^2 + (178.42)^2] \times \frac{10}{2}$$

$$= \frac{2.0536}{2}\text{MW} = 1.0468\ \text{MW}$$

$$\text{Efficiency of the line} = \frac{P_o}{P_o + \text{Total losses}} \times 100\%$$

$$= \frac{35}{35 + 1.0468} = 97.10\% \quad \textbf{Ans.}$$

$$\%\ \text{Voltage regulation} = \frac{V_r' - V_r}{V_r} \times 100\%$$

$$= \frac{81444.26 - 76210}{76210} \times 100\% = 6.87\% \quad \textbf{Ans.}$$

(ii) Nominal-π method:

The circuit diagram for nominal π method can be drawn as below

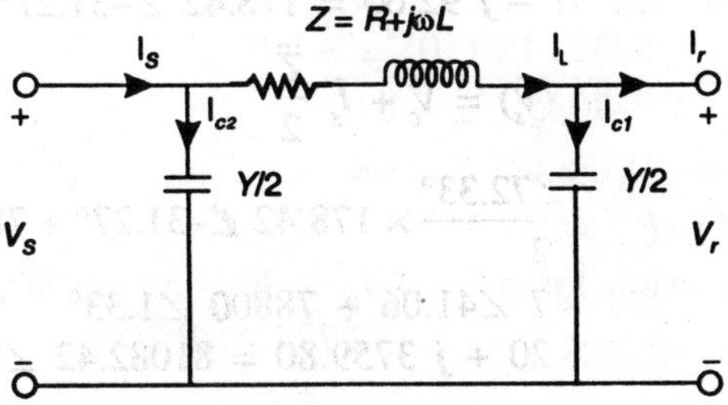

Fig. P. 4.3 (b)

Let us take V_r as reference phasor

$$\bar{V}_r = 76210 \angle 0° \text{ V}$$

$$\bar{I}_r = 191.36 \angle -36.9°$$

Voltage across shunt admittance at the receiving end

$$(V_{c1}) = V_r = 76210 \angle 0° \text{ V}$$

Current through the shunt admittance at the receiving end

$$(\bar{I}_{c1}) = \bar{V}_r \times \frac{\text{Y}}{2} = 76210 \angle 0° \times \frac{2.83 \times 10^{-4}}{2} \angle 90°$$

$$= 10.784 \angle 90° \text{ A}$$

Current through series impedance $(\bar{I}_L) = \bar{I}_{c1} + \bar{I}_r$

$$= 191.36 \angle -36.9 + 10.784 \angle 90°$$

$$= 153.03 - j\,104.11 = 185.09 \angle -34.23 \text{ A}$$

Voltage across shunt admittance of the sending end

$$(\bar{V}_{c2}) = \bar{I}_L \bar{Z} \times \bar{V}_{c1}$$

$$= 185.09 \angle -34.23° \times 32.95 \angle 72.33° + 76210 \angle 0°$$

$$= 6098.72 \angle 38.1° + 76210 \angle 0°$$

$$= 81009.3 + j\,3763.13$$

$$= 81096.66 \angle 2.66° \text{ V}$$

Sending end voltage (V_s) = 81096.66 V per phase

No-load receiving end voltage (V_r') = $\dfrac{|V_s|\left(-\dfrac{2j}{\omega c}\right)}{R + j\omega L - \dfrac{2j}{\omega c}}$

$$= \frac{81096.66\left(-\dfrac{2j}{2.83\times10^{-4}}\right)}{10 + j31.4 - \dfrac{2j}{2.83\times10^{-4}}}$$

$$= \frac{57312.12\times10^{-4}\angle -90^\circ}{10 - j7035.74}$$

$$= \frac{57312.12\times10^{-4}\angle -90^\circ}{7035.75\angle -89.92^\circ}$$

$$= 81458.44 \angle -0.08^\circ \text{ V}$$

Power output (P_o) = 35 MW

Total transmission losses = $3\, I_L^2 \times R = 3 \times (185.09)^2 \times 10$

= 1.02775 MW

% Efficiency of the line $= \dfrac{P_o}{P_o + \text{Total losses}} \times 100\%$

$= \dfrac{35}{35 + 1.02775} = 97.15\%$ **Ans.**

% Voltage regulation $= \dfrac{V_r' - V_r}{V_r} \times 100$

$= \dfrac{81458.44 - 76210}{76210} \times 100$

= 6.89% **Ans.**

Problem 4.4. A short 3-phase transmission line has a series line impedance per phase of (20 + j 50) ohm. The line delivers a load of 50 MW at 0.7 p.f. lag. Determine the regulation of the line and the A.B.C.D. parameters of the line. If the same load is delivered at 0.7 p.f. lead, determine the regulation of the line. System voltage 220 kV.

Solution: Given,

Series impedance (Z) of the line = (20 + j 50) Ω/phase

R = 20 Ω/phase, X = 50 Ω/phase, P_r = 50 MW, V_{Lr} = 220 kV

Receiving voltage per phase $(V_r) = \dfrac{220 \times 1000}{\sqrt{3}} = 127000$ V

Receiving end current $(I_r) = \dfrac{P_r}{\sqrt{3} V_{Lr} \times \cos\phi_r}$

$= \dfrac{50 \times 10^6}{\sqrt{3} \times 220 \times 10^3 \times 0.7}$

$= 187.45$ A

For short transmission line $I_s = I_r$ ....(i)

The circuit diagram of short-transmission line per phase can be drawn as below

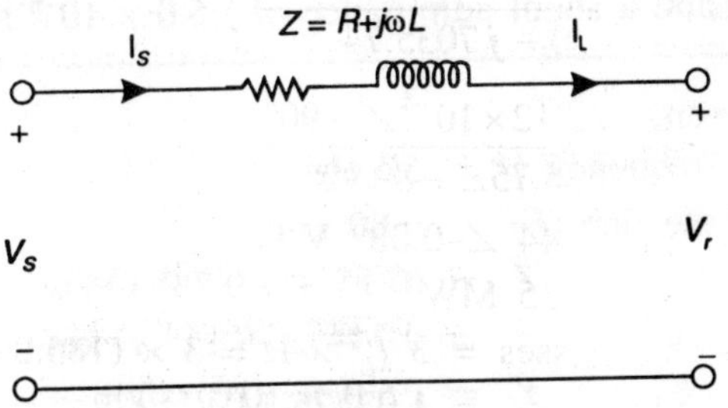

Fig. P. 4.4

Sending end voltage $(V_s) = V_r + I_r R \cos\phi_r + I_r X \sin\phi_r$

$\bar{V}_s = \bar{V}_r + \bar{I}_r \bar{Z}$ ......(ii)

We know that % regulation of transmission line

$$= \left(\frac{I_r R \cos\phi_r}{V_r} + \frac{I_r X \sin\phi_r}{V_r}\right) 100$$

$$= \left(\frac{187.48 \times 20 \times 0.7}{127000} + \frac{187.45 \times 50 \times 0.714}{127000}\right) 100$$

$= 7.34\%$ **Ans.**

For transmission line

$V_s = AV_r + BI_r$ ....(iii)

$I_s = CV_r + DI_r$ ....(iv)

On comparing equation (i) and (ii) with (iv) and (iii) respectively

A = 1.0 = 49.7 Ω

B = Z = 20 + *j* 50 = 53.85 ∠68.20°

C = 0.0 and D = 1.0

For leading p.f. % voltage regulation

$$= \left(\frac{I_r R \cos\phi_r}{V_r} - \frac{I_r X \sin\phi_r}{V_r}\right) \times 100\%$$

$$= \frac{1}{V_r}(I_r R \cos\phi_r - I_r \sin\phi_r X) \times 100\%$$

$$= \frac{I_r}{V_r}(R \cos\phi_r - X \sin\phi_r) \times 100\%$$

$$= \frac{187.45}{127000}(20 \times 0.7 - 50 \times 0.714) \times 100\%$$

$$= -3.2\% \quad \textbf{Ans.}$$

Problem 4.5. Find the A, B, C, D parameters of a 3-phase, 80 km, 50 Hz transmission line with series impedance of $(0.15 + j\,0.78)$ ohm per km and a shunt admittance of $j\,5.0 \times 10^{-6}$ mho per km.

Solution: Given,

Operating frequency $(f) = 50$ Hz

Length of the line $(L) = 80$ km

$$Z = (0.15 + j\,0.78)\ \Omega/\text{km}$$
$$= 0.744\ \angle 79.11°\ \Omega/\text{km}$$
$$Y = j\,5.0 \times 10^{-6}\ \mho/\text{km}$$
$$= 5.0 \times 10^{-6}\ \angle 90°\ \mho/\text{km}$$

For long transmission line the sending end voltage for any length l is given by

$$V_s = V_r \cos h\gamma l + I_r Z_c \sin h\gamma l \quad \text{........(1)}$$

The sending end current for any length l is given by

$$I_s = V_r \frac{\text{Sin}\,h\,\gamma l}{Z_C} + I_r \cos h\gamma l \quad \text{........(2)}$$

Now, characteristic impedance of the line $(Z_c) = \sqrt{\frac{Z}{Y}}$

$$= \sqrt{\frac{0.794\angle 79.11°}{5.0 \times 10^{-6}\angle 90°}} = \sqrt{1588\angle 79.11° - 90°}$$

$$= 39.85\ \angle -5.45°$$

Propagation constant $(\gamma) = \sqrt{Yz}$

$$= \sqrt{5 \times 10^{-6}\angle 90° \times 0.794\angle 79.11°}$$
$$= 1.99\ \angle 84.56° \times 10^{-3}$$

Now, $\gamma l = 1.99 \times 80 \times 10^{-3}\ \angle 84.56$

$$= 0.51 + j\,0.1585$$

The sending end voltage and current in terms of A, B, C and D are given by

$$V_s = AV_r + BI_r \qquad \text{.................(3)}$$

and $$I_s = CV_r + DI_r \qquad \text{.................(4)}$$

Comparing equations (1) and (2) with equations (3) and (4) respectively, we get

$$A = \text{Cos } h\gamma l = \cos h\ (0.0151 + j\ 0.1585)$$
$$= \text{Cos } h\ 0.0151 \cos 0.1585 + j \sin h\ 0.0151 \sin 0.1585$$
$$= 0.9876 + j\ 0.0024 = 0.9876\ \angle 0.14° \quad \textbf{Ans.}$$

$$B = Z_c \sin h\gamma l = Z_c \sin h\ (0.0151 + j\ 0.1585)$$
$$= Z_c (\sin h\ 0.015 \cos 0.1585 + j \cos h\ 0.015\ l \sin 0.1585)$$
$$= Z_c (0.0144 + j\ 0.15786)$$
$$= 398.5\ \angle{-5.45} \times 0.1586\ \angle 84.61°$$
$$= 63.20\ \angle 79.16° \quad \textbf{Ans.}$$

$$C = \frac{\text{Sin } h\gamma l}{Z_C} = \frac{0.1586\angle 84.61°}{398.5\angle -5.45°}$$
$$= 3.98 \times 10^{-4}\ \angle 90.06° \quad \textbf{Ans.}$$

$$D = \cos h\gamma l = 0.9876\ \angle 0.14° \quad \textbf{Ans.}$$

Problem 4.6. Determine both *T* and π equivalents for the line of problem 4.5. Also determine the propagation consistant and the surge impedance of the line.

Solution: We have

$Z_c = 39.85\ \angle{-5.45}$, $Y = 1.99\ \angle 84.56 \times 10^{-3}$

Sin $h\gamma l = 0.1586\ \angle 84.61°$ and cos $h\gamma l = 0.9876\ \angle 0.14$

The equivalent T-diagram for this problem can be respresented as below

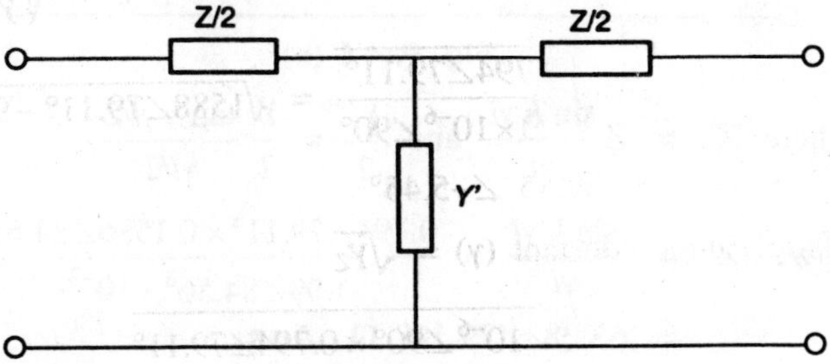

Fig. P. 4.6 (a)

Where $$\frac{Z'}{2} = \frac{Z}{2}\ \frac{\tan h\,\gamma l/2}{\gamma l/2} \text{ and } \gamma' = \gamma\ \frac{\sin h\,\gamma l}{\gamma l}$$

$$\text{Now, } \tan h\gamma l/2 = \frac{\sin h\,\gamma l/2}{\cos h\,\gamma l/2} = \frac{\sin h\,(0.0151 + j\,0.1585)/2}{\cos h\,(0.0151 + j\,0.1585)/2}$$

$$= \frac{\sin h\,(0.0076 + j\,0.0793)}{\cos h\,(0.0076 + j\,0.0793)}$$

$$= \frac{\sin h\,0.0076\,\cos 0.076 + j\,\cos h\,0.0076\,\sin 0.0793}{\cos h\,0.0076\,\cos 0.0793 + j\,\sin h\,0.0076\,\sin h\,0.0793}$$

$$= \frac{0.0076 + j\,0.0792}{0.997 + j\,0.0006} = \frac{0.0796\angle 84.52°}{0.997\angle .03°}$$

$$= 0.0798\ \angle 84.49°$$

$$\therefore\ \frac{Z'}{2} = 2Z\,\frac{\tan h\,\gamma l/2}{\gamma l} = \frac{2\times 0.794\angle 79.11°\times 0.0798\angle 84.49°}{2\times 1.99\times 10^{-3}\angle 84.56°}$$

$$= \frac{63.68}{2}\angle 79.4° = 31.84\ \angle 79.4°\quad \textbf{Ans.}$$

$$\text{and}\quad Y' = Y\,\frac{\sin h\,\gamma l}{\gamma l} = \frac{5.0\times 10^{-6}\angle 90°\times .1586\angle 84.61°}{1.99\angle 84.56°\times 10^{-3}}$$

$$= 3.98\times 10^{-4}\ \angle 90.5°\ \mho\quad \textbf{Ans.}$$

The equivalent π-network of this problem can be represented by the figure given below

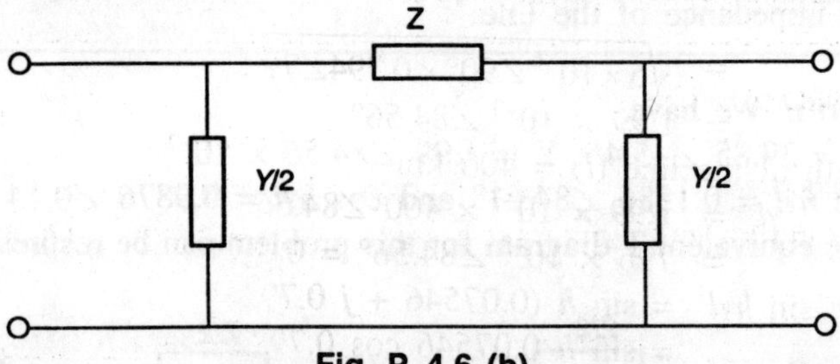

Fig. P. 4.6 (b)

$$\text{Where } Z' = Z\frac{\sin h\,\gamma l}{\gamma l}\ \text{ and }\ \frac{Y'}{2} = \frac{Y}{2}\,\frac{\tan h\,\gamma l/2}{\gamma l/2}$$

$$\text{Now, } Z' = \frac{\sin h\,\gamma l}{\gamma l} = \frac{0.794\angle 79.11°\times 0.1586\angle 84.61°}{1.99\angle 84.56°\times 10^{-3}}$$

$$= 63.28\ \angle 79.16°\ \Omega\quad \textbf{Ans.}$$

$$\frac{Y'}{2} = \frac{Y}{2}\,\frac{\tan h\,\gamma l/2}{\gamma l/2} = \frac{5.0\times 10^{-6}\angle 90°\times 0.0.0798\angle 84.44°}{1.99\angle 84.56°\times 10^{-3}}$$

$$= 2.01\times 10^{-4}\ \angle 89°.93°\ \mho\quad \textbf{Ans.}$$

Now, we have propagation constant (γ) = $1.99 \angle 84.50° \times 10^{-3}$
$= (1.890 + j\ 19.810) \times 10^{-4}$

To calculate surge impedance of the line, apply the relation.

$$\text{Surge impedance} = \sqrt{\frac{L}{C}} = \sqrt{\frac{\frac{X_L}{2\pi}}{\frac{Y}{2\pi}}} = \sqrt{\frac{X_L}{Y}} = \sqrt{\frac{0.78}{5 \times 10^{-6}}}$$

$= 394.97\ \Omega$ **Ans.**

> **Problem 4.7.** Determine A, B, C, D parameters of the line 400 km long having per unit impedance and admittance as in problem 4.5 assuming (i) the line could be represented by nominal-T or nominal-π and (ii) the exact representation

Solution: Given,

$Z = (0.15 + 0.78)\ \Omega/\text{km} = 0.794 \angle 79.11°\ \Omega/\text{km}$

$Y = j\ 5.0 \times 10^{-6}\ \mho/\text{km} = 5.0 \times 10^{-6} \angle 90°\ \mho/\text{km}$

$$\text{Characteristic impedance } (Z_C) = \sqrt{\frac{z}{Y}} = \sqrt{\frac{0.794 \angle 79.11°}{5.0 \times 10^{-6} \angle 90°}}\ \Omega$$

$= 398.5 \angle 5.45°\ \Omega$

$$\text{Propagation constant } (\gamma) = \sqrt{Yz}$$

$$= \sqrt{5 \times 10^{-6} \angle 90° \times 0.794 \angle 79.11°}$$

$= 1.99 \times 10^{-3} \angle 84.56°$

Length of the line (l) = 400 km

$\gamma l = 1.99 \times 10^{-3} \times 400 \angle 84.56°$

$= 796 \times 10^{-3} \angle 84.56° = 0.07546 + j\ 0.7924$

Now, $\sin h\gamma l = \sin h\ (0.07546 + j\ 0.7924)$

$= \sin h\ 0.07546 \cos 0.7924 + \cos h\ 0.07546 \sin 0.7924$

$= 0.0530 + j\ 0.7141 = 0.7161 \angle 85.76°$

$\text{Cos } h\gamma l = \cos h\ (0.07546 + j\ 0.7924)$

$= \cos h\ 0.07546 \cos 0.7924 + j \sin h\ 0.07546 \sin 0.7924$

$= 0.7041 + j\ 0.0538 = 0.7062 \angle 4.37°$

$A = \cos h\gamma l = 0.7062 \angle 4.37°$

$B = Z_c \sin h\gamma l = 398.5 \angle -5.45 \times 0.7161 \angle 85.76°$

$= 285.37 \angle 80.31°$

$$C = \frac{\sin h\ \gamma l}{Z_C} = \frac{0.7161 \angle 85.76°}{398.5 \angle -5.45°} = 17.97 \times 10^{-4} \angle 91.21°\ \mho$$

$D = A = \cos h\gamma l = 0.7062 \angle 4.37°$

Series Impedance of 400 km long line (Z) = Zl

$= 0.794 \times 400 \angle 79.11° \ \Omega = 317.6 \angle 79.11° \ \Omega$

Shunt admittance of 400 km long line (γ) = Yl

$= 5 \times 10^{-6} \angle 90° \times 400 \ ℧ = 2.0 \times 10^{-3} \angle 90° \ ℧$

(i) Nominal-T representation:

The nominal-T diagram can be represented as given below

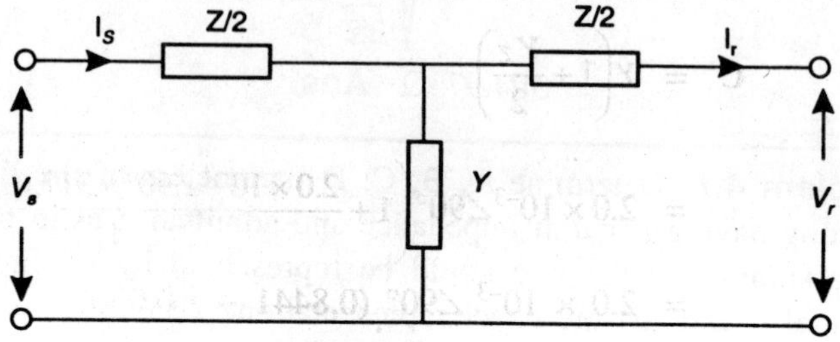

Fig. P. 4.7 (a)

For nominal-T representation

$$A = 1 + \sqrt{\frac{Yz}{2}} = 1 + \frac{2.0 \times 10^{-3} \angle 90° \times 317.6 \angle 79.11°}{2}$$

$$= 0.6881 + j\,0.060 = 0.69071 \angle 4.98 \quad \textbf{Ans.}$$

$$B = Z\left(1 + \frac{Yz}{4}\right)$$

$$= 317.6 \angle 79.11 \left(1 + \frac{2.0 \times 10^{-3} \angle 90° \times 317.6 \angle 79.11°}{4}\right)$$

$$= 317.6 \angle 79.11\ (0.8441 + j\,0.030)$$

$$= 317.6 \angle 79.11 \times 0.8446 \angle 2.04°$$

$$= 268.24 \angle 81.15° \ \Omega \quad \textbf{Ans.}$$

$$C = Y = 20 \times 10^{-4} \angle 90° \ ℧ \quad \textbf{Ans.}$$

$$D = 1 + \frac{Yz}{2} = 0.69071 \angle 4.98° \quad \textbf{Ans.}$$

(ii) Nominal-representation:

The nominal-π circuit can be represented as given below

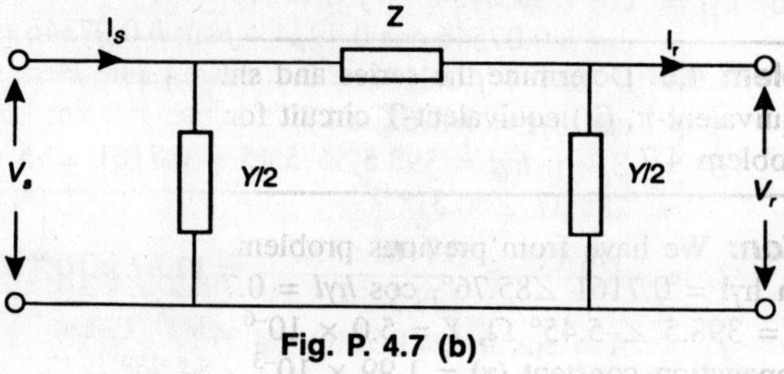

Fig. P. 4.7 (b)

For nominal-π circuit,

$$A = 1 + \frac{Yz}{2} = 1 + \frac{317.6\angle 79.11^\circ \times 2.0 \times 10^{-3}\angle 90^\circ}{2}$$

$$= 0.6881 + j\,0.060 = 0.69071\ \angle 4.98^\circ\quad \textbf{Ans.}$$

$$B = Z = 317.6\ \angle 79.11\ \Omega\quad \textbf{Ans.}$$

$$C = Y\left(1 + \frac{Yz}{2}\right)$$

$$= 2.0 \times 10^{-3}\angle 90^\circ\left(1 + \frac{2.0 \times 10^{-3}\angle 90^\circ \times 317.6\angle 79.11^\circ}{4}\right)$$

$$= 2.0 \times 10^{-3}\ \angle 90^\circ\ (0.8441 + j\,0.030)$$

$$D = 1 + \frac{Yz}{2}$$

$$= 2.0 \times 10^{-3}\ \angle 90^\circ \times 0.8446\ \angle 2.04$$

$$= 0.1689 \times 10^{-4}\ \angle 92.04\ \mho$$

$$\cong 0.1689 \times 10^{-4}\ \angle 90^\circ\ \mho\quad \textbf{Ans.}$$

$$= 1 + \frac{317.6\angle 79.11^\circ \times 2.0 \times 10^{-3}\angle 90^\circ}{2}$$

$$= 0.69071\ \angle 4.98^\circ\quad \textbf{Ans.}$$

(iii) Exact solution method

By exact solution method

$$A = \cos h\gamma l = 0.7062\ \angle 4.37^\circ\quad \textbf{Ans.}$$

(As we have calculated earlier.)

$$B = Z_c \sin h\gamma l = 398.5\ \angle{-5.45^\circ} \times 0.7161\ \angle 85.76^\circ$$

$$= 285.37\ \angle 80.31^\circ$$

$$C = \frac{\sin h\gamma l}{Z_C} = \frac{0.7161\angle 85.76^\circ}{398.5\angle -5.45^\circ}$$

$$= 17.97 \times 10^{-4}\ \angle 90\ \mho\quad \textbf{Ans.}$$

$$D = \cos h\gamma l = 0.7062\ \angle 4.37^\circ\quad \textbf{Ans.}$$

Problem 4.8. Determine the series and shunt parameters for the (i) equivalent-π, (ii) equivalent-T circuit for the 400 km long line of problem 4.7.

Solution: We have from previous problem.

Sin hγl = 0.7161 ∠85.76°, cos $h\gamma l$ = 0.7062 ∠4.37°

Z_c = 398.5 ∠–5.45° Ω, $Y = 5.0 \times 10^{-6}$ ∠90° ℧/km

propagation constant (γ) = 1.99×10^{-3} ∠84.56°

(i) Equivalent-π circuit:

The equivalent–circuit can be represented as shown in figure given below.

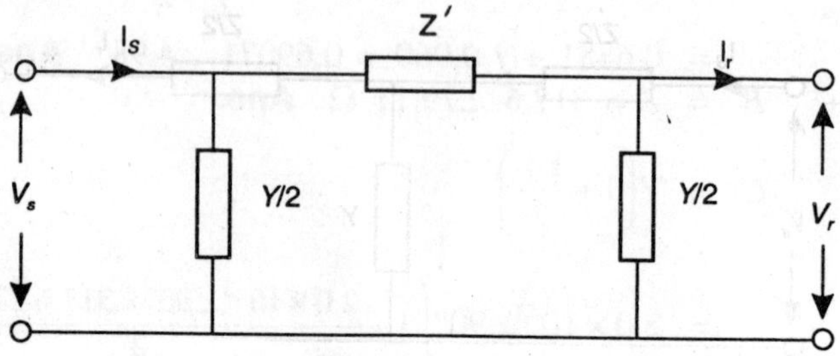

Fig. P. 4.8 (a)

$$\text{where } Z' = Z\frac{\sin h\,\gamma l}{\gamma l} \text{ and } \frac{Y'}{2} = \frac{Y}{2}\frac{\tan h\,\gamma l/2}{\gamma l/2}$$

$$\text{Now, series impedance } (Z') = Zl\frac{\sin h\,\gamma l}{\gamma l}$$

$$= \frac{0.794\angle 79.11° \times 0.7161\angle 85.76°}{1.99\times 10^{-3}\angle 84.56°}$$

$$= 285.72\ \angle 80.31°\ \Omega \quad \textbf{Ans.}$$

Shunt admittance

$$\frac{Y'}{2} = \frac{Y}{2}\frac{\tan h\,\gamma l/2}{\gamma l/2} = Y\times l\ \frac{\tan h\,\gamma l/2}{\gamma l}$$

$$= \frac{Y}{\gamma}\tan h\ \gamma l/2 = \frac{Y}{\gamma}\frac{\cos h\,\gamma l - 1}{\sin h\,\gamma l}$$

$$= \frac{5.0\times 10^{-6}\angle 90°}{1.99\times 10^{-3}\angle 84.56°}\times\frac{0.7062\ \angle 4.37° - 1}{0.7161\ \angle 85.76°}$$

$$= 3.5088\ \angle{-80.32}\times(-0.2956 + j\ 0.0538)$$

$$= 3.5088\ \angle{-80.32}\times 0.3005\ \angle 169.68°$$

$$= 1.0544\ \angle 89.36°\times 10^{-3}\ \mho$$

$$= 10.544\times 10^{-4}\ \angle 90°\ \mho \quad \textbf{Ans.}$$

(ii) Equivelent-T circuit:

This can be represented as shown in figure given below

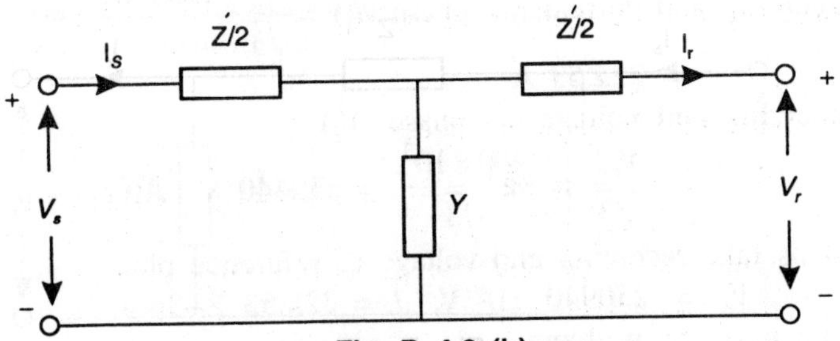

Fig. P. 4.8 (b)

where $\frac{Z'}{2} = \frac{Z}{2} \frac{\tan h\,\gamma l/2}{\gamma l/2}$ and $Y' = Y \frac{\sin h\,\gamma l}{\gamma l}$

Now, series impedance

$$\frac{Z'}{2} = \frac{Z}{2} \frac{\tan h\,\gamma l/2}{\gamma l/2} = Z \times \frac{\tan h\,\gamma l/2}{\gamma l}$$

$$= \frac{Z}{\gamma} \frac{\cos h\,\gamma l - 1}{\sin h\,\gamma l}$$

$$= \frac{0.794\,\angle 79.11°}{1.99 \times 10^{-3}\ \angle 84.56°} \times \frac{(-0.2956 + j\,0.0538)}{0.7161 \angle 85.76°}$$

$$= \frac{0.794\,\angle 79.11° \times 0.3005\,\angle 169.68°}{1.99 \times 10^{-3} \angle 84.56° \times 0.7161 \angle 84.76°}$$

$$= 167.432\ \angle 78.47°\ \Omega \quad \textbf{Ans.}$$

Shunt admittance (Y')

$$= \frac{Y}{2} \frac{\sin h\,\gamma l}{\gamma l/2} = \frac{Y}{\gamma}\ \sin\,h\gamma l$$

$$= \frac{5.0 \times 10^{-6} \angle 90°}{1.99 \times 10^{-3} \angle 84.56°}\ 0.7161\ \angle 85.76°$$

$$= 1.799 \times 10^{-3}\ \angle 91.2°\ ℧ \quad \textbf{Ans.}$$

$$= 17.99 \times 10^{-4}\ \angle 90°\ ℧ \quad \textbf{Ans.}$$

Problem 4.9. Determine the efficiency and regulation of the line of problem 4.7 when it delivers a load of 125 MW at 0.8 p.f. lag and 400 kV.

Solution: Given

P_r = 125 MW, cos ϕ_r = 0.8 p.f. (lagging) V_{Lr} = 400 kV

Receiving end current per phase (I_r) $= \dfrac{125 \times 10^6}{\sqrt{3} \times 400 \times 10^3 \times 0.8}$

= 225.53 A

Receiving end voltage per phase (V_r)

$= \dfrac{V_{Lr}}{\sqrt{3}} = \dfrac{400 \times 10^3}{\sqrt{3}}$ = 230440 V **Ans.**

Let us take receiving end voltage as reference phasor.

$\bar{V}_r$ = 230940 ∠0° V, $\bar{I}_r$ = 225.53 ∠–36.87°

From previous problem

A = 0.7062 ∠4.37°, B = 285.37 ∠80.31° Ω

C = 17.97 × 10⁻⁴ ∠90° ℧, D = 0.7062 ∠4.37°

Now, The sending end voltage is given by

$\bar{V}_s = A\bar{V}_r + B\bar{I}_r$

$\bar{V}_s$ = 0.7062 ∠4.37° × 230940 ∠0° + 285.37 ∠80.31° × 225.53 ∠–36.87°

= 163089.83 ∠4.37° + 64359.5 ∠43.44°

= 209346.792 + j 56680.193

= 216884.09 ∠15.15° V per phase

The receiving end current is given by

$\bar{I}_s = C\bar{V}_r + D\bar{I}_r$

= 17.97 × 10⁻⁴ ∠90° × 230940 ∠0° + 0.7062 ∠4.37° × 225.53 ∠–36.87°

= 415 ∠90° + 159.27 ∠–32.50

= 134 + j 329.42 = 355.631 ∠67.86° A

Regulation of the line

$$= \frac{\dfrac{V_S}{|A|} - V_r}{V_r} 100 = \left(\frac{\dfrac{216884.09}{0.7062} - 230940}{230940} \right) \times 100\ \%$$

= 32.98% **Ans.**

Phase difference between sending voltage and sending end current

= 67.86° –15.15° = 52.71°

Now, Power input to the line = 3 × V_s × I_s cos ϕ_s

= 3 × 216884.09 × 355.631 × cos 52.71° = 140 MW

Efficiency of the line

$$= \frac{\text{Power output}}{\text{Power input}} 100\% = \frac{125}{140} \times 100\%$$

= 89.29% **Ans.**

Problem 4.10. An 18-km 60 Hz single circuit three-phase line is composed of Partridge conductors equilaterally spaced with 1.6m. between centers. The line delivers 2500 kW at 11 kV to a balanced load. What must be the sending-end voltage when the power factor is (a) 80% lagging, (b) unity, and (c) 90% leading? Assume a wire temperature of 50°C.

Solution: From Table A1' for patridge conductor,

Resistance of each conductor at 50°C = 0.3792 Ω/mile

Now, mutual GMD of the given circuit is given by

$$D_m = \sqrt[3]{1.6 \times 1.6 \times 1.6} = 1.6 \text{ m} = \frac{1.6}{0.3048} \text{ ft}$$

$$= 5.25 \text{ ft}$$

∴ Inductive reactance of the line per phase

$$= 0.465 + 0.2794 \log 5.25 \; \Omega/\text{mile}$$

$$= 0.6662 \; \Omega/\text{mile}$$

Capacitive reactance of the line per phase

$$= 0.06831 \log 5.25 + 0.1074 \times 10^6$$

$$= 0.1566 \times 10^6 \; \Omega/\text{mile}$$

$$= \frac{0.1566 \times 10^6}{1.609} \; \Omega/\text{km} = 0.0973 \times 10^6 \; \Omega/\text{km}$$

∴ Series impedance of the line (Z)

$$= 0.3792 + j\,0.6662 \; \Omega/\text{mile}$$

$$= \frac{0.3792 + j\,0.6612}{1.609} \; \Omega/\text{km} = 0.4764 \angle 60.35 \; \Omega/\text{km}$$

Shunt admittance of the line

$$Y = j\frac{1}{X_C} = \frac{j}{0.0973 \times 10^6} \; \mho/\text{km} = j\,10.28 \times 10^{-6} \; \mho/\text{km}$$

$$= 10.28 \times 10^{-6} \angle 90° \; \Omega/\text{km}$$

Characteristic impedance $(Z_c) = \sqrt{\frac{Z}{Y}}$

$$= \sqrt{\frac{0.4764\angle 60.35°}{10.28 \times 10^{-6} \angle 90°}} = 215.27 \angle -14.83° \; \Omega$$

Propagation constant $(\gamma) = \sqrt{Yz}$

$$= \sqrt{0.4764\angle 60.35° \times 10.28 \times 10^{-6} \angle 90°}$$

$$= 2.213 \times 10^{-3} \angle 75.18°$$

$$\therefore \quad \gamma l = 2.213 \times 10^{-3} \times 18 \angle 75.18°$$

$$= 0.0398 \angle 75.18° = 0.0102 + j\,0.0385$$

Now,

$$\text{Sin } h\gamma l = \sin h\,(0.01.2 + j\,0.0385)$$
$$= \sin h\,0.0102 \cos 0.0385 + j \cos h\,0.0102 \sin 0.0385$$
$$= 0.0102 + j\,0.0385 = 0.0398 \angle 75.18°$$
$$\text{Cos h}\gamma\text{l} = \cos h\,(0.0102 + j\,0.0385)$$
$$= \cos h\,0.0102 \cos 0.0385 + j \sin h\,0.0102 \sin 0.0385$$
$$= 0.9993 + j\,0.0487 = 1.0004 \angle 2.79°$$

(a) For 80% lagging p.f.

Receiving end current per phase (I_r)

$$= \frac{25000 \times 10^6}{\sqrt{3} \times 11 \times 10^3 \times 0.8} = 164.02 \text{ A}$$

Receiving end voltage per phase (V_r)

$$= \frac{11 \times 1000}{\sqrt{3}} = 6351 \text{ V}$$

Let us take receiving end voltage as reference phasor.

$$\bar{V}_r = 6351 \angle 0° \text{ V}, \bar{I}_r = 164.02 \angle -36.87° \text{ A}$$

Now, Sending end voltage (V_s) = $V_r \cos h\gamma l + I_r \times Z_c \sin h\gamma l$

$$= 6351 \angle 0° \times 1.0004 \angle 2.79° + 164.02 \angle -36.87°$$
$$\times 215.274 \angle -14.83° \times 0.0398 \angle 75.18°$$
$$= 6353.54 \angle 2.79° + 1405.28 \angle 23.48$$
$$= 7632.79 + j\,869.166 = 7682.12 \angle 6.5 \text{ V}$$

Sending end line voltage (V_{Ls}) = $\sqrt{3} \times V_s$

$$= \sqrt{3} \times 7682.12 \text{ V} = 13.31 \text{ kV}$$ **Ans.**

(b) For unity power factor, $\cos \phi_r = 1$,

$$\text{Receiving end current } (I_r) = \frac{2500 \times 10^3}{\sqrt{3} \times 11 \times 10^3 \times 1} \text{ A} = 131.22 \text{ A}$$

Let us take receiving end voltage as reference phasor,

$$\bar{V}_r = 6351 \angle 0° \text{ V to neutral}$$
$$\bar{I}_r = 131.22 \angle 0° \text{ A per phase}$$

Now, Sending end voltage ($\bar{V}_s$) = V_r cos hγl + I_r Z_c sin hγl

$$= 6351 \angle 0° \times 1.0004 \angle 2.79° + 131.22 \angle 0° \times 215.27$$
$$\angle -14.83° \times 0.0398 \angle 75.18°$$
$$= 6353.54 \angle 2.79° + 1405.28 \angle 60.35°$$
$$= 7041.20 + j\,1530.54 = 7205.63 \angle 12.26° \text{ V}$$

Sending end line voltage = $\sqrt{3} \times 7205.63 = 12.48$ kV **Ans.**

Problem 4.11. A 100-mile, three-phase transmission line delivers 55 MVA at 0.8 power factor lagging to the load at 132 kV. The line is composed of Drake conductors with flat horizontal spacing of 11.9 ft between adjacent conductors. Determine the sending-end voltage, current, and power. Assume a wire temperature of 50°C.

Solution: Given,

Length of the line (l) = 100 mile

Receiving end p.f. ($\cos \phi_r$) = 0.8 (lagging)

Distance between the conductors = 11.9 ft

Mutual GMD of these conductors (D_m)

$$= \sqrt[3]{11.9 \times 11.9 \times 23.8} = 14.99 \text{ ft}$$

For drake conductors. From Table A.1,

Resistance of each conductor at 50° C = 0.1284 Ω/mi

Inductive reactance of the line per phase (X_L)

$$= 0.399 + 0.2974 \log D_m$$

$$= 0.399 + 0.2974 \log 14.99 = 0.7487 \ \Omega/\text{mi}$$

Capacitive reactance of the line per phase (X_C)

$$= 0.0912 \times 10^6 + 0.06831 \log 14.99 \times 10^6$$

$$= 0.1715 \times 10^6 \ \Omega/\text{mi}$$

Series impedance of the line per mile (Z)

$$= 0.1284 + j\ 0.7487 \ \Omega/\text{mi} = 0.7596 \angle 80.27° \ \Omega/\text{mi}$$

For 90% leading power factor

$$\cos \phi_r = 0.9,\ \phi_r = \cos^{-1} 0.9 = 25.84°$$

Receiving end current per phase

$$= \frac{2500 \times 10^3}{\sqrt{3} \times 11 \times 10^3 \times 0.9} \text{ A} = 145.80 \text{ A}$$

Let us take receiving end voltage as reference phasor

$$\bar{V}_r = 6351 \angle 0° \text{ V}$$

$\bar{I}_r$ = 145.80 ∠25.84°. Here +ve sign is for leading p.f.

Now, Sending end voltage ($\bar{V}_s$) = $\bar{V}_r \cos h\gamma l + \bar{I}_r \bar{Z}_C \sin h\gamma l$

$$= 6351 \angle 0° \times 1.0004 \angle 2.79° + 145.80 \angle 25.84 \times 215.27 \angle -14.83° \times 0.0398 \angle 75.18$$

$$= 6353.54 \angle 2.79° + 1249.18 \angle 86.19°$$

$$= 6429.01 + j\ 1555.68 = 6614.55 \angle 13.60 \text{ V}$$

Sending end line voltage = 6614.55 × $\sqrt{3}$ V = 11.5 kV **Ans.**

Shunt admittance of the line per phase (Y) = $\dfrac{j}{X_C}$

$$= \frac{j}{0.1715 \times 10^6} \ \mho/\text{mi} = 5.831 \times 10^{-6} \angle 90° \ \mho/\text{mi}$$

Characteristic impedance (Z_c)

$$= \sqrt{\frac{Z}{Y}} = \sqrt{\frac{0.7596\angle 80.27^\circ}{5.831\times 10^{-6}\angle 90^\circ}}\ \Omega$$

$$= 360.93\ \angle 4.87^\circ\ \Omega$$

$$\gamma l = \sqrt{YZ} = \sqrt{0.7596\angle 80.27^\circ \times 5.831\times 10^{-6}\angle 90^\circ \times 100}$$

$$= 0.2105\ \angle 85.14^\circ = 0.0178 + j\ 0.2097$$

Now,

$$\begin{aligned}
\text{Sin } h\gamma l &= \text{Sin } h\ (0.0178 + j\ 0.2097)\\
&= \text{Sin } h\, 0.0178 \cos 0.2097 + j \cos h\, 0.0178 \sin 0.2097\\
&= 0.0178 \times 0.9781 + j\ 1.0002 \times 0.2082\\
&= 0.2090\ \angle 85.22^\circ\\
\text{Cos } h\gamma l &= \cos h\ (0.017 + j\ 0.2097)\\
&= \cos h\, 0.0178 \cos 0.2097 + j \sin h\, 0.0178 \sin 0.2097\\
&= 1.0002 \times 0.9781 + j\ 0.0178 \times 0.2082\\
&= 0.9783\ \angle 0.22^\circ
\end{aligned}$$

Receiving end voltage to neutral ($\bar{V}_r$) $= \dfrac{132\times 10^3}{\sqrt{3}} = 76210\text{V}$

Receiving end current per phase ($\bar{I}_r$) $= \dfrac{55\times 10^6}{\sqrt{3}\times 132\times 10^3} = 240.56$ A

Let us take receiving end voltage as reference phasor.

$$\therefore \quad \bar{V}_r = 76210\ \angle 0^\circ\text{V}$$

$$\bar{I}_r = 240.56\ \angle{-\cos^{-1}}\ (0.8) = 240.56\ \angle{-36.87^\circ}\ \text{A}$$

Sending end voltage of the transmission line is given by

$$\begin{aligned}
\bar{V}_s &= \bar{V}_r \cos h\gamma l + \bar{I}_r Z_c \sin h\gamma l\\
&= 76210\ \angle 0^\circ \times 0.9783\ \angle 0.22^\circ + 240.50\ \angle{-36.87^\circ}\\
&\quad \times 360.93\ \angle 4.87^\circ \times 0.2090\ \angle 85.22^\circ\\
&= 74556\ \angle 0.22^\circ + 18146.49\ \angle 53.22\\
&= 85420.553 + j\ 14820.53 = 86696.71\ \angle 9.84\ \text{V}
\end{aligned}$$

Sending end line voltage = 86696.71 × $\sqrt{3}$ V = 150.161 kV **Ans.**

The sending end current of the transmission line is given by

$$\begin{aligned}
\bar{I}_s &= \frac{\sin h\,\gamma l}{Z_C} \times \bar{V}_r + \cos h\gamma l\ \bar{I}_r\\
&= \frac{0.2090\angle 85.22^\circ}{360.93\angle 4.87^\circ} \times 76210\ \angle 0^\circ + 0.9783\ \angle 0.22^\circ\\
&\quad \times 240.56\ \angle{-36.87^\circ}\\
&= 44.13\ \angle 80.35^\circ + 235.34\ \angle{-36.65^\circ}\\
&= 196.21 - j\ 96.975 = 218.87\ \angle{-26.3^\circ}\ \text{A}
\end{aligned}$$

AT sending end, line current = 218.87 A

Power factor at sending end ($\cos \phi_s$)

$$= \cos (26.3 + 9.84) = 0.8076$$

$$\text{Power input} = 3 \times V_s I_s \cos \phi_s$$

$$= 3 \times 86696.71 \times 218.87 \times 0.8076$$

$$= 46 \text{ MW} \quad \textbf{Ans.}$$

Problem 4.12. Find the ABCD constants of a circuit having a 600-Ω resistor for the shunt branch at the sending end, a 1-kΩ resistor for the shunt branch at the receiving end, and an 80-Ω resistor for the series branch.

Solution: Given,

Series resistance (R) = 80 Ω

Shunt branch at receiving end = 1 kΩ

Shunt conductance at receiving end (g_r)

$$= \frac{1}{1 \times 10^3} \mho = 1 \times 10^{-3} \mho$$

Shunt branch at sending end = 600 Ω

Shunt conductance at sending end (g_s)

$$= \frac{1}{6 \times 10^2} \mho = 0.167 \times 10^{-2} \mho$$

Let, V_s = sending end voltage

V_r = receiving end voltage

I_s = sending end current

I_r = receiving end current

The π-circuit of the given problem can be represented as shown in figure given below.

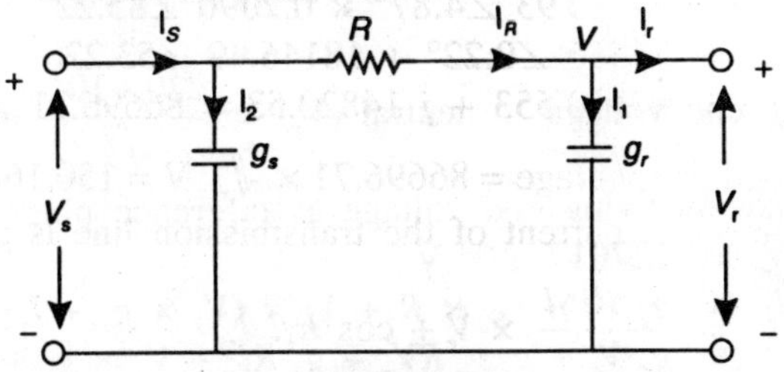

Fig. P. 4.12

Current through conductance at receiving and (I_1)

$$= V_r \times g_r$$

Current through series resistance (I_R) = $I_1 + I_r$

$$= V_r \times g_r + I_r$$

Voltage across the conductance at sending end (V_s)

$$= I_R \times R + V_r$$
$$= V_r + RI_R$$

Current through conductance at sending end (I_s)

$$= I_R = g_s (V_r + RI_R)$$
$$I_s = g_s V_r + (g_s R + 1) I_R$$
$$D = g_s R + 1 = 0.167 \times 10^{-2} \times 80 + 1$$
$$= 1.134 \quad \textbf{Ans.}$$

Problem 4.13. The ABCD constants of a three-phase transmission line are

$$A = D = 0.936 + j\,0.016 = 0.936 \angle 0.98°$$
$$B = 33.5 + j\,138 = 142.0 \angle 76.4°\ \Omega$$
$$C = (-5.18 + j\,914) \times 10^{-6}\ \mho$$

The load at the receiving end is 50 MW at 220kV with a power factor of 0.9 lagging. Find the magnitude of the sending-end voltage and the voltage regulation. Assume the magnitude of the sending-end voltage remains constant.

Solution: Given,

$$A = D = 0.936 + j\,0.016 = 0.936 \angle 0.98°$$
$$B = 33.5 + j\,138 = 142.0 \angle 76.4°\ \Omega$$
$$C = -5.18 + j\,914 \times 10^{-6}\ \mho$$
$$= 914.015 \times 10^{-6} \angle 90.32\ \mho$$

Power at load (P_r) = 50 MW

Receiving end line voltage = 220 kV

p.f. angle at receiving end = $\cos^{-1}(0.9)$ = 25.84°

$$\text{Receiving end current } (I_r) = \frac{50 \times 10^6}{\sqrt{3} \times 220 \times 10^3 \times 0.9} = 145.80 \text{ A}$$

$$\text{Receiving end voltage to neutral } (V_r) = \frac{220 \times 10^3}{\sqrt{3}} = 127017 \text{ V}$$

Let us take receiving end voltage as reference phasor

$$\bar{V}_r = 127017 \angle 0° \text{ V}$$
$$I_s = g_s Vr + (g_s \times R + 1) \times (V_r \times g_r + I_r)$$
$$= g_s V_r + g_s g_r RV_r + g_s RI_r + V_r \times g_r + I_r$$
$$I_s = (g_s + g_s g_r R + g_r) V_r + (g_s R + 1) I_r \quad \text{.........(1)}$$

Now, Sending end voltage

$$(V_s) = V_r + RI_R,\ V_s = V_r + R(V_r g_r + I_r)$$
$$V_s = V_r + Rg_r V_r + RI_r$$
$$V_s = (Rg_r + 1) V_r + RI_r \quad \text{.................(2)}$$

The sending end voltage and current in term of A, B, C and D contants are given by

$$V_s = AV_r + BI_r \quad \text{......(3)}$$

and $$I_s = CV_r + DI_r \quad \text{......(4)}$$

Now, comparing the equations (1) and (2) with equations (4) and (3) respectively, we get

$$A = Rg_r + 1 = 80 \times 1 \times 10^{-3} + 1 = 1.08 \quad \textbf{Ans.}$$

$$B = R = 80\ \Omega \quad \textbf{Ans.}$$

$$C = g_s + g_r + g_s g_v R$$
$$= 0.167 \times 10^{-2} + 1 \times 10^{-3} + 0.167 \times 10^{-2} \times 1 \times 10^{-3} \times 80$$
$$= 2.804 \times 10^{-3}\ \mho \quad \textbf{Ans.}$$

> **Problem 4.14.** A three-phase 60-Hz transmission line is 250 mile long. The voltage at the sending end is 220 kV. The parameters of the line are R –0.2 Ω/mi, X = 0.8 Ω/mi, and Y = 5.3 ℧/mi. Find the sending end current when there is no load on the line.

Solution: Given,

Length of the line (l) = 250 mi, R = 0.2 Ω/mi, X = 0.8 Ω/mi

$Y = 5.3 \times 10^{-6}$ ℧/mi

Series impedance of the line (Z)

$$= (0.2 + j\ 0.8)\ \Omega/\text{mi} = 0.8246\ \angle 75.96\ \Omega/\text{mi}$$

Shunt admittance of the line (Y)

$$= j\ 5.3 \times 10^{-6}\ \mho/\text{mi} = 5.3 \times 10^{-6}\ \angle 90°\ \mho/\text{mi}$$

$$\text{Propagation constant } (\gamma) = \sqrt{YZ}$$
$$= \sqrt{5.3 \times 10^{-6} \angle 90° \times 0.8246 \angle 75.96°}$$
$$= 2.091 \times 10^{-3}\ \angle 82.98°$$
$$\gamma l = 2.091 \times 10^{-3}\ \angle 82.98 \times 250 = 0.5228\ \angle 82.98$$
$$= 0.06339 + j\ 0.5189$$
$$\bar{I}_r = 145.80\ \angle -25.84°\ \text{A}$$

Now, The sending end voltage of a transmission line in terms of transmission constant is given by

$$V_s = AV_r + BI_r$$
$$= 0.936\ \angle 0.098° \times 127017\ \angle 0° + 142.0 \angle 76.4 \times 145.80\ \angle -25.84°$$
$$= 118887.912\ \angle 0.98° + 20703.6\ \angle 50.56°$$
$$= 132022.89 + j\ 18022.56 = 133247.35\ \angle 7.77°\ \text{V}$$

At sending end, Line voltage

$$(V_L) = 133247.35 \times \sqrt{3}\ \text{V} = 230.8\ \text{kV} \quad \textbf{Ans.}$$

At No-load, $I_r = 0$

No-load receiving end voltage (V_r^1)

$$= \frac{V_S}{A} = \frac{133247.35}{0.936} = 142358.28 \text{ V}$$

% Voltage regulation

$$= \frac{V_r' - V_r}{V_r} 100\% = \frac{\frac{V_S}{|A|} - V_r}{V_r} 100\%$$

$$= \frac{142358.28 - 133247.35}{133247.35} 100\% = 6.84\% \quad \textbf{Ans.}$$

Now, cos $h\gamma l$

$$= \cos h\,(0.0639 + j\,0.5189)$$
$$= \cos h\,0.0639 \cos 0.5189 + j \sin h\,0.0639 \sin 0.5189$$
$$= 0.8701 + j\,0.0317 = 0.8707\ \angle 2.09°$$

The sending end voltage of a long transmission line is given by

$$V_s = V_r \cos h\gamma l + I_r Z_c \sin h\gamma l$$

At No-load, $I_r = 0$

$$V_r = \frac{V_S}{\text{Cos } h\gamma l} = \frac{220}{0.8707} \text{ kV} = 252.67 \text{ kV (line to line)}$$

Receiving end voltage to neutral

$$= \frac{252.67}{\sqrt{3}} \text{ kV} = 145879 \text{ V}$$

Characteristic impedance (Z_c) $= \sqrt{\frac{Z}{Y}}$

$$= \sqrt{\frac{0.8246\angle 75.96°}{5.3 \times 10^{-6} \angle 90°}} = 394.44\ \angle -7.02°\ \Omega$$

$$\text{Sin } h\gamma l = \text{Sin } h\,(0.0639 + j\,0.5189)$$
$$= \text{Sin } h\,0.0639 \cos 0.5189 + j \cos h\,0.0639 \sin 0.5189$$
$$= 0.0555 + j\,0.497 = 0.5001\ \angle 83.63°$$

The sending end current in terms of transmission constant is given by

$$I_s = \frac{\sin h\,\gamma l}{Z_C} V_r + \cos h\gamma l\ I_r$$

At No-load $I_r = 0$

$$I_s = \frac{\sin h\,\gamma l}{Z_C} \times V_r = \frac{0.5001\angle 83.63^\circ}{394.44\angle -7.02^\circ} \times 145879$$

$$= 184.92\ \angle 90.65^\circ \text{ A}$$

Sending end line current (I_L) = 184.92 A **Ans.**

Problem 4.15. A three-phase transmission line is 300 mi long and serves load of 400 MVA. 0.8 lagging-power factor at 345 kV. The ABCD constants of the line are

$$A = D = 0.8180\ \angle 1.3^\circ$$
$$B = 172.2\ \angle 84.2^\circ\ \Omega$$
$$C = 0.001933\ \angle 90.4^\circ\ \mho$$

(a) Determine the sending-end line-to-neutral voltage, the sending-end current and the percent voltage drop at full load.

(b) Determine the receiving-end line-to-neutral at no load, the sending-end current at no load, and the voltage regulation.

Solution: Given,

Length of the line (l) = 300 mi

$$A = D = 0.8180\ \angle 1.3^\circ$$
$$B = 172.2\ \angle 84.2^\circ\ \Omega$$
$$C = 0.0019333\ \angle 90.4^\circ\ \mho$$

(a) At receiving end line voltage = 345 kV

Line voltage to neutral

$$= \frac{345}{\sqrt{3}} \times 1000 \text{ V} = 199185.84 \text{ V}$$

Receiving end current per phase

$$= \frac{400 \times 10^6}{\sqrt{3} \times 345 \times 10^3} = 669.39 \text{ A}$$

p.f. angle = $\cos^{-1} 0.8 = 36.87^\circ$

Let us take receiving end voltage as reference phasor

$$\bar{V}_r = 1999185.84\ \angle 0^\circ \text{ V}$$
$$\bar{I}_r = 669.39\ \angle -36.87^\circ \text{ A}$$

The sending end voltage line to neutral is given by

$$V_s = AV_r + BI_r$$
$$= 0.8180\ \angle 1.3^\circ \times 199185.84\ \angle 0^\circ + 172.2\ \angle 84.2^\circ \times 669.39\ \angle -36.87^\circ$$
$$= 162934.02\ \angle 1.3^\circ + 115268.96\ \angle 47.33^\circ$$
$$= 241018.476 + j\,88450.298 = 256735.98\ \angle 20.15 \text{ V}$$

Sending end line to neutral voltage = 256.74 kV **Ans.**

The sending end current of this transmission line is given by

$$I_s = CV_r + DI_r$$
$$= 0.001933\angle 90.4° \times 199185.84\angle 0° + 0.8180\angle 1.3° \times 669.39\angle -36.87°$$
$$= 385.03\angle 90.4° + 547.50\angle -35.57°$$
$$= 442.7 + j\,66.51 = 447.67\angle 8.54°$$

Sending end current (I_s) = 447.67 A **Ans.**

$$\%\text{ Voltage drop at full load} = \frac{V_{nL} - V_{FL}}{V_{FL}} \times 100\%$$

$$= \frac{256735.98 - 199185.84}{199185.84} \times 100\%$$

$$= 28.89\% \quad \textbf{Ans.}$$

Problem 4.16. If the load on the line described in Prob. 4.14 is 80 MW at 220 kV, with unity power factor, calculate the current, voltage, and power at the sending end. Assume that the sending-end voltage is held constant, and calculate the voltage regulation of the line for the load specified above.

Solution: Given,

Length of the line (l) = 250 mi, R = 0.2 Ω/mi, X = 0.8 Ω/mi

$\gamma = 5.3 \times 10^{-6}$ ℧/mi

load power (P_r) = 80 MW

Receiving end line voltage (V_{Lr}) = 220 kV, $\cos\phi_r = 1.00$, $\phi_r = 0°$

Series impedance of the line (Z)

$$= (0.2 + j\,0.8)\ \Omega/\text{mi} = 0.8246\angle 75.96°\ \Omega/\text{mi}$$

Shunt admittance of the line (Y)

$$= j\,5.3 \times 10^{-6}\ \text{℧/mi} = 5.3 \times 10^{-6}\angle 90°\ \text{℧/mi}$$

Characteristic impedance (Z_c)

$$= \sqrt{\frac{Z}{Y}} = \sqrt{\frac{0.8246\angle 75.96°}{5.3\times 10^{-6}\angle 90°}} = 394.44\angle -7.02°\ \Omega$$

(b) For transmission line $V_s = AV_r + BI_r$ and $I_s = CV_r + DI_r$

At No-load

No-load receiving end voltage (V_r)

$$= \frac{V_S}{A} = \frac{256.74}{0.8180}\ \text{kV} = 313.86\ \text{kV}$$

No-load receiving end line to neutral voltage = 313.86 kV **Ans.**

Again, At No-load, $I_r = 0$

Sending end current at No-load $(I_{SN}) = CV_r$

$$= 0.001933 \angle 90.4° \times 256735.98 \angle 20.15°$$
$$= 496.27 \angle 110.55° \text{ A} = 496.27 \text{ A} \quad \textbf{Ans.}$$

$$\text{Voltage regulation} = \frac{\frac{V_S}{|A|} - V_r}{V_r} \times 100\%$$

$$= \left(\frac{\frac{256735.98}{0.8180} - 199185.84}{199185.84} \right) \times 100 \%$$

Propagation constant (γ)

$$= \sqrt{Yz} = \sqrt{5.3 \times 10^{-6} \angle 90° \times 0.8246 \angle 75.96°}$$
$$= 2.091 \times 10^{-3} \angle 82.98°$$
$$\gamma l = 2.091 \times 10^{-3} \angle 82.98° \times 250$$
$$= 0.5228 \angle 82.98° = 0.0639 + j\, 0.5189°$$
$$\cos h\gamma l = \cos h\,(0.0639 + j\, 0.5189)$$
$$= \cos h\, 0.0639 \cos 0.5189 + j \sin h\, 0.0639 \sin 0.5189$$
$$= 0.8701 + j\, 0.0317 = 0.8707 \angle 2.09°$$
$$\sin h\gamma l = \sin\,(0.0639 + j\, 0.5189)$$
$$= \sin h\,(0.0639) \cos 0.5189 + j \cos h\, 0.0639 \sin 0.5189$$
$$= 0.0555 + j\, 0.497 = 0.5001 \angle 83.63°$$

Receiving end current (I_r)

$$= \frac{80 \times 10^6}{\sqrt{3} \times 220 \times 10^3 \times 1} = 209.95\text{A}$$

Receiving end line to neutral voltage (V_r)

$$= \frac{220 \times 1000}{\sqrt{3}} = 127017 \text{ V}$$

Let us take receiving end voltage as reference phasor.

$$\bar{V}_r = 127017 \angle 0° \text{ V}, \bar{I}_r = 209.95 \angle 0° \text{ A}$$

At sending end,

$$I_s = \frac{\sin h\,\gamma l}{Z_C} V_r + I_r \cos h\gamma l$$

$$= \frac{127017 \angle 0°}{394.44 \angle -7.02°} \times 0.5001 \angle 83.63° + 0.8707 \angle 2.09° \times 209.95 \angle 0°$$

$= 161.04 \angle 90.67° + 182.80 \angle 2.09°$

$= 180.8 + j\,167.70 = 246.601 \angle 42.85°$

Sending end current (I_s) = 246.601 A

Sending end line to neutral voltage (V_s)

$= V_r \cos h\gamma l + Z_c \sin h\gamma l\, I_r$

$= 127017 \angle 0° \times 0.8707 \angle 2.09° + 394.44 \angle -7.02° \times 0.5001 \angle 83.63° \times 209.95 \angle 0°$

$= 110593.7 \angle 2.09° + 41414.62 \angle 66.61°$

$= 126961.23 + j\,42044.60 = 133741.92 \angle 18.32°$ V

Sending end line voltage

$= 133741.92 \times \sqrt{3}$ V = 231.65 kV **Ans.**

Sending end p.f. = cos (42.85°–18.32°) = 0.9097

Power input $= 3\, V_s I_s \cos \phi_r$

$= 3 \times 231.65 \times 10^3 \times 246.601 \times 0.9097$ W

= 155.9 MW **Ans.**

Now, Voltage regulation of the line

$$= \frac{\dfrac{V_S}{|A|} - V_r}{V_r} \times 100\%$$

$$= \frac{\dfrac{V_S}{\cos h\gamma l} - V_r}{V_r} \times 100\% = \frac{\dfrac{231.65}{0.8707} - 220}{220} \times 100\%$$

= 20.93% **Ans.**

Problem 4.17. A series capacitor bank is to be installed at the midpoint of the 300 mile line of Prob. 4.15. The ABCD constants for 150 mile of line are

A = D = 0.9534 ∠0.3°, B = 90.33 ∠84.1° Ω

C = 0.001014 ∠90.1° ℧

The *ABCD* constants of the series capacitor bank are

A = D = 1 ∠0°, *B* = 146.6 ∠–90° Ω, *C* = 0

(a) Determine the equivalent ABCD constants of the series combination of the line-capacitor-line.

(b) Rework Prob. 4.15 using these equivalent *ABCD* constants.

Solution: For 150 mile transmission line

A = D = 0.9534 ∠0.3°, B = 90.33 ∠84.1°

C = 0.001014 ∠90.1° ℧

For series capacitor bank A = D = 1 ∠0°, B = 146.6 ∠–90° Ω

C = 0

(a) To calculate equivalent ABCD constants of the series combination of the line-capacitor-line, apply the relation

$$A_{eq} = A_1 A_2 + B_1 C_2$$
$$B_{eq} = A_1 B_2 + B_1 D_2$$
$$C_{eq} = A_2 C_1 + C_2 D_1$$
$$D = B_2 C_1 + D_1 D_2$$
$$A_{eq} = 0.9534 \angle 0.3° \times 1 \angle 0° + 90.33 \angle 84.1° \times 0$$
$$= 0.9534 \angle 0.3 \quad \textbf{Ans.}$$
$$B_{eq} = 0.9534 \angle 0.3° \times 146.6 \angle -90° + 90.33 \angle 84.1° \times 1 \angle 0°$$
$$= 139.77 \angle -89.7° + 90.33 \angle 84.1° = 10.02 - j\,49.92$$
$$= 50.92 \angle -78.65° \ \Omega \quad \textbf{Ans.}$$
$$C_{eq} = 1 \angle 0° \times 0.001014 \angle 90.1° + 0 \times 0.9534 \angle 0.3$$
$$= 0.001014 \angle 90.1° = 10.14 \times 10^{-4} \angle 90.1° \ \mho \quad \textbf{Ans.}$$
$$D_{eq} = 146.6 \angle -90° \times 10.14 \times 10^{-4} \angle 90.1° + 0.9534 \angle 0.3° \times 1 \angle 0°$$
$$= 0.1487 \angle 0.1° + 0.9534 \angle 0.3°$$
$$= 1.1021 + j\,0.0053 = 1.1021 \angle 0.3° \quad \textbf{Ans.}$$

(b) Given, Load volt-ampere = 400 MVA

Receiving end line voltage = 345 kV

$$\text{Receiving end current } (I_r) = \frac{400 \times 10^6}{\sqrt{3} \times 345 \times 10^3} = 669.39\text{A}$$

Load p.f angle = $\text{Cos}^{-1}\ 0.8 = 36.87°$

Receiving end line to neutral voltage

$$= \frac{345 \times 1000}{\sqrt{3}} = 199185.84 \text{ V}$$

Let us take receiving end voltage as reference phasor.

$$\bar{V}_r = 199185.84 \angle 0° \text{ V}, \ \bar{I}_r = 669.39 \angle -36.87° \text{ A}$$

The sending end line to neutral voltage is given by

$$V_s = A_{eq} \bar{V}_r + B_{eq} \bar{I}_r$$
$$= 0.9534 \angle 0.3° \times 199185.84 \angle 0° + 50.92 \angle -78.65° \times 669.39 \angle -36.87$$
$$= 189903.78 \angle 0.3 + 34085.34 \angle -115.52$$
$$= 175215.54 - j\,29765.47 = 177725.82 \angle -9.64° \text{ V}$$

At sending end, Line to neutral voltage = 177.73 kV **Ans.**

The sending end current of this line is given by

$$I_s = C_{eq} V_r + D_{eq} I_r$$
$$= 10.14 \times 10^{-4} \angle 90.1° \times 199185.84 \angle 0° + 1.1021 \angle 0.3° \times 669.39 \angle -36.87°$$
$$= 201.97 \angle 90.1° + 737.73 \angle -36.84°$$
$$= 590.06 + j\,939.07 = 1109.06 \angle 57.86°$$

At sending end, Line current (I_s) = 1109.06 A **Ans.**

Voltage drop at full load

$$= \frac{V_S - V_r}{V_r} 100\ \% = \frac{177.73 - 199.19}{199.19} 100\%$$

$$= -10.77\% \quad \textbf{Ans.}$$

At No-load $I_r = 0$

$$\text{No-load receiving end voltage } (V_r') = \frac{V_S}{A_{eq}} = \frac{177.73}{0.9534}$$

$$= 186.42 \text{ kV (line to neutral)} \quad \textbf{Ans.}$$

No-load sending end current.

$$I_{sn} = CV_r = 10.14 \times 10^{-4} \angle 90.1° \times 199185.84 \angle 0°$$

$$= 201.97 \angle 90.1° \text{ A} = 201.97 \text{ A}$$

% Voltage regulation

$$= \frac{V_r' - V_r}{V_r} \times 100\% = \frac{\dfrac{V_S}{A_{eq}} - V_r}{V_r} \times 100\%$$

$$= \frac{186.24 - 199.185}{199.185} \times 100\% = -6.5\% \quad \textbf{Ans.}$$

Problem 4.18. The admittance of a 300-mile transmission line is

$$y_c = 0 + j\ 6.87 \times 10^{-6}\ ℧/\text{mi}$$

Determine the ABCD constants of a shunt reactor that will compensate for 60% of the total shunt admittance.

Solution: Given,

Length of the line (l) = 300-mi

Shunt admittance of the line (Y_c) = 0 + j 6.87 × 10^{-6} ℧/mi

For 300-mile long line shunt admittance (B_c)

$$= j\ 6.87 \times 10^{-6} \times 300 = 0.002061 \angle 90°\ ℧$$

For 60% compensation, shunt admittance of the compensated line (B_L)

$$= -j\ 0.6 \times 0.002061 = 0.001237 \angle -90°\ ℧$$

From the table, the ABCD constants of the shunt reactor is given by

$$\text{A} = \text{D} = 1, \text{B} = 0 \text{ and C} = -j\ 0.001237\ \Omega \quad \textbf{Ans.}$$

Solution: Given,

Length of the line (l) = 175 mi

Total series impedance of the line (Z)

$$= 35 + j\,140\ \Omega = 144.31\ \angle 75.96^\circ\ \Omega$$

Total series impedance of the line (γ) = $930 \times 10^{-6}\ \angle 90^\circ\ \Omega$

Characteristic impedance of the line (Z_C)

$$= \sqrt{\frac{\frac{Z}{l}}{\frac{\gamma}{l}}} = \sqrt{\frac{Z}{\gamma}} = \sqrt{\frac{144.31\angle 75.96^\circ}{930\times 10^{-6}\angle 90^\circ}}$$

$$= 393.92\ \angle -7.02^\circ\ \Omega$$

Propagation constant (γ)

$$= \sqrt{yz} = \sqrt{\frac{\gamma z}{l2}} = \sqrt{\frac{\gamma z}{l}}$$

$$\therefore \quad \gamma l = \sqrt{\gamma z} = \sqrt{930\times 10^{-6}\angle 90^\circ \times 144.31\angle 75.96^\circ}$$

$$= 0.3603\ \angle 82.98^\circ = 0.0448 + j\,0.3636$$

Now,

$$\text{Cos}\ h\gamma l = \cos h\,(0.0448 + j\,0.3636)$$

$$= \cos h\ 0.0448 \times \cos 0.3636 + j \sin h\ 0.0448\ \sin 0.3636$$

$$= 0.9356 + j\,0.0159 = 0.9357\ \angle 0.97^\circ$$

$$\text{Sin}\ h\gamma l = \sin h\,(0.0448 + j\,0.3636)$$

$$= \sin h\,0.0448 \cos 0.3636 + j \cos \text{h}\ 0.0448 \sin 0.3636$$

$$= 0.04189 + j\,0.3560 = 0.3585\ \angle 83.29^\circ$$

Again, Given,

Load power (P_r) = 40 MW

Receiving and line voltage = 220 kV

Load p.f. = 90% = 0.9 (lagging)

Receiving end p.f. angle (ϕ_r) = $\cos^{-1} 0.9 = 25.84^\circ$

Receiving end current (I_r)

$$= \frac{40\times 10^6}{\sqrt{3}\times 220\times 10^3 \times 0.9} = 116.64\ \text{A}$$

Receiving end line to neutral voltage (V_r)

$$= \frac{220\times 10^3}{\sqrt{3}} = 127.017\ \text{V}$$

Let us receiving end voltage as reference phasor.

$$\therefore \quad V_r = 127017\ \angle 0^\circ\ \text{V},\ I_r = 116.64\ \angle -25.84^\circ\ \text{A}$$

(a) For short line approximation. Let us assume shunt admittance as vegligible

Sending end voltage $(V_s) = V_r + I_r Z$

$= 127017 \angle 0° + 116.64 \angle -25.84 \times 144.31 \angle 75.96°$

$= 12701 \angle 0° + 16832.32 \angle 50.12°$

$= 137809.58 + j\ 12916.94 = 138413.61 \angle 5.35°$ V

At sending end, line voltage (V_L)

$= 138413.61 \times \sqrt{3} = 239.74$ kV **Ans.**

At sending end, line voltage (V_L)

$= 130159.50° \times \sqrt{3}$ V $= 225.44$ kV **Ans.**

> **Problem 4.19.** Use per-unit values on a base of 230 kV, 100 MVA to find voltage, current, power and power factor at the sending end of a transmission line delivering a load of 60 MW at 230 kV with 0.8 power-factor lagging. The three-phase line is arranged in flat, horizontal spacing with 15 ft between adjacent ostrich conductors. Line length 175 mile. Assume a wire temperature of 50° C. Note that base admittance must be the reciprocal of base impedance.

Solution: Given,

Base volt-ampere = 100 MVA

Base voltage (*V*) = 230 kV

Base impedance (Z_b)

$$= \frac{(\text{Base voltage})^2}{\text{Base voltage}-\text{ampere}} = \frac{230 \times 230}{100} = 529\ \Omega$$

$$\text{Base admittance } (Y_b) = \frac{1}{Z_b} = \frac{1}{529}\ \mho = 0.00189\ \mho$$

For ostrich conductors, from Table A.1,

Resistance of each conductor at 50° = 0.3372 Ω/mi

Mutual GMD of this three-phase circuit

$$D_m = \sqrt[3]{15 \times 15 \times 30} = 18.90 \text{ ft}$$

For the nominal-π approximation,

$$A = 1 + \frac{YZ}{2} = 1 + \frac{930 \times 10^{-6} \angle 90° \times 144.31 \angle 75.96°}{2}$$

$= 1 + 0.0671 \angle 165.96.2$

$= 0.9344 + j\ 0.0163 = 0.9350 \angle 1°$

$B = Z = 144.31 \angle 75.96°$

Now, Sending end voltage $(V_s) = Av_r + BI_r$

$= 0.9350 \angle 1° \times 127017 \angle 0° + 144.31 \angle 75.96°$

$\times\ 116.64 \angle -25.48$

$= 118760.9 \angle 1° + 16832.32 \angle 50.12°$

$= 129535.39 + j\ 14989.6 = 130399.8 \angle 6.6°\text{V}$

At sending end, line voltage (V_L)

$= 130399.8 \times \sqrt{3} = 225.86$ kV **Ans.**

(c) For long line equation.

The sending end voltage is given by

$V_s = \cos h\gamma l\ V_r + Z_c \sin h\gamma l\ I_r$

$= 127017 \angle 0° \times 0.9357 \angle 0.97° + 393.92 \angle -7.02°$

$\times 0.3585 \angle 83.29° \times 116.64 \angle -25.84°$

$= 118849.81 \angle 0.97° + 16471.94 \angle 50.43°$

$= 129325.74 + j\ 14709.34 = 130159.56 \angle 6.49°\text{V}$

From Table A.2 Indicative reactance of the line per conductor (X_L)

$= 0.458 + 0.2794 \log D_m$

$= 0.458 + 0.2794 \log 18.9$

$= 0.8146\ \Omega/\text{mi}$

Capacitive reactance of the line per conductor

$(X_c) = 0.1057 + 0.06831 \log D_m$

$= (0.1057 + 0.06831 \log 18.9) \times 10^6\ \Omega$

$= 0.1929 \times 10^6\ \Omega/\text{mi}$

Shunt admittance of the line (y)

$$= \frac{j}{0.1929 \times 10^6}\ \mho/\text{mile} = 5.184 \times 10^{-6} \angle 90°\ \mho/\text{mi}$$

Series impedance of the line (Z)

$= (0.3372 + j\ 0.8146)\ \Omega/\text{mile}$

$= 0.8816 \angle 67.51°\ \Omega/\text{mile}$

Characteristic impedance (Z_C)

$$= \sqrt{\frac{Z}{Y}} = \sqrt{\frac{0.8816 \angle 67.51°}{5.18 \times 10^{-6} \angle 90°}}\ \Omega = 412.39 \angle -11.25°\ \Omega$$

Characteristic impedance in p.u

$$Z_{cp.u} = \frac{Z_{actual}}{Z_b} = \frac{412.39 \angle -11.25°}{529}$$

$= 0.7796 \angle -11.25°$ per unit

Propagation constant (γ)

$$= \sqrt{YZ} = \sqrt{5.184 \times 10^{-6} \angle 90° \times 0.8816 \angle 67.51°}$$

$= 2.13 \times 10^{-3} \angle 78.76°$

$\gamma l = 2.138 \times 10^{-3} \angle 78.76 \times 175$

$= 0.3742 \angle 78.76° = 0.0729 + j\ 0.3670$

Now,

$$\begin{aligned}
\text{Cos } h\gamma l &= \cos h\ (0.0729 + j\ 0.3670) \\
&= \cos h\ 0.0729 \cos 0.3670 + j \sin h\ 0.0729 \sin 0.3670 \\
&= 0.9359 + j\ 0.0262 = 0.9363\ \angle 1.6^\circ \\
\text{Sin } h\gamma l &= \sin h\ (0.0729 + j\ 0.3670) \\
&= \sin h\ 0.0729 \cos 0.3670 + j \cos h\ 0.0729 \sin 0.3670 \\
&= 0.0681 + j\ 0.3598 = 0.3662\ \angle 79.28^\circ
\end{aligned}$$

Base current

$$= \frac{\text{Base voltage} - \text{ampere}}{\text{Base voltage}} = \frac{100 \times 10^6}{230 \times 10^3} = 434.78 \text{ A}$$

Again, Given

Receiving end power (P_r) = 60 MW

p.f. = 0.8 (logging)

p.f. angle = $\cos^{-1} 0.8 = 36.87^\circ$

Receiving end voltage = 230 kV

Receiving end current (I_r)

$$= \frac{60 \times 10^6}{\sqrt{3} \times 230 \times 10^3 \times 0.8} = 188.27 \text{ A}$$

$$\text{Receiving end voltage in p.u.} = \frac{230}{V_b} = \frac{230}{230} = 1.0 \text{ p.u.}$$

$$\text{Receiving end current in p.u} = \frac{188.27}{I_b} = \frac{188.27}{434.78} = 0.4330 \text{ p.u}$$

Let us take receiving end voltage as reference phasor,

$$\bar{V}_r = 1.0\ \angle 0^\circ \text{ p.u}, \ \bar{I}_r = 0.4330\ \angle{-36.87^\circ} \text{ p.u.}$$

The sending end voltage of their line is given by

$$\begin{aligned}
\bar{V}_s &= V_r \cos h\gamma l + Z_{\text{cp.u}} \sin h\gamma l\ I_r \\
&= 1.0\ \angle 0^\circ \times 0.9363\ \angle 1.6^\circ + 0.7796\ \angle{-11.25^\circ} \\
&\quad \times 0.3662\ \angle 79.28^\circ \times 0.4330\ \angle{-36.87^\circ} \\
&= 0.9363\ \angle 1.6^\circ + 0.1236\ \angle 31.16^\circ \\
&= 1.042 + j\ 0.0901 = 1.046\ \angle 4.94^\circ \text{ p.u.}
\end{aligned}$$

At sending end, line voltage = 1.046 × 230 kV = 240.58 kV **Ans.**

Shunt admittance in p.u. ($\gamma_{\text{p.u}}$)

$$= \frac{Y}{Y_b} = \frac{5.184 \times 10^{-6} \times 175\ \angle 90^\circ}{0.00189} = 0.480\ \angle 90^\circ \text{ p.u}$$

At sending end, Sending current (I_s)

$$= \frac{\sin h\,\gamma l}{Z_{Cp.u.}} V_r + \cos h\gamma l\ I_r$$

$$= \frac{0.3662 \angle 79.28° \times 1.0 \angle 0°}{0.7796 \angle -11.25°} + 0.9363 \angle 1.6° \times 0.4330 \angle -36.87°$$

$$= 0.4697 \angle 90.53° + 0.4054 \angle -35.27°$$

$$= 0.3266 + j\, 0.2356 = 0.4027 \angle 35.81°$$

Sending end current

$$= 0.4027 \times I_b = 0.4027 \times 434.78 \text{ A} = 175.09 \text{ A}$$ **Ans.**

Power factor at the sending end

$$= \cos (35.81 - 4.94°) = 0.8583$$ **Ans.**

Power input at sending end

$$= \sqrt{3}\ V_L I_L \cos \phi_s$$

$$= \sqrt{3} \times 175.09 \times 240.58 \times 10^3 \times 0.8583$$

$$= 62.62 \text{ MW}$$ **Ans.**

Chapter 5

High Voltage D.C. Transmission

Problem 5.1. A bridge connected rectifier in fed from 238 kV/110 kV transformer from 238 kV supply. Calculate the direct voltage output when the commutation angle is 20° and delay angle (a) 0°, (b) 30°, (c) 60°.

Solution: Given,

Commutation angle (γ) = 20°, V_L = 110 kV

The d.c. output voltage in given by

$$V_d = \frac{3\sqrt{2} \times V_L}{2 \times \pi} [\cos \alpha + \cos (\alpha + \gamma)]$$

where α is delay angle.

(a) For delay angle (α) = 0°

D.C. out voltage

$$= \frac{3\sqrt{2} \times 110}{2\pi} [\cos 0° + \cos (0° + 20°)]$$

= 144.07 kV **Ans.**

(b) For delay angle (α) = 30°

D.C. output voltage

$$= \frac{3\sqrt{2} \times 110}{2\pi} [\cos 30° + \cos (30° + 20°)]$$

$$= \frac{3\sqrt{2} \times 110}{2\pi} [\cos 30° + \cos 50°]$$

= 112.07 kV **Ans.**

(c) For delay angle (α) = 60°

D. C. output voltage (V_d)

$$= \frac{3\sqrt{2}\times 110}{2\pi}[\cos 60° + \cos (60° + 20°)]$$

$$= \frac{3\sqrt{2}\times 110}{2\pi}[\cos 60° + \cos 80°] = 50.04 \text{ kV} \quad \textbf{Ans.}$$

Problem 5.2. It is required to obtain a direct voltage of 100 kV form a bridge-connected rectifier operating with α = 30° and γ = 15°. Calculate the necessary line secondary voltage of the rectifier transformer which is normally rated at 345 kV/150 kV. Calculate the tap ratio required.

Solution: Given,

V_d = 100 kV,

Delay angle (α) = 30°

Commutation angle = 15°

Rating of rectifier transformer = 345 kV/150 kV = V_p/V_s

We know that for three phase,

$$\text{D.C. output voltage } (V_d) = \frac{3\sqrt{2}\times V_L}{2\times\pi}[\cos\alpha + \cos(\gamma+\alpha)]$$

$$V_L = \frac{2\pi\, V_d}{3\sqrt{2}[\cos\alpha + \cos(\alpha+\gamma)]}$$

$$= \frac{2\times\pi\times 100\times 10^3}{3\sqrt{2}[\cos 30° + \cos(30° + 15°)]}$$

$$= \frac{2\times\pi\times 100\times 1000}{3\sqrt{2}[\cos 30° + \cos 45°]} = 94.14 \text{ kV}$$

$\therefore$ As we know that

$$\text{Top ratio} = \frac{V_L}{V_S} = \frac{94.14}{150} = 0.63 \quad \textbf{Ans.}$$

Problem 5.3. If the rectifier in Problem 5.1 delivers a current of 500 amps, determine the effective reactance per phase.

Solution: Given,

I_d = 500A

Delay angle (α) = 30°

Commutation angle (γ) = 15°

Rating of rectifier transformer (V_p/V_s) = 345 kV/150 kV

Let us calculate first V_L, applying the relation

$$V_d = \frac{3\sqrt{2} \times V_L}{2\times\pi}[\cos\alpha + \cos(\alpha + \gamma)]$$

$$\text{or} \quad 100 \times 10^3 = \frac{3\sqrt{2} \times V_L}{\pi}[\cos 30° + \cos 45°]$$

$$V_L = 94.14 \text{ kV}$$

To calculate reactance (X) per phase, apply the relation,

$$V_d = \frac{3\sqrt{2} \times V_L}{\pi} \cos\alpha - \frac{3XI_d}{\pi}$$

$$\text{or} \quad 100 \times 1000 = \frac{3\sqrt{2} \times 94.14 \times 1000 \times \cos 30°}{\pi} - \frac{3 \times 500 \times X}{\pi}$$

This yields X = 21.155 Ω **Ans.**

> **Problem 5.4.** A d.c. link has a loop resistance of 5 ohm and is connected to transformers giving secondary voltage of 110 kV at each end. The bridge connected converters operate as follows :
>
> Rectifier: α = 15°, Inverter: δ_0 = 10°, X = 10 Ω, γ = 15°
>
> Allow 5° margin on δ_0 for δ
>
> X = 10 Ω
>
> Determine the direct current delivered if the inverter operates on constant.

Solution: Given,

Delay angle (α) = 15°, reactance (X) = 10 Ω

100 persistence (R) = 5 Ω

$V_L = V_P$ = 110 kV, δ_0 = 10°, γ = 15°

Rectifier output voltage in given by

$$V_{or} = \frac{3\sqrt{2} \times V_L}{\pi} \cos\alpha - \frac{3 \times I_d}{\pi}$$

$$= \frac{3\sqrt{2} \times 110 \times 10^3}{\pi} \cos 15° - \frac{3 \times I_d \times 10}{\pi}$$

$$= 143.49 \times 10^3 - \frac{3 \times I_d}{\pi}$$

Inverter d.c. output voltage is given by

$$V_{oi} = \frac{3\sqrt{2}V_L}{\pi} \cos(\delta + \gamma) + \frac{3\,I_d X}{\pi}$$

$$= \frac{3\sqrt{2}\times 110\times 10^3}{\pi} \cos(10 + 15) + \frac{3\times I_d \times 10}{\pi}$$

$$= 134.63 \times 10^3 + \frac{30\,I_d}{\pi}$$

To calculate direct current delivered, apply the relation

$$RI_d = V_{or} - V_{oi}$$

or $$5\,I_d = 143.49 \times 10^3 - \frac{30\,I_d}{\pi} - \left(134.63\times 10^3 + \frac{30\,I_d}{\pi}\right)$$

or $$5\,I_d = 8.86 \times 103 - \frac{60\,I_d}{\pi}$$

or $$I_d\left(5+\frac{60}{\pi}\right) = 8.86 \times 10^3$$

$$I_d = 367.66 \text{ A}$$

Direct curent delivered (I_d) = 367.66 A **Ans.**

Problem 5.5. A bridge connected rectifier in fed from 330 kV/ 220 kV transformer with primary connected to 330 kV. Determine the d.c. output voltage when commutation angle in 10° and the delay angle (a) 15° (b) 20° (c) 30°.

Solution: Given

V_L = 220 kV, Commutation angle (γ) = 10°

The d.c. output voltage of the rectifier is given by

$$V_d = \frac{3\sqrt{2}V_L}{2\times\pi}[\cos\alpha + \cos(\alpha + \gamma)] \text{ V}$$

(a) For delay angle (α) = 15°

D.C. output voltage (V_d)

$$= \frac{3\sqrt{2}\times 220}{2\times\pi}[\cos 15^\circ + \cos(15 + 10^\circ)] \text{ kV}$$

$$= \frac{3\sqrt{2}\times 220}{2\times\pi}[\cos 15^\circ + \cos 25^\circ] = 278.13 \text{ kV}$$

(b) For delay angle (α) = 20°

D.C. output voltage (V_d)

$$= \frac{3\sqrt{2} \times 220}{2 \times \pi} [\cos 20° + \cos (20° + 10°] \text{ kV}$$

$$= \frac{3\sqrt{2} \times 220}{2 \times \pi} [\cos 20° + \cos 30°] = 268.25 \text{ kV}$$ **Ans.**

(c) For delay angle (α) = 30°

$$\text{D.C. output voltage} = \frac{3\sqrt{2}}{2\pi} V_L [\cos 30° + \cos (30° + 10°)]$$

$$= \frac{3\sqrt{2}}{2\pi} \times 220 [\cos 30° + \cos 40°] \text{ kV}$$

$$= 242.45 \text{ kV}$$ **Ans.**

Problem 5.6. A bridge connected rectifier operates with delay angle 30° and commutation angle 15°. Determine the necessary line secondary voltage of the rectifier transformer which in normally rated at 440/220 kV, if it is required to obtain a d.c. output voltage of 200 kV. Also determine the tap ratio required.

Solution: Given,

Delay angle (α) = 30°, Commutation angle (α) = 15°

V_d = 220 kV, V_L = 2

To calculate secondary voltage, apply the relation

$$V_d = \frac{3\sqrt{2}V_L}{2\pi} [\cos \alpha + \cos (\gamma + \alpha)]$$

$$V_L = \frac{V_d \times \pi \times 2}{3\sqrt{2}(\cos 30° + \cos 45°)} = 188.28 \text{ kV}$$

Secondary line voltage (V_L) = 188.28 kV

The tap ratio

$$= \frac{V_L}{V_S} = \frac{188.28}{220} = 0.86$$

Problem 5.7. A bridge connected rectifier operates with (α) = 30° and commutation angle (γ) = 15°. Determine the necessary line secondary voltage of the rectifier transformer which is normally rated at 220/210 kV, if it is required to obtain a d.c. output voltage of 100 kV. Also determine the effective reactance per phase if the rectifier delivers a current of 500 A.

Solution: Given,

Delay angle (α) = 30°

Commutation angle (α) = 15°

V_d = 100 kV, I_d = 500 A

To calculate necessary secondary line voltage, apply the relation.

$$V_d = \frac{3\sqrt{2}V_L}{2\pi}[\cos\alpha + \cos(\alpha+\gamma)]$$

or,
$$V_L = \frac{V_d \times \pi \times 2}{3\sqrt{2}\times[\cos\alpha + \cos(\alpha+\gamma)]}$$

$$= \frac{100\times10^3\times\pi\times2}{3\sqrt{2}[\cos 30° + \cos 45°]} = 94.14 \text{ kV}$$

Necessary secondary line voltage (V_L) = 94.14 kV

To calculate reactance, apply the relation.

$$V_d = \frac{3\sqrt{2}V_L}{2\pi}\cos\alpha - \frac{3\,I_d X}{\pi}$$

$$\text{or } V_d\times\pi = 3\sqrt{2}\,V_L\cos\alpha - 3\,I_d\,X$$

$$X = \frac{3\sqrt{2}V_L\cos\alpha - \pi\times V_d}{3I_d}$$

$$= \frac{3\sqrt{2}\times94.14\times100\cos 30° - \pi\times100\times1000}{3\times500}$$

$= 21.16\ \Omega$ **Ans.**

Problem 5.8. A d.c. link has a loop resistance of 15 Ω and is connected to transformers giving secondary voltage of 140 kV at each end. The bridge connected converter operates as follows

Rectifier: α = 20°, X = 20 Ω, Inverter = δ_0 = 15°, γ = 20°, X = 20 Ω.

Allow 5° Margin on δ_0 for δ.

Calculate the direct current delivered if inverter operates on constant β control.

Solution: Given

$R = 15\ \alpha$, delay angle $(\alpha) = 20°$, reactance $(X) = 20\ \Omega$,

$V_L = V_P = 140$ kV, $\delta_0 = 15°$, $\gamma = 20°$

The rectifier output voltage in given by

$$V_{or} = \frac{3\sqrt{2}V_L}{\pi}.\cos\alpha - \frac{3\,I_d X}{\pi}$$

$$= \frac{3\sqrt{2}\times 140\times 10^3}{\pi}.\cos 20° - \frac{3\times I_d\times 10}{\pi}$$

$$= \frac{558.149\times 10^3}{\pi} - \frac{60\,I_d}{\pi}$$

Inverter d.c. output voltage is given by

$$V_{oi} = \frac{3\sqrt{2}V_L}{\pi}\cos(\delta+\gamma) + \frac{3\,I_d X}{\pi}$$

$$= \frac{3\sqrt{2}\times 140\times 10^3\cos 45°}{\pi} + \frac{3\times 20\times I_d}{\pi}$$

$$= \frac{420\times 10^3}{\pi} + \frac{60\,I_d}{\pi}$$

To calculate direct current delivered, apply the relation

$$R_{I_d} = V_{or} - V_{oi}$$

$$15\,I_d = \frac{558.149\times 10^3}{\pi} - \frac{60\,I_d}{\pi} - \left[\frac{420\times 10^3}{\pi} + \frac{60\,I_d}{\pi}\right]$$

$$15\,I_d = \frac{138.149\times 10^3}{\pi} - \frac{120\,I_d}{\pi}$$

or $$I_d\left[15+\frac{120}{\pi}\right] = \frac{138.149\times 10^3}{\pi},\ I_d = 826.63\text{ A}$$

$\therefore$ d.c. current = 826.03 A **Ans.**

Chapter 6

Corona

Problem 6.1. Determine the corona characteristics of a 3-phase, 50 Hz, 132 kV transmission line 100 km long running through terrain at an altitude of 600 metres, temp. of 30ºC and barometric pressure 74 cm. The conductors are 1.5 cm diameter and spaced with equilateral spacing of 2.75 metres. Assume surface irregularity factor of 0.9 and m_v = 0.75.

Solution: Given

f = 50 Hz, V = 132 kV, length of the line (l) = 100 km

t = 30° C, b = 74 cm.

Diameter of each conductor = 1.5 cm

Radius of each conductor = 0.74 cm, $r = 0.75 \times 10^{-2}$ m

Surface irregularity factor (m) = 0.9, m_v = 0.75, d = 2.75 m

The air density factor (δ) at 30° C and 74 cm pressure is given by

$$\delta = \frac{3.92b}{273+t} = \frac{3.92 \times 74}{30+273} = 0.9574$$

The critical disruptive voltage is given by

$$V_d = 21.1 \text{ m } \delta r \ln \frac{d}{r} \text{ kV to neutral}$$

$$= 21.1 \times 0.9 \times 0.9574 \times 0.75 \ln \frac{2.75 \times 100}{0.75} \text{ kV}$$

to neutral

= 80.51 kV to neutral

Critical disruptive line-to-line voltage

$= 80.51 \times \sqrt{3}$ kV = 139.45 kV **Ans.**

The visual disruptive voltage is given by

$$V_v = 21\ m_v\ \delta r \left(1+\frac{0.3}{\sqrt{r\delta}}\right) \ln \frac{d}{r} \text{ kV to neutral}$$

Now, $\sqrt{r\delta} = \sqrt{0.75 \times 0.9574} = 0.8473$

$$V_v = 21.1 \times 0.75 \times 0.9574 \times 0.75 \left(1+\frac{0.3}{0.8473}\right)$$

$$ln \frac{2.75 \times 100}{0.75} \text{ kV to neutral} = 90.85 \text{ kV to neutal}$$

Visual disruptive line to line voltage

$$= \sqrt{3} \times 90.85 \text{ kV} = 157.36 \text{ kV} \quad \textbf{Ans.}$$

Problem 6.2. A 3-phase, 50 Hz, 132 kV transmission line consists of conductors of 1.17 cm dia and spaced equilaterally at a distance of 3 metres. The line conductors have smooth surface with value for m = 0.96. The barometric pressure is 72 cm of Hg and temperature cf 20° C. Determine the fair and foul weather corona loss per km per phase.

Solution: Given

$f = 50$ Hz, $V_p = 132$ kV, $d = 3$ m

Diameter of each conductor = 1.17 cm

$$\text{Radius of each conductor} = \frac{1.17}{2} \text{ cm} = 0.585 \text{ cm}$$

$$= 0.585 \times 10^{-2} \text{ m}$$

$m = 0.96$, $b = 72$ cm of Hg, $t = 20°$ C

The air density correction factor is given by

$$\delta = \frac{3.92\,b}{273+t} = \frac{3.92 \times 72}{273+20} = 0.9633$$

The critical disruptive nuptive voltage is given by

$$V_d = 21.1\ m\delta r \ln \frac{d}{r} \text{ kV to neutral}$$

$$= 21.1 \times 0.9633 \times 0.96 \times 0.585 \ln \frac{3 \times 100}{0.585} \text{ kV}$$

$$= 71.23 \text{ kV}$$

$$\text{Operating voltage per phase } (V_p) = \frac{132}{\sqrt{3}} = 76.21 \text{ kV}$$

For fair weather condition, corona loss

$$= 241 \times 10^{-5} \left(\frac{f+25}{\delta}\right) \sqrt{\frac{r}{d}} \left(V_p - V_d\right)^2 \text{ kW/km/phase}$$

$$= 241 \times 10^{-5} \left(\frac{50+25}{0.9633}\right) \times \sqrt{\frac{0.585 \times 10^{-2}}{3}} \times$$

$$(76.21 - 71.23)^2 \text{ kW/km/phase}$$

$$= 0.2055 \text{ kW/phase/km}$$

For foul weather condition, critical disruptive voltage

$$= 0.8 \times V_d$$

$$V_d' = 0.8 \times 71.23$$

$$V_d' = 56.98 \text{ kV}$$

In this condition, Corona loss

$$= 241 \times 10^{-5} \left(\frac{f+25}{\delta}\right) \sqrt{\frac{r}{d}} \left(V_p - V_d\right)^2 \text{ kW/km/phase}$$

$$= 241 \times 10^{-5} \left(\frac{50 \times 25}{0.9633}\right) \times \sqrt{\frac{0.585 \times 10^{-2}}{3}}$$

$$(76.21 - 56.98)^2 \text{ kW/km/phase}$$

$$= 3.064 \text{ kW/km/phase} \quad \textbf{Ans.}$$

Problem 6.3. A 3-phase, 50 Hz, 138 kV transmission, line has conductors in equilateral formation spaced 2.5 metres apart. The conductor diameter is 1.04 cm and the surface factor is 0.85. The air pressure and temperature are 74 cm of Hg and 21° C respectively. Determine the critical visual voltage for corona and the corona loss per km per phase of the line, $m_v = 0.72$

Solution: Given,

Operating voltage (V_p) = 138 kV,

Operating frequency (f) = 50 Hz, d = 2.5 m

Diameter of each conductor = 1.04 cm

Radius of each conductor (r) = 0.52 cm = 0.52×10^{-2} m

$m = 0.85$, $b = 74$ cm, $t = 21°$ C, $m_v = 0.72$

The air density connection factor is given by

$$\delta = \frac{3.92b}{273+t} = \frac{3.92 \times 74}{273+21} = 0.9867$$

The critical visual voltage is given by

$$V_v = 21.1\ m_v\ \delta r \left(1+\frac{0.3}{\sqrt{r\delta}}\right) \ln \frac{d}{r} \text{ kV to neutral}$$

$$= 21.1 \times 0.72 \times 0.9867 \times 0.52 \left(1+\frac{0.3}{\sqrt{0.52 \times 0.9867}}\right) \ln \left(\frac{2.5 \times 100}{0.52}\right) \text{ kV}$$

$$= 21.1 \times 0.72 \times 0.9867 \times 0.52 \left(1+\frac{0.3}{0.7163}\right) \ln \left(\frac{250}{0.52}\right) \text{ kV} = 68.30 \text{ kV to neutral}$$

Critical visual line to line voltage

$$= \sqrt{3} \times 68.30 \text{ kV} = 118.30 \text{ kV}$$ **Ans.**

The critical disruptive voltage is given by

$$V_d = 21.1\ m\delta r \ln \frac{d}{r} \text{ kV}$$

$$= 21.1 \times 0.85 \times 0.9867 \times 0.52 \ln \left(\frac{2.5 \times 100}{0.52}\right) \text{kV}$$

$$= 56.83 \text{ kV}$$

$$\text{Operating voltage per phase} = \frac{138}{\sqrt{3}} \text{ kV} = 79.67 \text{ kV}$$

Now, corona loss

$$= 241 \times 10^{-5} \left(\frac{f+25}{\delta}\right) \sqrt{\frac{r}{d}}\ (V_p - V_d)^2 \text{ kW/km/phase}$$

$$= 241 \times 10^{-5} \left(\frac{50+25}{0.9867}\right) \sqrt{\frac{0.52 \times 10^{-2}}{2.5}}\ (79.67 - 56.83)^3 \text{ kW/km/phase}$$

$$= 4.358 \text{ kW/km/phase}$$ **Ans.**

Problem 6.4. A single phase transmission line has conductors of diameter 1.25 cm and spaced 2.5 metres apart. Derive an expression for the potential gradient at any point on a line joining the centres of the conductors if the operating voltage of line is 60 kV. Calculate the voltage at which corona will start.

Solution: Given,

Operating voltage (V) = 60 kV, d = 2.5 m

Diameter of each conductor = 1.25 cm

$$\text{Radius of each conductor } (r) = \frac{1.25}{2} = 0.625 \text{ cm}$$

$$= 0.625 \times 10^{-2} \text{ m}$$

Expression for Potential Gradient at any point

Let *a* and *b* be the conductors of single phase transmission line and d be the separation between these as shown in Fig. P. 6.4.

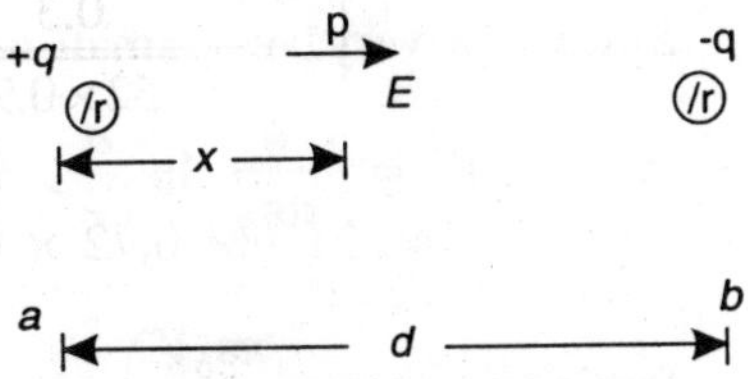

Fig. P. 6.4

Let $+q$ be the charge per unit length on the conductor *a* and $-q$ be the charge per unit length on the conductor *b*.

Let *p* be any point at a distance *x* from the conductor a on the line joining the centres of the conductors.

The electric field intensity at point p due to conductor a is given by

$$E_a = \frac{q}{2\pi\varepsilon_o x}$$

Again, the electric field intensity at point p due to conductor b is given by

$$E_b = \frac{q}{2\pi\varepsilon_o (d-x)}$$

The resultant electric field intensity at point p is given by

$$E_x = E_a + E_b = \frac{q}{2\pi\varepsilon_o x} + \frac{q}{2\pi\varepsilon_o (d-x)}$$

$$= \frac{q}{2\pi\varepsilon_o}\left[\frac{1}{x} + \frac{1}{d-x}\right] \qquad \text{....(1)}$$

The potential difference between the conductors is given by

$$V = -\int_{d-r}^{r} E_x dx = \int_{r}^{d-r} E_x dx$$

$$= \int_{r}^{d-r} E_x dx \, \frac{q}{2\pi\varepsilon_o}\left[\frac{1}{x} - \frac{1}{d-x}\right]$$

$$= \frac{q}{2\pi\varepsilon_o}\left[\ln x - \ln (d-x)\right]_r^{d-r}$$

$$= \frac{q}{2\pi\varepsilon_o} 2 [\ln (d - r) - \ln r]$$

$$= \frac{q}{\pi\varepsilon_o} \ln \frac{d-r}{r}$$

Since r is very very small as compared to d, $d - r \cong d$

$$\therefore \quad V = \frac{q}{\pi\varepsilon_o} \ln \frac{d}{r}, \ q = \frac{\pi\varepsilon_o V}{2\pi\varepsilon_o \ln \frac{d}{r}}$$

Putting $q = \dfrac{\pi\varepsilon_o V}{2\pi\varepsilon_o \ln \frac{d}{r}}$ in equation (1), we get

$$E_x = \frac{\pi\varepsilon_o V}{2\pi\varepsilon_o \ln \frac{d}{r}} \left[\frac{1}{x} + \frac{1}{d-x}\right]$$

or
$$E_x = \frac{V}{2 \ln \frac{d}{r}} \left[\frac{d-x+x}{x(d-x)}\right] = \frac{V_d}{2x(d-x) \ln \frac{d}{r}}$$

Thus the potential gradient at any point on line joining the centres of the conductors is given by

$$E_x = \frac{V_d}{2x(d-x) \ln \frac{d}{r}}$$ This is the required expression.

Problem 6.5. Find the critical disruptive voltage and the critical voltages for local and general corona or a 3-phase overhead transmission line, consisting of three-stranded copper conductors spaced 2 m apart at the corners of an equilateral triangle. Air temperature and pressure are 20° C and 72 cm Hg respectively. The conductor dia, irregularity factor and surface factors are 10.6 mm, 0.87, 0.7 and 0.8 respectively

Solution: Given,

$d = 2$ m, $t = 20°$ C, $b = 72$ cm

Diameter of each conductor = 10.6 mm

Radius of each conductor = 5.3 mm = 0.53 cm

$= 0.53 \times 10^{-2}$ m

$m = 0.87$, m_v = 0.7 for local corona

= 0.8 for general corona

The air-density correction factor is given by

$$\delta = \frac{3.92b}{273+t} = \frac{3.92\times72}{273+20} = 0.9633$$

The critical disruptive voltage is given by

$$V_d = 21.1 \text{ m } \delta r \ln\frac{d}{r} \text{ kV to neutal}$$

$$= 21.1 \times 0.87 \times 0.9633 \times 0.53 \ln\frac{2\times100}{0.53} \text{ kV}$$

$$= 55.61 \text{ kV}$$

Critical disruptive line to line voltage

$$= 55.61 \times \sqrt{3} = 96.32 \text{ kV}$$ **Ans.**

$$\text{Now, } \sqrt{r\delta} = \sqrt{0.53\times0.9633} = 0.7145$$

The visual critical voltage is given by

$$V_v = 21.1\, m_v\, \delta r\left(1+\frac{0.3}{\sqrt{r\delta}}\right)\ln\frac{d}{r} \text{ kV to neutral}$$

For local corona, $m_v = 0.7$

$$V_v = 21.1 \times 0.7 \times 0.9633 \times 0.53\left(1+\frac{0.3}{0.7145}\right)$$

$$\ln\frac{2\times100}{0.53} \text{ kV} = 63.53 \text{ kV}$$

Visual critical line to line voltage

$$= 63.53 \times \sqrt{3} = 110.04 \text{ kV}$$ **Ans.**

The visual critical voltage for general corona is given by

$$V_v = \frac{0.8}{0.7}\times V_v \text{ for local corona}$$

$$= \frac{0.8}{0.7}\times 110.04 \text{ kV} = 125.76 \text{ kV}$$ **Ans.**

Problem 6.6. A conductor with dia 3 cm is passed centrally through a porcelain bushing $\varepsilon_r = 5$ having internal diameter and external diameter of 4 cm and 10 cm respectively. The voltage between the conductor and an earthed clamp surrounding the porcelain is 22 kV/cm. Determine whether the corona will be present in the air space round the conductor.

Solution: Given,

Diameter of the conductor = 3 cm
Radius of the conductor (r) = 1.8 cm
Relative permilivity of the porcelain (ε_r) = 5
Internal diameter of the porcelain = 4 cm
$\therefore$ Internal radius of the porcelain (r_i) = 2 cm
External diameter of the porcelain = 10 cm
$\therefore$ External radius of the porcelain (r_o) = 5 cm, V_{cp} = 22 kV

Let E_{1max} be the maximum gradient on the surface of the conductor and E_{2max} be the maximum gradient on the inner surface of the porcelain.

$$E_{1max} = \frac{q}{2\pi\varepsilon_o r} \quad .....(1)$$

where q is the charge per unit length on the conductor.

$$\text{and } E_{2max} = \frac{q}{2\pi\varepsilon_o\varepsilon_r r_2} \quad .....(2)$$

From Eqs. 1 and 2

$$E_{1max} \times r = E_{2max} \times \varepsilon_r \times r_i,\ E_{1max} \times 1.5 = E_{2max} \times 5 \times 2$$

$$E_{2max} = 0.15\ E_{1max}$$

Now, p.d. between conductor and earthed clamp

$$= E_{1max}\ r \ln\frac{r_i}{r} + E_{2max}\ r_i \ln\frac{r_o}{r}$$

$$\text{or} \quad 22 = E_{1max} \times 1.5 \times \frac{2}{1.5} + E_{2max} \times 2 \times \ln\frac{5}{2}$$

$$\text{or} \quad 22 = 1.5\ E_{1max} \ln\frac{2}{1.5} + 0.15\ E_{1max} \times 2 \times \ln\frac{5}{2}$$

$$\text{or} \quad 22 = 0.7064\ E_{1max}$$

$$\therefore \quad E_{1max} = \frac{22}{0.7064} = 31.14 \text{ kV/cm}$$

Since the surface gradient exceeds 21.1 kV/cm. The corona will be present.

Problem 6.7: Determine the critical disruptive voltage and corona loss for a 3-phase line operating at 120 kV which has conductor of 1.5 cm dia arranged in a 3 m detta. Assume air density factor of 1.06 and the dielectric strength of air to be 21.1 kV/cm. Take m = 1

Solution: Given,

Operating voltage (v_o) = 120 kV
Diameter of each conductor = 1.5 cm
Radius of each conductor (r) = 0.75 cm
d = 3 δ, = 1.06, g_0 = 21.1 kV/cm
The critical disruptive voltage is given by

$$V_d = g_o \, m \, \delta r \ln \frac{d}{r} \text{ kV to neutral.}$$

$$= 21.1 \times 1 \times 1.06 \times 0.75 \times \ln \frac{3 \times 100}{0.75} \text{ kV}$$

$$= 100.5 \text{ kV}$$

Critical disruptive line to line voltage

$$= 180.5 \times \sqrt{3} \text{ kV} = 174.07 \text{ kV} \quad \textbf{Ans.}$$

Since the operating voltage is 120 kV, the corona loss will be absent. Hence corona loss = 0 W

Problem 6.8. A single phase overhead line has two conductors of diameter 1.2 cm with a spacing of 1.5 m between the centres. If the dielectric strength of air is 21.1 kV/cm, determine the line voltage for which corona will commence on the line. Take m = 1 and δ = 1.01.

Solution: The corona will commence on the line when the line voltage becomes equal to the line, when the line voltage becomes equal to the critical disruptive voltage.

Given, g_o = 21.1 kV/cm, d = 1.5 m
Diameter of each conductor = 1.2 cm

Radius of each conductor (r) = $\frac{1.2}{2}$ = 0.6 cm = 0.6 × 10^{-2} m

The critical disruptive voltage of the line is given by

$$V_d = g_o \, \text{m} \, \delta r \ln \frac{d}{r} \text{ kV}$$

$$= 21.1 \times 1 \times 1.01 \times 0.6 \ln \frac{1.5 \times 100}{0.6} \text{ kV}$$

$$= 70.60 \text{ kV} \quad \textbf{Ans.}$$

Problem 6.9. A certain three-phase equilaterally spaced transmission line has a total corona loss of 53 kW at 106 kV and a loss of 98 kW at 110.9 kV. What is the disruptive critical voltage between lines' and what is the corona loss at 113 kV?

Solution: Given

$$P_1 = 98 \text{ kW at } 110.9 \text{ kV} \qquad .....(1)$$
$$P_2 = 53 \text{ kW at } 106 \text{ kV} \qquad .....(2)$$

The corona loss of a line is given by

$$P = 241 \times 10^{-5} \left(\frac{f+25}{\delta}\right)\sqrt{\frac{r}{d}}\,(V_p - V_d)^2$$

Let V_d be the disruptive critical voltage between the lines.
For the constant values of f, r, d and δ
Corona loss α $(V_p - V_d)^2$ this yield

$$\frac{P_1}{P_2} = \left(\frac{Vp_1 - V_d}{Vp_2 - V_d}\right)^2 \text{ or } \frac{Vp_1 - V_d}{Vp_2 - V_d} = \sqrt{\frac{P_1}{P_2}}$$

For $V_1 = 110.9$ kV, $P_1 = 98$ kW
$V_2 = 106$ kV, $P_2 = 53$ kW

or $$\frac{110.9 - V_d}{106 - V_d} = \sqrt{\frac{98}{53}} \text{ or } \frac{110.9 - V_d}{106 - V_d} = 1.36$$

or $$110.9 - V_d = 1.36 \times 106 - 1.36\ V_d$$

or $$1.36\ V_d - V_d = 1.36 \times 106 - 110.9$$

or $$0.36\ V_d = 33.26$$

$$\therefore \quad V_d = \frac{33.26}{0.36} = 92.39 \text{ kV}$$

Thus, critical disruptive voltage between the lines
$= 92.36$ kV **Ans.**

At $V_{p3} = 113$ kV, $P_3 = ?$

To calculate P_3, apply the relation.

$$\frac{P_3}{P_2} = \left(\frac{Vp_3 - V_d}{Vp_2 - V_d}\right)^2$$

$$\therefore \quad P_3 = \left(\frac{Vp_3 - V_d}{Vp_2 - V_d}\right)^2 \times P_2$$

$$= \left(\frac{113 - 92.36}{106 - 92.36}\right)^2 \times 53 \text{ kW}$$

$= 121.36$ kW **Ans.**

Chapter 7

Mechanical Design of Transmission Lines

Problem 7.1. An overhead line has a conductor of cross-section 2.5 cm^2 hard drawn copper and a span length of 150 metres. Determine the sag which must be allowed if the tension is not to exceed one-fifth of the ultimate strength of 4175 kg/cm^2 (a) in still air, and (b) with a wind pressure of 1.3 kg/metre and an ice coating of 1.25 cm. Determine also the vertical sag in the latter.

Solution: Given,

Area of crosssection of the conductor (A) $= 2.5 \text{ cm}^2$
$= 2.5 \times 10^{-4} \text{ m}^2$

Span length (l) = 150 m

Ultimate strength (u) = 4175 kg/cm^2

(*a*) In still air, working tension (T)

$$= \frac{1}{5} \times u \times A = \frac{1}{5} \times 4175 \times 2.5 = 2087.5 \text{ kg}$$

As we know that density of the copper is 8850kg/m^3

Weight per unit length (w) = density of copper × area

$= 8850 \times 2.5 \times 10^{-4}$ kg/m = 2.2125 kg/m

To calculate sag, apply the relation

$$d = \frac{We^2}{8T} = \frac{2.2125 \times (150)^2}{8 \times 2087.5} = 2.98 \text{ m}$$

$\therefore$ Sag = 2.98 m **Ans.**

(*b*) In this case, w_a = 1.3 kg/m, ice-coating (t) = 1.25 cm

Let D be the diameter of the conductor

Given, Area of the conductor = 2.5 cm^2,

$$\text{or} \quad \frac{\pi D^2}{4} = 2.5, \ D = 1.78 \text{ cm}$$

The overall diameter of the conductor with ice-coating

$$D_o = D + 2t = 1.78 + 1.25 \times 2 = 1.78 + 2.50$$
$$= 4.28 \text{ cm}$$

Area of section of the ice

$$= \frac{\pi}{4}(D_o^2 - D^2)$$

$$A_i = \frac{\pi}{4} = \left[(4.28)^2 - (1.78)^2\right] = 11.90 \text{ sq. cm}$$

We know that density of ice (ρ_i) is 0.916 g/cm³

Weight of ice per unit length of the conductor (W_i)

$$= A_i \times \delta_i = 11.90 \times 0.916 = 10.9 \text{ g/cm}$$
$$= 1.09 \text{ kg/m}$$

Total loading on the conductor per unit length

$$(W) = \sqrt{(w + w_L) + w_a^2}$$

$$= \sqrt{(2.2125 + 1.09)^2 + (1.3)^2} = 3.549 \text{ kg/m}$$

The sag in the conductor is given by

$$d = \frac{WC^2}{8T} = \frac{3.549 \times (150)^2}{8 \times 2087.5} = 4.78 \text{ m}$$

Here $\theta = \tan^{-1}\left(\frac{w_a}{w + w_L}\right) = \tan^{-1}\left(\frac{1.3}{2.2125 + 1.09}\right)$

$$= 21.49 \text{ to the vertical}$$

The required vertical sag = 4.78 cos 21.49 = 4.45 m **Ans.**

Problem 7.2. An overhead conductor consists of 7 strands of silicon–bronze having an ultimate strength of 10,000 kg/cm² and an area of 2.5 cm² when erected between supports 650 metres apart and having a 20 metre difference in level, determine the vertical sag which must be allowed so that the factor of safety shall be 5. Assume the wire weighs 2 kg/metre, ice loading 1 kg/metre and wind loading is 1.75 kg/metre.

Solution: Given,

Ultimate strength (u) = 10,000 kg/cm², Span length (l) = 650 m

Area of the conductor (A) = 2.5 cm²

Difference in levels (h) = 20 m, Factor of safety (Fos) = 5

Weight of the wire permit length (w) = 2 kg/m

Ice loading (w_i) = 1 kg/m, Wind loading (w_a) = 1.75 kg/m

Now, Working stress (f) $= \dfrac{U}{Fos} = \dfrac{10{,}000}{5} = 2000$ kg/cm^2

$= 10 \times 10^6$ kg/m^2

Total loading on the conductor

$$(W) = \sqrt{(w+w_i)^2 + w_a^2} = \sqrt{(2+1)^2 + (1.75)^2}$$

$= 3.473$ kg/m

Loading factor (q) $= \dfrac{W}{w_i} = \dfrac{3.473}{2} = 1.74$

Density of the conductor (δ) $= \dfrac{w}{A} = \dfrac{2}{2.5 \times 10^{-4}} = 8000$ kg/m^2

To calculate sag, apply the relations

$$d = \frac{q\rho L^2}{8f} + \frac{fh^2}{298\,L^2} - \frac{h}{2}$$

$$= \frac{1.74 \times 8000 \times (650)^2}{8 \times 2000 \times 10^4} + \frac{2000 \times 10^4 \times (20)^2}{2 \times 1.74 \times 8000 \times (650)^2} - \frac{20}{2}$$

$= 36.76 + 0.68 - 10 = 37.44 - 10 = 27.49$ m

Angle made by the conductor with vertical

$$(\theta) = \tan^{-1} \frac{w_a}{w_i + w} = \tan^{-1}\left(\frac{1.75}{2+1}\right) = \tan^{-1}\left(\frac{1.75}{3}\right)$$

$= 30.26°$

The required vertical sag in given by

$d_r = d \cos \theta = 27.44 \cos 30.26° = 23.70$ m

$\therefore$ vertical sag = 23.70 m **Ans.**

Problem 7.3. An overhead line has the following data : Span length 185 metres, difference in levels of supports 6.5 metres, conductor dia 1.82 cm, weight per unit length of conductor 2.5 kg/metre, wind pressure 49 kg/m^2 of projected area. Maximum tensile stress of the conductor 4250 kg/cm^2. Factor of safety 5. Calculate the allowable sag in metres at the lower support.

Solution: Given,

Span length (l) = 185 m,

Difference in levels of supports (h) = 6.5 m

Diameter of the conductor (D) = 1.82 cm

Weight per unit length of the conductor (w) = 2.5 kg/m

Wind pressure (P_i) = 49 kg/m^2 of projected area
Ultimate strength (u) = 4250 kg/cm^2
Factor of safety (Fos) = 5

Now, Working stress (f) = $\dfrac{u}{Fos} = \dfrac{4250}{5}$ kg/cm^2 = 850 kg/cm^2

Projected area per metre length of the conductor (A_p)
$= 1.82 \times 10^{-2} \times 1 = 1.82 \times 10^{-2}$ sq.m

Winding loading per unit length of the conductor (w_a)
$= A_p \times p_i = 49 \times 1.82 \times 10^{-2}$ kg/m
$= 0.8918$ kg/m

Total loading on the wire per unit length

$$(W) = \sqrt{w^2 + w_a^2} = \sqrt{(2.5)^2 + (0.8918)^2} = 2.654 \text{ kg/m}$$

Area of the conductor = $\dfrac{\pi D^2}{4}$

$$\therefore \text{ Working tension } (T) = f \times \frac{\pi D^2}{4} = 850 \times \frac{\pi \times (1.82)^2}{4}$$
$$= 2211.32 \text{ kg}$$

To calculate the distance of the lower support from origin. Apply the relation,

$$a = \frac{l}{2} - \frac{Ch}{l}$$
$$= \frac{l}{2} - \frac{Th}{Wl}$$
$$= \frac{185}{2} - \frac{2211.32 \times 6.5}{2.654 \times 185}$$
$$= 92.5 - 29.27$$
$$= 63.23 \text{ m}$$

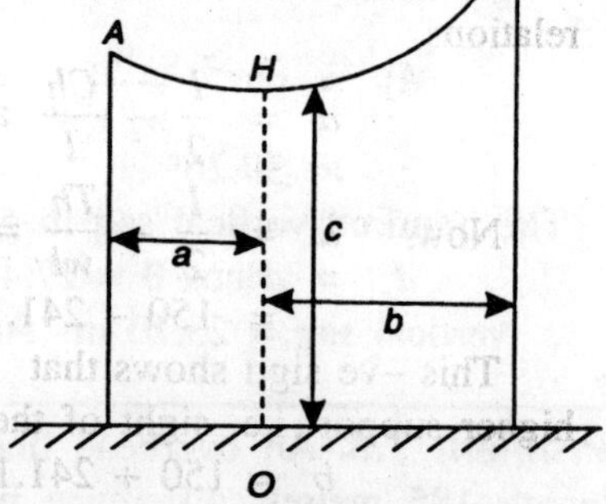

Fig. P. 7.3

The allowable sag at the lower support is given by

$$d = \frac{a^2}{2C} = \frac{wa^2}{2T} = \frac{2.654 \times (63.23)^2}{2 \times 2211.32} = 2.40 \text{ m}$$

$\therefore$ Allowable sag at lower support = 2.40 m **Ans.**

Problem 7.4. Transmission line conductor at a river crossing is supported from two towers at heights of 45 metres and 75 metres above water level. The span length is 300 metres. Weight of the conductor 0.85 kg/metre. Determine the clearance between the conductor and water at a point midway between towers if the tension in the conductor is 2050 kg.

Solution: Given,

h_1 = 45 m, h_2 = 75 m, span-length (l) = 300 m

weight of the conductor permit length (w) = 0.85 kg/m

working tension in the wire (T) = 2050 kg.

The situation can be represented as shown in Fig. P. 7.4.

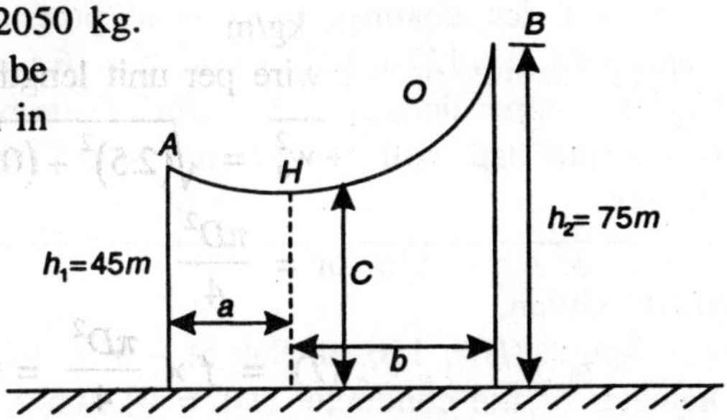

Fig. P. 7.4

Difference in support levels = $h_2 - h_1$

$$h = 75 - 45 = 30\text{m}$$

The distances of supports from origin can be calculated by the relations

$$a = \frac{l}{2} - \frac{Ch}{l} \text{ and } b = \frac{l}{2} + \frac{Ch}{l}$$

Now, $$a = \frac{l}{2} - \frac{Th}{wl} = \frac{300}{2} - \frac{2050 \times 30}{0.85 \times 300}$$

$$= 150 - 241.18 = -91.18 \text{ m}$$

This –ve sign shows that lower support lies on the same side of higher support, i.e. sight of the origin as shown in figure.

$$b = 150 + 241.18 = 391.18 \text{ m}$$

Now, Mid-way distance from origin

$$= \frac{391.18 + 91.18}{2} = 241.18 \text{ m}$$

The sag at lower support is given by

$$d_1 = \frac{a^2}{2c} = \frac{wa^2}{2T} = \frac{0.85 \times (91.18)^2}{2 \times 2050} = 1.72 \text{ m}$$

The sag at 241.18 m is given by

$$d_2 = \frac{w \times (241.18)^2}{2T} = \frac{0.85 \times (241.18)^2}{2 \times 2050} = 12.06 \text{ m}$$

Therefore, height of mid-point with respect to *A* as shown in Fig. P. 7.4 is given by

$$d = d_2 - d_1 = 12.06 - 1.072 = 10.34 \text{ m}$$

Therefore, the clearance between the conductor and water at a point mid-way between the tower will $d + h_1$

Clearance $= d + h_1 = 44 + 10.34 = 55.34$ m **Ans.**

Problem 5: An overhead line having a conductor of dia 10 mm and a span length of 150 metres has a sag of 3.5 metres at –5° with 10 mm thick ice coating and a wind pressure of 40 kg/m^2 of projected area. E = 127 × 104 kg/cm^2, $\alpha = 16.6 \times 10^{-6}$/C, ice density 910 kg/m^3, copper density 8850 kg/m^3. Determine the temperature at which the sag will remain the same under fair weather conditions.

Solution: Given,

Span length (l) = 150 m, Sag at – 5° C (d_1) = 3.5 m

Diameter of the conductor (D) = 10 mm = $10^{-3} \times 10$ m

Thickness of ice coating (t) = 10 mm

Wind pressure (p_i) = 40 kg/m^2

$$E = 127 \times 10^4 \text{ kg/m}^2 = 1.27 \times 10^4 \text{ kg/mm}^2$$

$$\alpha = 16.6 \times 10^{-6} \text{ /°C}, \ \rho_i = 910 \text{ kg/m}^3$$

$$\rho_e = 8850 \text{ kg/m}^3 = 8.850 \times 10^{-3} \text{ kg/m/sq.mm}$$

Now, The overall dia of the conductor with ice coating

$$(D_o) = 10 + 2 \times 10 = 30 \text{ mm}$$

The projected area per unit length of the conductor

$$(A_p) = 30 \times 10^{-3} \times 1 \text{ sq. m} = 30 \times 10^{-3} \text{ sq.m}$$

Wind loading per unit length of the conductor

$$(w_a) = P_i \times A_p = 40 \times 30 \times 10^{-3} \text{ kg/m} = 1.20 \text{ kg/m}$$

Area of section of the ice

$$(A_i) = \frac{\pi}{4}(D_o^2 - D^2) = \frac{\pi}{4}[(30)^2 - (10)^2] \times 10^{-6} \text{ sq.m}$$

$$= 628.32 \times 10^{-6} \text{ sq.m.}$$

Area of the conductor

$$(A) = \frac{\pi}{4} D^2 = \frac{\pi}{4} \times (10 \times 10^{-3})^2 \text{ sq.m}$$

$$= 78.54 \times 10^{-6} \text{ sq. m} = 0.7854 \text{ sq.cm.}$$

Ice loading per unit length of the conductor

$$(w_i) = \rho_i \times A_i = 910 \times 628.32 \times 10^{-6} \text{ kg/m}$$

$$= 0.5718 \text{ kg/m}$$

Weight of the conductor per unit length

$$(w) = \rho_c \times A = 8850 \times 78.54 \times 10^{-6} \text{ kg/m}$$

$$= 0.6951 \text{ kg/m}$$

Total loading on the conductor per unit length

$$(w) = \sqrt{(w+w_i)+w_a^2}$$

$$= \sqrt{(0.6951+0.5718)^2+(1.20)^2} = 1.719 \text{ kg/m}$$

Loading factor $(q_1) = \frac{W}{w} = \frac{1.719}{0.6951} = 2.47$

Let T_1 be the working temperature at –5° C

Given, Sag = 3.5 m

$$\therefore \quad \frac{Wl^2}{8T_1} = 3.5 \text{ or } T_1 = \frac{Wl^2}{8\times 3.5} = \frac{1.719\times(150)^2}{8\times 3.5}$$

$$= 1381.34 \text{ kg}$$

Working stress at –5°

$$(f_1) = \frac{T_1}{A} = \frac{1381.34}{0.7854} \text{ kg/cm}^2 = 1757.5 \text{ kg/cm}^2$$

$$= 17.58 \text{ kg/mm}^2$$

Let T_2 be the working tension at t° C.

According to question, the sag remains same at both the temperature.

$$\therefore \quad \text{Sag} = \frac{wl^2}{8T_2}$$

$$\therefore \quad T_2 = \frac{wl^2}{8\times Sag} = \frac{0.6951\times(150)^2}{8\times 3.5} = 558.56 \text{ kg.}$$

Working stress (f_2) at t°C $= \frac{T_2}{A} = \frac{558.56}{0.7854}$ kg/cm²

$$= 711.2 \text{ kg/cm}^2 = 7.112 \text{ kg/mm}^2$$

To calculate t_2, apply the relation,

$$f_1 - \frac{L^2\delta_c^2 q_1^2 E}{24 f_1^2} = f_2 - \frac{L^2\delta_c^2 q_2^2 E}{24 f_2^2} + \alpha\Delta t E$$

For fair weather and higher temperature, $q_2 = 1$

$$= 17.58 - \frac{(150)^2\times(8.850)^2\times 10^{-6}\times(2.47)^2\times 1.27\times 10^4}{24\times(17.58)^2}$$

$$= 7.112 - \frac{(150)^2\times(8.850)^2\times 10^{-6}\times 1\times 1.27\times 10^4}{24\times(7.112)^2} + \alpha\Delta t E$$

or $17.58 - 18.41 = 7.112 - 18.44 + \alpha\Delta t E$

or $\alpha\Delta t E = 17.58 - 18.41 + 18.44 - 7.112$

or $\alpha \Delta t E = 10.498$ or $\Delta t = \dfrac{10.498}{\alpha E}$

or $\Delta t = \dfrac{10.498}{16.6 \times 10^{-6} \times 1.27 \times 10^4}$ or $t_2 - t_1 = 49.80$

or $t_2 = 49.80 + t_1 = 49.80 - 5°\text{ C} = 44.80°\text{ C}$ **Ans.**

> **Problem 7.6.** An overhead line has the following data: Span length 150 meters, conductor dia 0.90 cm, weight per unit length of the conductor 0.60 kg/metre. Ultimate stress 4200 kg/cm^2, wind pressure 40 kg/cm^2 of projected area. Factor of safety 5. Calculate the sag?

Solution: Given,

Span length (l) = 150 m

Diameter of the conductor (D) = 0.90 cm.

Area of the conductor $= \dfrac{\pi D^2}{4} = \dfrac{\pi \times (0.9)^2}{4}$ sq.cm

$= 0.636$ sq. cm.

$w = 0.60$ kg/m

Ultimate stress = 4200 kg/cm^2, P = 40 kg/cm^2

Factor of safety = 5

Now, Working stress (f)

$$= \frac{\text{Ultimate Stress}}{\text{Factor of safety}} = \frac{4200 \text{ kg/cm}^2}{5} = 840 \text{ kg/cm}^2$$

Winding loading per unit length of the conductor

$(w_a) = P \times A = 40 \times 0.636 \times 10^{-2}$ kg/m = 0.254 kg/m.

Total loading $(W) = \sqrt{w^2 + w_a^2} = \sqrt{(0.60)^2 + (0.254)^2}$

$= 0.652$ kg/m

Working tension (T) = $f \times A$ = 840 × 0.636 kg. = 534.24 kg

To calculate sag, apply the relation

$$d = \frac{Wl^2}{8T} = \frac{0.652 \times (150)2}{8 \times 534.24} = 3.43 \text{ m}$$

$\therefore$ Sag = 3.43 m **Ans.**

> **Problem 7.7.** Transmission line conductor having a dia of 19.0 mm weighs 0.80 kg/m. The span is 270 metres. The wind pressure is 38 kg/m^2 of projected area with ice coating 12 mm. The ultimate stress of the conductor is 8100 kg. Calculate the maximum sag if factor of safety is 3 and ice weights 910 kg/m^3.

Solution: Given

W = 0.80 kg/m,

Diameter of the conductor (*D*) = 19 mm = 19×10^{-3} m

$$\text{Area of the conductor } (A) = \frac{\pi D^2}{4} = \frac{\pi\left(19\times10^{-3}\right)^2}{4} \text{ sq.m}$$

$$= 283.53 \times 10^{-3} \text{ sq.m}$$

Span length (*l*) = 270 m, *P* = 38 kg/m^2

t = 12 mm, = 12×10^{-3} m

Ultimate strength (*u*) = 8100 kg

Factor of safety = 3

W_c = 910 kg/m^3

Working tension of the conductor (*T*)

$$= \frac{\text{Ultimate stress}}{\text{Factor of safety}} = \frac{8100}{3} = 2700 \text{ kg}$$

Area of section of *ice* = $\pi\ [(r + t)^2 - r^2]$

$A_i = \pi\ [(9.5 + 12)^2 - (9.5)^2] \times 10^{-6}$ sq.m

$A_i = 11.69 \times 10^{-3}$ sq.m

Overall dia of the conductor with ice coating

$D_o = 19 + 24 = 43 \text{ mm} = 43 \times 10^{-3}$ m

Protected area per metre length of the conductor

$A_p = D_o \times 1 = 43 \times 10^{-3}$ sq.m

Wind load per meter length (w_a) = $A_p \times P$

$= 38 \times 43 \times 10^{-3}$ kg/m = 1.634 kg/m

Ice loading per metre length (w_i) = $A_i \times W_c$

$= 1.169 \times 10^{-3} \times 910$ kg/m = 1.064 kg/m

$$\text{Total loading } (W) = \sqrt{(w + w_1)^2 + w_a^2}$$

$$= \sqrt{(1.06 + 0.80)^2 + (1.634)^2} = 2.4788 \text{ kg/m}$$

To calculate maximum sag, apply the relation.

$$\text{Maximum sag} = \frac{Wl^2}{8T} = \frac{2.4788 \times (270)^2}{8 \times 2700} \text{ m} = 8.37 \text{ m}$$

$\therefore$ Maximum sag = 8.37 m **Ans.**

Problem 7.8. Determine the sag at 32° and 60° C in an 8 SWG copper conductor erected on a 45 m span length. The wind pressure is 48 kg/m^2 of projected area at a temperature of 4° C, weight of wire is 0.1160 kg/m, the working stress shall not exceeds one-half the ultimate tensile strength.

Modulus of elasticity (E) = 1.25×10^4 kg/mm^2
Coefficient of linear expansion (α) = 16.5×10^{-6} per ºC.
Ultimate stress of the conductor = 40 kg/mm^2
Dia of the conductor = 4 mm

Solution: Given,

$t_1 = 4°$ C, $t_2 = 32°$ C, $t_3 = 60°$ C

Span length (l) = 45 m

$P_i = 48$ kg/m^2, $w = 0.1160$ kg/m,

$E = 1.25 \times 10^4$ kg/mm^2, $\alpha = 16.5 \times 10^{-6}$ /ºC

$U = 40$ kg/mm^2

Diameter of the conductor (D) = 4 mm = 4×10^{-3} m

Cross-sectional area of the conductor = $\dfrac{\pi D^2}{4}$

$$(A) = \frac{\pi(4)^2}{4} \text{ sq.mm} = 12.57 \text{ sq.mm}$$

Projected area of the conductor per unit length

$$A_p = D \times 1 = 4 \times 10^{-3} \times 1 = 4 \times 10^{-3} \text{ sq.m}$$

Winding loading per meter length = $A_p \times p_i$

$$w_a = 4 \times 10^{-3} \times 48 \text{ kg/m} = 0.192 \text{ kg/m}$$

$$\text{Total loading (W)} = \sqrt{w^2 + w_a^2} = \sqrt{(0.1160)^2 + (0.192)^2}$$

$$= 0.2243 \text{ kg/m}$$

$$K = f_1 - \frac{l^2 \delta^2 q_1^2 E}{24 f_1^2}$$

$$= 20 - \frac{(45)^2 \times (9.228 \times 10^{-3})^2 \times (1.934)^2 \times 1.25 \times 10^4}{24 \times (20)^2}$$

$$= 20 - 0.83 = 19.17$$

$$M = \frac{l^2 \delta^2 q_2^2 E}{24}$$

At higher temperature, $q_2 = 1$

$$\therefore \quad M = \frac{(45)^2 \times (9.228 \times 10^{-3})^2 \times (1)^2 \times 1.25 \times 10^4}{24} = 89.81$$

$$N = K - \alpha \Delta t E = 19.17 - 5.775 = 13.4$$

To calculate f_2, apply the relation

$$f_2^2 (f_2 - N) = M \text{ or } f_2^2 (f_2 - 13.4) = 89.81$$

By inspection $f_2 = 13.87$ kg/mm^2

Working tension $(T) = f_2 \times A = 13.87 \times 12.57 = 174.35$ kg
According to the question, working stress (f_1)
$= U \times ½ = ½ \times 40 = 20$ kg/mm²
Working tension $(T) = f_1 \times A = 20 \times 12.57$ kg $= 251.4$ kg
To calculate sag at 4° C, apply the relations,

$$d = \frac{Wl^2}{8T} = \frac{0.2243 \times (45)^2}{8 \times 251.4} = 22.58 \text{ cm} \quad \textbf{Ans.}$$

SAG at 32° C

Here, $\Delta t = t_2 - t_1 = 32 - 4 = 28°$ C
$\alpha \Delta t E = 16.5 \times 10^{-6} \times 28 \times 1.25 \times 10^4 = 5.775$

$$\text{Loading factor } (q_1) = \frac{W}{w} = \frac{0.2243}{0.1160} = 1.934$$

$$\delta = \frac{w}{A} = \frac{0.1160}{12.57}$$

$$= 9.228 \times 10^{-3} \text{ kg/m/sq.mm}$$

$$\text{Sag} = \frac{wl^2}{8\,T} = \frac{0.1160 \times (45)^2}{8 \times 174.35} = 16.84 \text{ cm} \quad \textbf{Ans.}$$

Sag at 60°C

Here, $\Delta t = t_3 - t_1 = 60 - 4 = 56°$ C
$\alpha \Delta t E = 16.5 \times 10^{-6} \times 56 \times 1.25 \times 10^4 = 11.55$
$\therefore \quad N = K - \alpha \Delta t E = 19.17 - 11.54 = 7.62$
To Calculate f_3, apply the relation
$f_3^2\,(f_3 - N) = M$ or $f_3^2\,(f_3 - 7.62) = 89.81$
By hit and trial method. $f_3 = 8.79$ kg/mm²
In this case, working tension $(T) = f_3 \times A$
$= 8.79 \times 12.57 = 110.49$ kg

$$\therefore \quad \text{Sag} = \frac{wl^2}{8\,T} = \frac{0.1160 \times (45)^2}{8 \times 110.49} = 26.57 \text{ cm} \quad \textbf{Ans.}$$

Problem 7.9. An overhead line at a river crossing is supported from two towers of heights 40 m and 100 m above water level with a span of 400 m. The weight of the conductor is 2 kg/m and the working tension is 4400 kg. Determine the clearance between the conductor and water level mid way between the towers.

Solution: Given

$h_1 = 40$ m, $h_2 = 100$ m, $l = 400$ m, $w = 2$ kg/m, $T = 4400$ kg

Height of the lowest point of the conductor above sea-level

$$(c) = \frac{T}{w} = \frac{4400}{2} = 2200 \text{ m}$$

and $h = h_2 - h_1 = 100 - 40 = 60$ m

To calculate a and b, apply the relations

$$a = \frac{l}{2} - \frac{Ch}{l} \text{ and } b = \frac{l}{2} + \frac{Ch}{l}$$

Now,

$$a = \frac{400}{2} - \frac{2200 \times 60}{400} = -130 \text{ m}$$

$$b = \frac{400}{2} + \frac{2200 \times 60}{400} = 200 + 330 = 530 \text{ m}$$

Now, The sag at 130 m is given by

$$d_1 = \frac{a^2}{2c} = \frac{(130)^2}{2 \times 2200} = 3.84 \text{ m}$$

Here, $\dfrac{|a| + b}{2} = \dfrac{130 + 530}{2} = 330$ m

Here, The sag at 330 m is given by

$$d_2 = \frac{(330)^2}{2c} = \frac{(330)^2}{2 \times 2200} = 24.75 \text{ m}$$

The diagram can be represented as shown in Fig. P. 7.9.

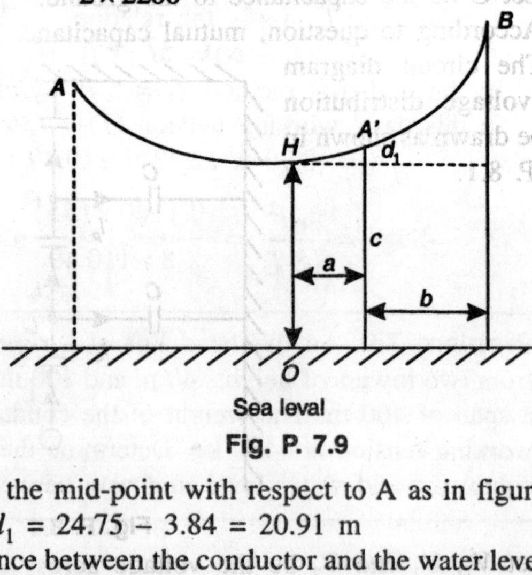

Fig. P. 7.9

Therefore, height of the mid-point with respect to A as in figure

$$d = d_2 - d_1 = 24.75 - 3.84 = 20.91 \text{ m}$$

Therefore, the clearance between the conductor and the water level midway between the towers will be h_1 + d *ie* 40 + 20.91 = 60.91 m.

Chapter 8

Insulators

Problem 8.1. Each conductor of a 33 kV, 3-phase system is suspended by a string of three similar insulators, the capacitance of each disc is nine times the capacitance to ground. Calculate the voltage across each insulator. Determine the string efficiency also.

Solution: Given,

$$\text{Operating voltage } (V) = \frac{33}{\sqrt{3}} \text{ kV to neutral}$$

Let C be the capacitance to the ground.

According to question, mutual capacitance = 9 C

The circuit diagram and voltage distribution can be drawn as shown in Fig. P. 8.1.

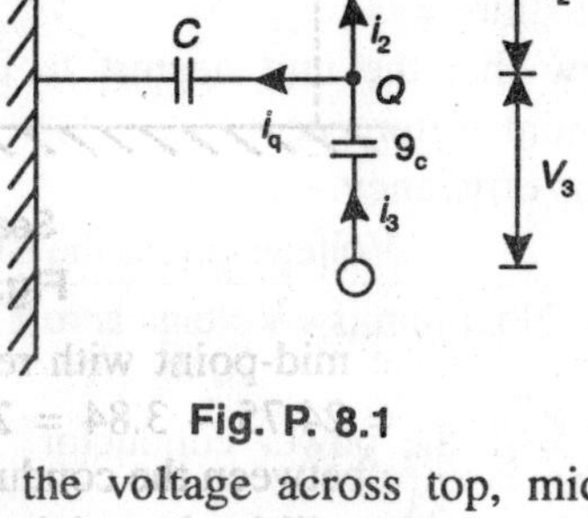

Fig. P. 8.1

Let V_1, V_2 and V_3 be the voltage across top, mid and bottom insulators, respectively.

From circuit diagram, at node P, $i_2 = i_p + i_1$

$V_2\omega \times 9\text{ C} = V_1\omega \times C + V_1\,\omega 9\text{ C}$ or $9\,V_2 = 10\,V_1$

$\therefore \quad V_2 = \frac{10}{9} V_1$ .... (1)

Again, At node Q, $i_3 = i_2 + i_q$

$V_3\omega \times 9\,C = V_2\omega \times 9\,C + (V_1 + V_2)\,\omega c$

or $\quad 9\,V_3 = 9\,V_2 + V_1 + V_2$ or $9\,V_3 = 10\,V_2 + V_1$

or $\quad 9\,V_3 = 10 \times \frac{10}{9} V_1 + V_1$

or $\quad V_3 = \frac{1}{9}\left[\frac{100}{9} V_1 + V_1\right]$

$\therefore \quad V_3 = \frac{109}{81} V_1$ ... (ii)

Now, Voltage across the spring $(V) = V_1 + V_2 + V_2$

or $\quad V = V_1 + \frac{10}{9} V_1 + \frac{109}{81} V_1$

or $\quad V = V_1\left[1 + \frac{10}{9} + \frac{109}{81}\right]$

or $\quad V = V_1\left[\frac{81 + 90 + 109}{81}\right]$ or

$\therefore \quad V = \frac{280}{81}, V_1 = \frac{81}{280} \times \frac{33}{\sqrt{3}} = 5.51$ kV

Thus voltage across top-most insulator = 5.51 kV **Ans.**

Voltage across mid-insulator (V_2)

$$= \frac{10}{9} V_1 = \frac{10}{9} \times 5.51 \text{ kV} = 6.12 \text{ kV} \quad \textbf{Ans.}$$

Voltage across the bottom insulator (V_3)

$$= \frac{109}{81} V_1 = \frac{109}{81} \times 5.51 \text{ kV} = 7.41 \text{ kV} \quad \textbf{Ans.}$$

Here, no of units = 3

As we know that the unit nearest to the power conductor has maximum potential difference.

$\therefore$ % String efficiency

$$= \frac{\text{Voltage across the string}}{\text{No. of units} \times \text{Voltage across unit nearest to the power conductor}} \times 100$$

× to the power conductor

$$= \frac{33}{\sqrt{3} \times 3 \times 7.41} \times 100\% = 85.71\%$$

Thus, the string efficiency = 85.71% **Ans.**

Problem 8.2. A string of eight suspension insulators is to be graded to obtain uniform distribution of voltage across the string. If the capacitance of the top unit is 10 times the capacitance to ground each unit, determine the capacitance of the remaining seven units.

Solution: Let C be the capacitance to the ground.

The circuit diagram of this string can be drawn as given below

According to the question,

$C_1 = 10\text{ C}$

Let V be the voltage across this string.

For equal voltage distribution across each unit

$$V_1 = V_2 = V_3 = V_4$$

$$= V_5 = V_6 = V_7 = V_8 = \frac{V}{8}$$

Applying KC_1 at node g,

$$i_g + i_1 = i_2,$$

$$V_1 \times \omega \times C + V_1 \times \omega \times C_1 = V_2 \times \omega \times C_2$$

$$\frac{V}{8} \times C + \frac{V}{8} \times 10\text{ C} = \frac{V}{8}\ C_2$$

$$C_2 = (10 + 1)\ C = 11\text{ C}$$

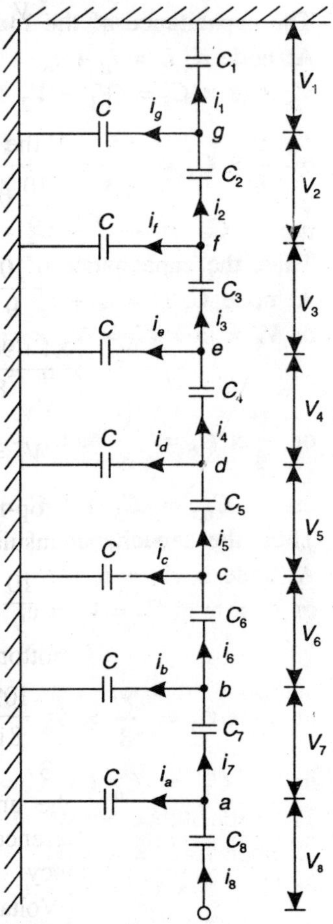

Fig. P. 8.2

Thus, capacitance of the 2nd top most insulator = 11 C

Node f_1, $i_f + i_2 = i_3$

$$(V_1 + V_2) \times \omega \times C + V_2\omega \times C_2 = V_3 \times \omega \times C_3$$

$$\text{or}\quad \left(\frac{V}{8} + \frac{V}{8}\right) \times C + \frac{V}{8}\ C_2 = \frac{V}{8} \times C_3$$

$C_3 = 2C + C_2 = 2C + 11C$

The capacitance of the third unit $(C_3) = 13$ C **Ans.**

At node e, $i_4 = i_3 + i_e$

or $V_4 \times \omega \times C_4 = V_3 \times \omega \times C_3 + (V_1 + V_2 + V_3) \times \omega \times C$

or $\frac{V}{8} \times C_4 = \frac{V}{8} \times C_3 + \left(\frac{V}{8} + \frac{V}{8} + \frac{V}{8}\right) \times C$

or $C_4 = C_3 + 3C$ or $C_4 = 13C + 3C = 16C$

The capacitance of the fourth unit $(C_4) = 16$ C **Ans.**

At node d, $i_5 = i_d + i_4$,

$V_5 \times \omega \times C_5 = (V_1 + V_2 + V_3 + V_4) \times \omega \times C + 4 \times \omega \times C_4$

or $\frac{V}{8} \times C_5 = \left(\frac{V}{8} + \frac{V}{8} + \frac{V}{8} + \frac{V}{8}\right) \times C + \frac{V}{8} \times C_4$

or $C_5 = 4C + C_4 = 4C + 16C = 20C$

Thus, the capacitance of fifth unit $(C_5) = 20$ C **Ans.**

At node C, $i_6 = i_5 + i_c$

or $V_5 \times \omega \times C_6 = V_5 \times \omega \times C_5 + (V_1 + V_2 + V_3 + V_4 + V_5) \times \omega \times C$

or $\frac{V}{8} \times C_6 = \frac{V}{8} \times C_5 + \left(\frac{V}{8} + \frac{V}{8} + \frac{V}{8} + \frac{V}{8} + \frac{V}{8}\right) \times C$

or $C_6 = C_5 + 5C = 20C + 5C = 25C$

Thus, the capacitance of sixth unit $(C_6) = 25$ C **Ans.**

At node b, $A_7 = i_6 + i_b$

or $V_7 \times \omega \times C_7 = V_6 \times \omega \times C_6 + (V_1 + V_2 + V_3 + V_4 + V_5 + V_6) \times \omega \times C$

or $\frac{V}{8} \times C_7 = \frac{V}{8} \times C_6 + \left(\frac{V}{8} + \frac{V}{8} + \frac{V}{8} + \frac{V}{8} + \frac{V}{8} + \frac{V}{8}\right) \times C$

or $C_7 = C_6 + 6C = 25C + 6C = 31C$

The capacitance across seventh unit = 31 C **Ans.**

At node a, $i_8 = i_7 + i_a$

or $V_8 \times \omega \times C_8 = V_7 \times \omega \times C_7 + (V_1 + V_2 + V_3 + V_4 + V_5 + V_6 + V_7) \times \omega \times C$

or $\frac{V}{8} \times C_8 = \frac{V}{8} \times C_7 + \left(\frac{V}{8} + \frac{V}{8} + \frac{V}{8} + \frac{V}{8} + \frac{V}{8} + \frac{V}{8} + \frac{V}{8}\right) \times C$

or $C_8 = C_7 + 7C = 31C + 7C = 38C$

The capacitance of the bottom most insulator = 38 C **Ans.**

Problem 8.3. A string of six insulator unit has mutual capacitance 10 times the capacitance to ground. Determine the voltage across each unit as a fraction of the operating voltage. Also determine the string efficiency.

Solution: Let V be the operating voltage and C be the capacitance to the ground.

Given, No. of insulator unit (n) = 6

According to the question, Mutual capacitance

$$= 10 \times \text{capacitance to the ground} = 10\ C$$

$$\therefore \text{Here, } m = \frac{\text{Mutual capacitance}}{\text{Capacitance to the ground}} = \frac{10\ C}{C} = 10$$

Let V_1, V_2, V_3, V_4, V_5 and V_6 be the voltages across the different units as shown in Fig. P. 8.3.

To calculate voltage across different units, apply the relation

$$V_{n+1} = \frac{V_n}{m} + V_n$$

where n = 1, 2, 3, 4, 5 and 6

$$\text{Now, } V_2 = \frac{V_1}{m} + V_1$$

$$= \frac{V_1}{m} + V_1$$

$$= \left[\frac{1}{10} + 1\right] V_1$$

$$= \frac{11}{10}$$

$$= 1.1\ V_1$$

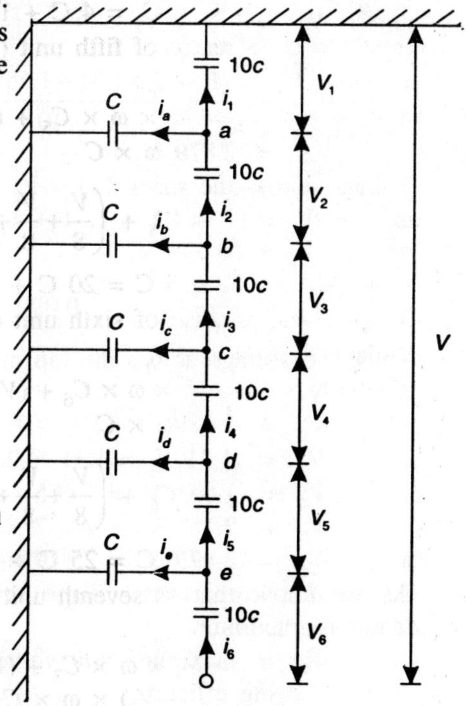

Fig. P. 8.3

The voltage across the third unit is given by

$$V_3 = \frac{V}{m} + V_2 = \frac{V_1 + V_2}{10} + V_2 = \frac{V_1}{10} + \frac{1}{10} V_2 + V_2$$

$$= \frac{V_1}{10} + \frac{11}{10} V_2 = \frac{V_1}{10} + \frac{11}{10} \times \frac{11}{10} V_1$$

$$= \left(\frac{1}{10} + \frac{121}{100}\right) V_1 = 1.31 \ V_1$$

Similarly,

$$V_4 = \frac{V}{m} + V_3 = \frac{V_1 + V_2 + V_3}{10} + V_3$$

$$= \frac{(1.1 + 1.31 + 1)V_1}{10} + 1.31 \ V_1 = 1.651 \ V_1$$

$$V_5 = \frac{V}{m} + V_4 = \frac{V_1 + V_2 + V_3 + V_4}{10} + V_4$$

$$= \frac{(1 + 1.1 + 1.31 + 1.651)V_1}{10} + 1.651 \ V_1 = 2.157 \ V_1$$

and $$V_6 = \frac{V}{m} + V_5 = \frac{V_1 + V_2 + V_3 + V_4 + V_5}{m} + V_5$$

$$= \frac{(1 + 1.1 + 1.31 + 1.651 + 2.157)V_1}{10} + 2.157 \ V_1$$

$$= 2.879 \ V_1$$

Voltage across the string $(V) = V_2 + V_2 + V_3 + V_4 + V_5 + V_6$

or $V = (1 + 1.1 + 1.31 + 1.651 + 2.157 + 2.879) \ V_1$

or $V = 10.097 \ V_1$

$$\therefore \quad V_1 = \frac{1}{10.097} V = 0.099 \ V$$

Thus, the voltage across the top most unit $(V_1) = 0.099$ V

Similarly,

$V_2 = 1.1 \ V_1 = 1.1 \times 0.99 \ V = 0.109 \ V$ **Ans.**

$V_3 = 1.31 \ V_1 = 1.31 \times 0.099 \ V = 0.130 \ V$ **Ans.**

$V_4 = 1.651 \ V_1 = 1.6951 \times 0.099 \ V = 0.163 \ V$ **Ans.**

$V_5 = 2.157 \ V_1 = 2.157 \times 0.099 \ V = 0.214 \ V$ **Ans.**

and $V_6 = 2.879 \ V_1 = 2.879 \times 0.099 \ V = 0.285 \ V$ **Ans.**

As we know that the voltage across the unit nearest to the conductor is maximum

$\therefore \quad V_6$ = maximum unit voltage = 0.285 V

Now, % string efficiency

$$= \frac{\text{operating voltage}}{n \times \text{maximum voltage across the unit nearest to the conductor}} \times 100\%$$

$$= \frac{V}{6 \times 0.285 \ V} \times 100\% = 58.48 \ \%$$ **Ans.**

Problem 8.4. Determine the maximum voltage that the string of the suspension insulators ás shown in figure can withstand if the maximum voltage per unit is 20 kV.

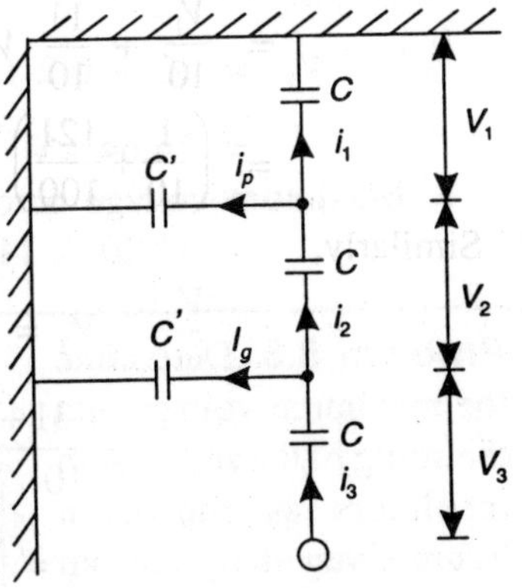

In figure, $C = 0.2$ C

Fig. P. 8.4

Solution: Let voltages across the various units be V_1, V_2 and V_3 such that $V = V_1 + V_2 + V_3$, where V is the desired withstand voltage of the string.

At point P, $i_2 = i_p + i_1$ or $V_2\ \omega C = V_1\omega C_1 + V_1\omega C$

$$\therefore \quad V_2 = (C + C')\ \frac{V_1}{C} \qquad \text{.......(1)}$$

At point q, $i_3 = i_\theta + i_2$ or $V_3\omega C = V_2\ \omega C + (V_1 + V_2)\ \omega C'$

$$\text{or} \quad V_3\ C = V_2\ C + V_2\ C' + V_1\ C'$$

$$\text{or} \quad V_3 \times C = V_2\ (C + C') + V_1\ C'$$

$$\text{or} \quad V_3 = \frac{1}{C}\left[(C+C')\times\frac{(C+C')}{C}V_1 + V_1C'\right]$$

$$\text{or} \quad V_3 = \frac{V_1}{C}\left[\frac{(C+C')^2}{C} + C'\right]$$

$$\text{or} \quad V_3 = \frac{V_1}{C}\left[\frac{(C+0.2C)^2}{C} + 0.26\right]$$

$$\text{or} \quad V_3 = \frac{V_1}{C}\left[\frac{1.44\ C^2}{C} + 0.2\ C\right]$$

$$\text{or} \quad V_3 = V_1\ (1.44 + 0.2)\ 1.64\ V_1 \qquad \text{.......(2)}$$

As we know that the maximum voltage exists across the capacitor nearest to the conductor.

$$\therefore \quad V_3 = V_{max} = 20\ \text{kV or}\ 1.64\ V_1 = 20\ \text{kV}$$

$$\therefore \quad V_1 = 12.20\ \text{kV}$$

From Eq. (1)

$$V_2 = (C + 0.2\,C)\,\frac{V_1}{C} = 1.2 \times 12.20 \text{ kV}$$

$$= 14.64 \text{ kV}$$

$\therefore$ Maximum voltage across the string $(V) = V_1 + V_2 + V_3$

$$= 12.20 + 14.64 + 20 = 28.84 \text{ kV} \quad \textbf{Ans.}$$

Problem 8.5. Determine the maximum voltage that the string of the suspension insulators as shown in figure given alongside can withstand if the minimum voltage per unit is 10 kV.

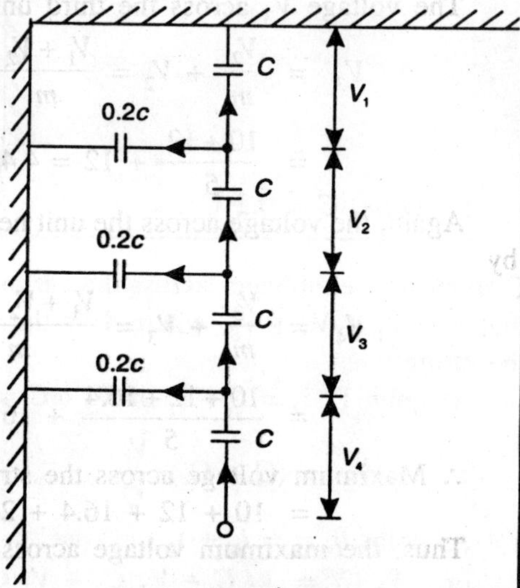

Fig. P. 8.5

Solution: Let voltages across the various units be V_1, V_2, V_3 and V_4 such that maximum voltage across the string

$$(V) = V_1 + V_2 + V_3 + V_4$$

Given, mutual capacitance = C, Capacitance to ground = 0.2 C

As we know that

$$m = \frac{\text{mutual capacitance}}{\text{capacitance to ground}} = \frac{C}{0.2\,C} = 5$$

As we know that the voltage across the unit nearest to the cross arm is minimum,

$$\therefore \quad V_1 = 10 \text{ kV}$$

To calculate voltage across various units, apply the formula

$$V_{n+1} = \frac{V_n}{m} + V_n$$

where n = 1, 2, 3

Now,

$$V_2 = \frac{V_1}{m} + V_1 = \frac{V_1}{m} + V_1$$

$$= \left(\frac{1}{m}+1\right) V_1 = \left(\frac{1}{5}+1\right) V_1 = 1.2 \times 10 = 12 \text{ kV}$$

$\therefore$ $V_2 = 12$ kV

The voltage V_3 across the third unit is given by

$$V_3 = \frac{V_2}{m} + V_2 = \frac{V_1 + V_2}{m} + V_2$$

$$= \frac{10+12}{5} + 12 = 4.4 + 12 = 16.4 \text{ kV}$$

Again, the voltage across the unit nearest to the conductor is given by

$$V_4 = \frac{V_3}{m} + V_3 = \frac{V_1 + V_2 + V_3}{m} + V_3$$

$$= \frac{10+12+16.4}{5} + 16.4 = 24.08 \text{ kV}$$

$\therefore$ Maximum voltage across the string $(V) = V_1 + V_2 + V_3 + V_4$
$= 10 + 12 + 16.4 + 24.08 = 62.48$ kV

Thus, the maximum voltage across the string will be 62.48 kV

Problem 8.6. Determine the voltage across each disc of the suspension insulators as a percentage of the line voltage to earth. The self and capacitance to the ground of each disc is C and 0.5 C respectively as shown in figure. The capacitance between the link pen and the guard ring is 0.1 C. If the capacitance to the line of lower link pin were increased to 0.2 C by means of guard ring, determine the redistribution of voltages. Also determine the string efficiency in each case.

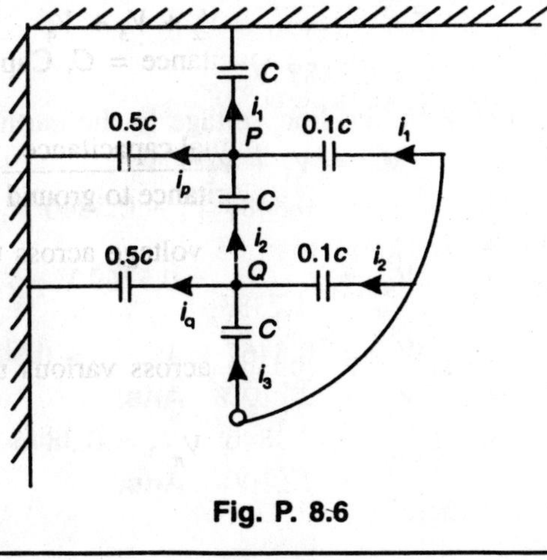

Fig. P. 8.6

Solution: Let V_1 be the voltage across the topmost unit

V_2 be the voltage across the middle unit and

V_3 be the voltage across the bottom unit.

Again Let I_1, I_2 and I_3 be the current through these three units.

At point P, $i_1' + i_2 = i_1 + i_p$

or $(V_3 + V_2)\ \omega \times 0.1\ C + V_2\omega\ C = V_1\ \omega C + V_1 \times \omega \times 0.5\ C$

or $\quad 0.1\ V_3 + 0.1\ V_2 + V_2 = V_1 + 0.5\ V_1$

or $\quad 1.5\ V_1 - 1.1\ V_2 - 0.1\ V_3 = 0$ ....(1)

At point Q, $i_3 + i_2 = i_q + i_2$

$V_3\omega C + V_3\omega \times 0.1\ C = (V_1 + V_2) \times \omega \times 0.5\ C + V_2\omega C$

or $V_3 + 0.1\ V_3 = 0.5\ V_1 + 0.5\ V_2 + V_2$

or $1.1\ V_3 = 0.5\ V_1 + 1.5\ V_2$

or $0.5\ V_1 + 1.5\ V_2 - 1.1\ V_3 = 0$ ....(2)

Let us calculate V_1 and V_2 in terms of V_3.

Multiplying Eq. (2) by 3 and subtracting from Eq. (1), we get

$-1.1\ V_2 - 0.1\ V_3 - 1.5\ V_2 + 1.1\ V_3 = 0$

or $-2.6\ V_2 + 1 \times V_3 = 0$

$$\therefore V_2 = \frac{1}{2.6} V_3$$

$\therefore \quad V_2 = 0.3846\ V_3$

Putting $V_2 = 0.3846\ V_3$ in equation (2), we get

$1.5\ V_1 - 1.1 \times 0.3846\ V_3 - 0.1\ V_3 = 0$

or $1.5\ V_1 - 0.5231\ V_3 = 0$ or

$$\therefore V_1 = \frac{0.5231\ V_3}{1.5}$$

$\therefore \quad V_1 = 0.3487\ V_3$

Let V be the line voltage to the earth, such that

$V = V_1 + V_2 + V_3$

Or, $\quad V = 0.3487\ V_3 + 0.3846\ V_3 + V_3$ Or, $V = 1.7333\ V_3$

$\therefore \quad V_3 = \dfrac{1}{1.7333} = 0.5769\ V = 57.69\%$ of $V = 57.69\%$ **Ans.**

$V_1 = 0.3467 \times V_3,\ V_1 = 0.3487 \times 57.69\%$

$\therefore \quad V_1 = 20.12\%$ **Ans.**

and $\quad V_2 = 0.3846 \times V_3 = 0.3846 \times 57.69\%$

$\therefore \quad V_2 = 22.19\%$ **Ans.**

As we know that String efficiency

$$= \frac{\text{Voltage across the string}}{\text{No. of units} \times \text{Voltage across unit nearest to the power conductor}}$$

$$= \frac{V}{3 \times V_3} \times 100\% = \frac{V}{3 \times \frac{1}{1.733} \times V_3} \times 100\%$$

$$= \frac{1.733}{3} \times 100\% = 57.77\% \quad \textbf{Ans.}$$

When the capacitance to the line of the lower pin is increased to 0.2 C. The equation (1) remains the same while equation (2) gets changed and is given by

$$i_3 + i_2' = i_q + i_2,$$

$$V_3\omega \times 0.2\ C + V_3\omega \times 1.0\ C = (V_1 + V_2) \times \omega \times 0.5\ C + V_2\omega C$$

or $1.2\ V_3 = 1.5\ V_2 + 0.5\ V_1$

$\therefore\ 0.5\ V_1 + 1.5\ V_2 - 1.2\ V_3 = 0$ ....(iii)

Again, Let us calculate V_1 and V_2 in terms of V_3. Multiplying equation (3) by 3 and subtracting it from equation (2), We get

$$-1.1\ V_2 - 0.1\ V_3 - 4.5\ V_2 + 3.6\ V_3 = 0$$

or $-5.6\ V_2 + 3.5\ V_3 = 0$ or $V_2 = \frac{3.5}{5.6}\ V_3$

$\therefore\quad V_2 = 0.625\ V_3$

Substituting $V_2 = 0.625\ V_3$ in equation (1), We get

$$1.5\ V_1 - 1.1 \times 0.025\ V_3 - 0.1\ V_3 = 0$$

or $1.5\ V_1 = 0.7875\ V_3$ or $V_1 = \frac{0.7578}{1.5}\ V_3$

$\therefore\quad V_1 = 0.525\ V_3$

Let V be the line voltage to the earth such that

$$V = V_1 + V_2 + V_3$$

or $\quad V = 0.252\ \text{V}_3 + 0.625\ \text{V}_3 + \text{V}_3$ or $V = 2.15\ \text{V}_3$

$\therefore\quad V_3 = \frac{1}{2.15}\ \text{V} = 0.4651\ \text{V} = 46.51\ \%$ of V

i.e. $\quad V_3 = 46.51\%$ **Ans.**

Now,

$$V_1 = 0.525 \times V_3 = 0.525 \times 46.51\ \% = 24.42\ \% \quad \textbf{Ans.}$$

$$V_2 = 0.625\ V_3 = 0.625 \times 46.51\% = 29.07\% \quad \textbf{Ans.}$$

$$\text{Now, \% String efficiency} = \frac{V}{3 \times V_3} \times 100\%$$

$$= \frac{V}{3 \times \dfrac{V}{2.15}} \times 100\% = \frac{2.15}{3} \times 100\% = 71.67\%$$

> **Problem 8.7.** A string of seven suspension insulators is to be fitted with a grading sign. If the pin to earth capacitances one all equal to C, find the values of line to pin capacitances that would give a uniform voltage distribution over the string.

Solution: Let us draw the circuit of this suspension insulators (Fig. P. 8.7).

Let V be the maximum voltage across the string.

$$V = V_1 + V_2 + V_3 + V_4 + V_5 + V_6 + V_7$$

For equal voltage distribution,

$$V_1 = V_2 = V_3 = V_4 = V_4 = V_5 = V_6 = V_7 = \frac{V}{7}$$

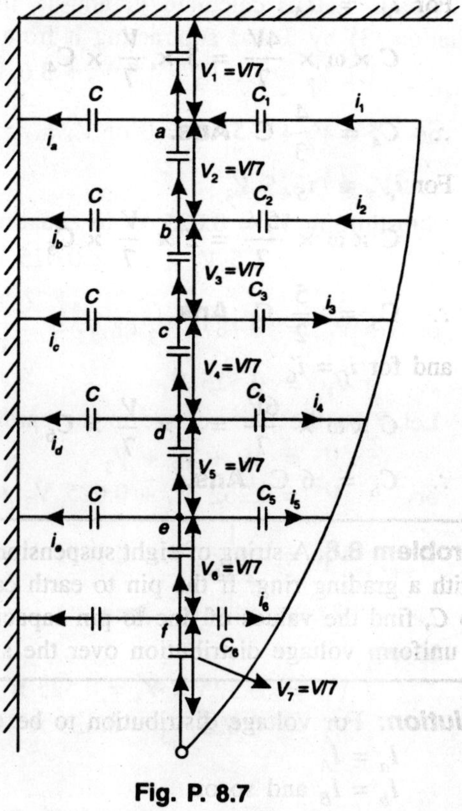

Fig. P. 8.7

$i_a = i'_1,\ i_b = i'_i$ and so on

For $i_a = i'_1$

$$C \times \omega \times \frac{V}{7} = 6 \times \frac{V}{7} \times C_1$$

$$\therefore \quad C_1 = \frac{1}{6}\,C \quad \textbf{Ans.}$$

For $i_b = i'_2$

$$C \times \omega \times \frac{2V}{7} = 5 \times \frac{V}{7} \times C_1$$

$$\therefore \quad C_2 = \frac{2}{5}\,C \quad \textbf{Ans.}$$

For $i_c = i'_3$

$$C \times \omega \times \frac{3V}{7} = 4 \times \frac{V}{7} \times C_3$$

$$\therefore \quad C_3 = \frac{3}{4}\,C \quad \textbf{Ans.}$$

For $i_d = i'_4$

$$C \times \omega \times \frac{4V}{7} = 3 \times \frac{V}{7} \times C_4$$

$$\therefore \quad C_4 = \frac{4}{3}\,C \quad \textbf{Ans.}$$

For $i_e = i'_5$

$$C \times \omega \times \frac{5V}{7} = 2 \times \frac{V}{7} \times C_5$$

$$\therefore \quad C_5 = \frac{5}{2}\,C \quad \textbf{Ans.}$$

and for $i_f = i'_6$

$$C \times \omega \times \frac{6V}{7} = 1 \times \frac{V}{7} \times C_6$$

$$\therefore \quad C_6 = 6\,C \quad \textbf{Ans.}$$

Problem 8.8. A string of eight suspension insulators is to be fitted with a grading ring. If the pin to earth capacitances are all equal to *C*, find the values of line to pin capacitances that would given a uniform voltage distribution over the string.

Solution: For voltage distribution to be uniform

$I_a = I_A$

$I_b = I_B$ and so on

Also,

$E_1 = E_2 = E_3 \text{ ..... } E_8$

$I_a = \omega C \dfrac{E}{8}$ and

$I_A = C_1 \omega \dfrac{7E}{8}$

and $\omega C \dfrac{E}{8} = \omega C_1 \times \dfrac{7E}{8}$

or $C_1 = \dfrac{C}{7}$

Similarly $I_b = I_B$

$I_b = \dfrac{2E}{8} \omega C$

and $I_B = \dfrac{6E}{8} \omega C$

$\omega C = 3C_2 \omega$

or $C_2 = \dfrac{C}{3}$

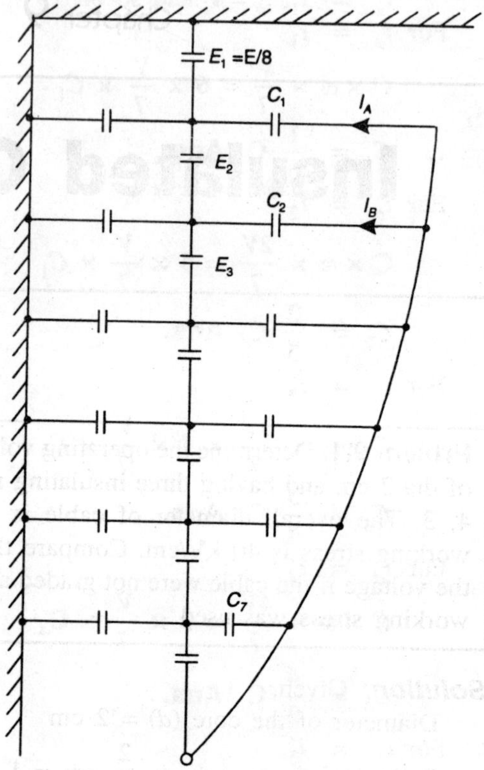

Fig. P. 8.8

Similarly other results can be obtained. It will be seen that the other values are $\dfrac{3C}{5}$, C, $\dfrac{5C}{3}$, $3C$ and $7C$, respectively.

Chapter 9

Insulated Cables

Prblem 9.1. Determine the operating voltage of a single core cable of dia 2 cm and having three insulating material of permitvities 5, 4, 3. The overall diameter of cable is 5 cm and the maximum working stress is 40 kV/cm. Compare the operating voltage with the voltage if the cable were not graded and the material with same working stress was used.

Solution: Given,

Diameter of the core (d) = 2 cm

Radius of the core $(r) = \frac{2}{2}$ cm = 1 cm

Let relative permitivties E_1, E_2 and E_3 be 5, 4, and 3 respectively.

Overall diameter of the cable (D) = 5 cm

Overall radius of the cable (R) = 2.5 cm

Maximum working stress (g_{max}) = 40 kV/cm

As we know that the dielectric with highest permitivity should be placed near the core, the dielectric distribution can be arranged as shown in Fig. P. 9.1.

Let λ be the charge per unit length of the cable.

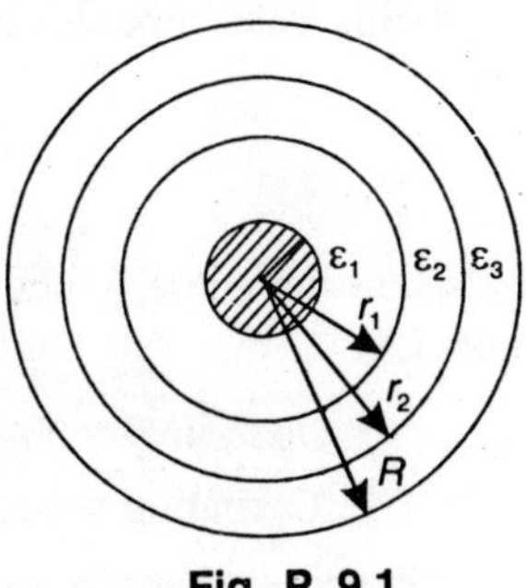

Fig. P. 9.1

The maximum electric stress is given by

$$g_{max} = \frac{\lambda}{2\pi\varepsilon_o\varepsilon_1 r} = \frac{\lambda}{2\pi\varepsilon_o\varepsilon_2 r_1} = \frac{\lambda}{2\pi\varepsilon_o\varepsilon_3 r_2}$$

This yields $\varepsilon_1 r = \varepsilon_2 r_1 = \varepsilon_3 r_2$

When $\varepsilon_1 r = \varepsilon_2 r_1$

$$\therefore \quad r_1 = \frac{\varepsilon_1}{\varepsilon_2} \times r = \frac{5}{4} \times 1 = 1.25 \text{ cm}$$

When $\varepsilon_1 r = \varepsilon_3 r_2$

$$\therefore \quad r_2 = \frac{\varepsilon_1}{\varepsilon_3} \times r = \frac{5}{3} \times 1 = 1.67 \text{ cm}$$

To calculate operating voltage, apply the equation.

$$V = g_{max}\left[r \ln\frac{r_1}{r} + r_1 \ln\frac{r_2}{r_1} + r_2 \ln\frac{R}{r_2}\right] \text{ kV}$$

$$= 40\left[1 \ln\frac{1.25}{1} + 1.25 \ln\frac{1.67}{1.23} + 1.67 \ln\frac{2.5}{1.67}\right] \text{ kV}$$

$$= 50.36 \text{ kV} \quad \textbf{Ans.}$$

Operating voltage (V) = 50.36 kV

When the cable is not graded, the situation is shown in Fig. P. 9.1 (a).

We have, R = 2.5 cm, r = 1 cm

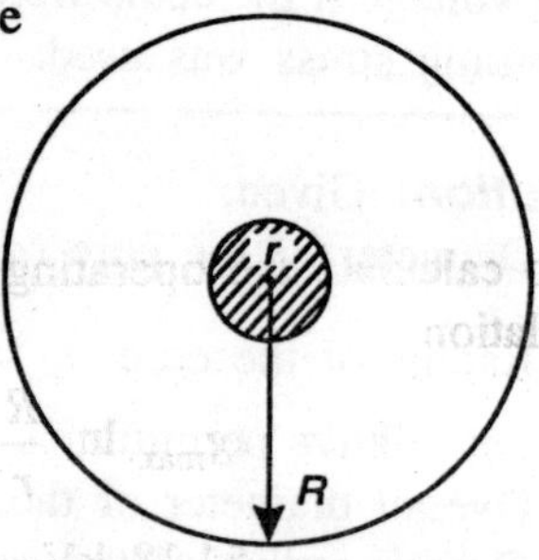

Fig. P. 9.1 (a)

In this case, operating voltage= $r g_{max} \times \ln \dfrac{R}{r}$

$$V' = 1 \times 40 \ln \frac{2.5}{1}$$

$$V' = 36.65 \text{ kV} \quad \textbf{Ans.}$$

When the cable is not graded, the operating voltage is reduced to 36.65 kV.

$$\text{Now,} \quad \frac{\text{Operating voltage of the graded cable}}{\text{Operating voltage of ungraded cable}} = \frac{50.36}{36.65} = 1.374$$

This shows that the operating voltage of the graded cable is 1.374 times that of the ungraded cable.

Problem 9.2. A single core cable has an inner diameter of 5 cms and a core diameter of 1.5 cm. Its paper dielectric has a working maximum dielectric stress of 60 kV/cm. Calculate the maximum permissible line voltage when such cables are used on a 3-phase power system.

Solution: Given,

Diameter of the core (d) = 1.5 cm

Radius of the core (r) = 0.75 cm

Inner diameter of the cable (d) = 5 cm

$\therefore$ Inner radius of the cable (R) = 2.5 cm

as shown in Fig. P. 9.2.

Maximum working dielectric, Stress (g_{max}) = 60 kV/core

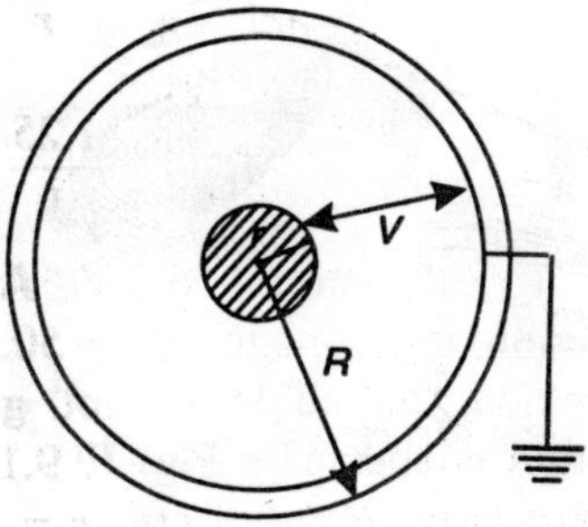

Fig. P. 9.2

To calculate the operating voltage of the cable per phase, apply the relation

$$V = rg_{max} \ln \frac{R}{r} \text{ kV} = 0.75 \times 60 \times \ln \frac{2.5}{0.75} \text{ kV}$$

$$= 54.18 \text{ kV}$$

When such cables are used on a 3-phase power system,

$\therefore$ operating line voltage (V_L)

$$= \sqrt{3} \times V = \sqrt{3} \times 54.18 \text{ kV} = 93.84 \text{ kV} \quad \textbf{Ans.}$$

Problem 9.3. A 1-phase concentric cable 5 km long has a capacitance of 0.2 μF per km, the relative permitivity of the dielectric being 3.5 The diameter of the inner conductor is 1.5 cm and the supply voltage is 66 kV at 50 Hz. Calculate the inner diameter of the outer conductor, the rms voltage gradient at the surface of the inner conductor and the rms value of the charging current.

Solution: Let d be the inner diameter of the conductor and D be the outer diameter of the conductor as shown in Fig. P. 9.3.

Given, d = 1.5 cm, Operating voltage (V) = 66 kV

Capacitance (C) = 0.2 μF/km = 0.2×10^{-9} F/m

Let us calculate charge per unit length of the cable.

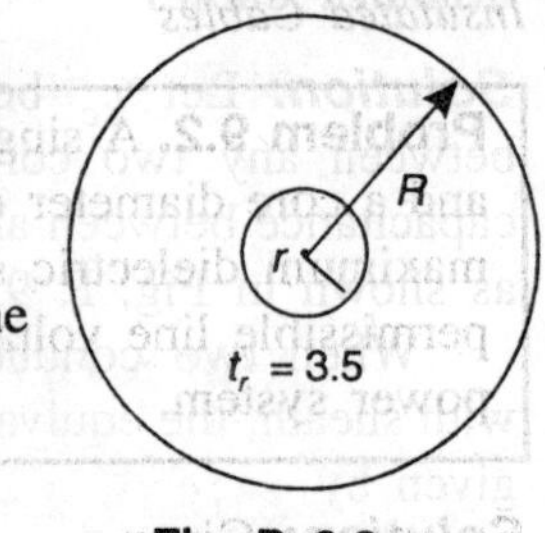

Fig. P. 9.3

We have, $q = CV$

$$\therefore \quad \lambda = C \times V = 0.2 \times 10^{-9} \times 66 \times 10^{3} \text{ C/m}$$

$$= 13.2 \times 10^{-6} \text{ μC/m} = 13.2 \times 10^{-8} \text{ μC/cm.}$$

The rms voltage gradient at the surface of inner conductor is given by

$$g_{max} = \frac{\lambda}{2\pi\varepsilon_o\varepsilon_r \times r} = \frac{13.2\times10^{-8}\times2}{2\pi\times8.854\times10^{-12}\times3.5\times1.5}$$

$$= 90.39 \text{ kV/cm} \quad \textbf{Ans.}$$

To calculate the inner diameter of the outer conductor, apply the relation,

$$V = rg_{max} \ln\frac{D}{d} \text{ or } \ln\frac{D}{d} = \frac{V}{r\times g_{max}}$$

$$\text{or } \ln\frac{D}{d} = \frac{66\times2}{1.5\times90.39} = 0.974$$

$$\therefore \quad D = d \times e^{0.974} = 3.97 \text{ cm}$$

Thus, inner diameter of the outer conductor = 3.97 cm **Ans.**

Total capacitance of the cable (C_T) = C × length of the cable

$$= 0.2 \text{ μF} \times 5 = 1 \text{ μF}$$

Now, The charging current of the cable in given by

$$= \frac{V}{X_c} = 2\,\pi FC \times V$$

$$= 2\pi\ 50 \times 1 \times 10^{-6} \times 66 \times 10^{3} \text{ A} = 20.73 \text{ A} \quad \textbf{Ans.}$$

Problem 9.4. The capacitances of a 3-core cable belted type are measured and found to be as follows :

(i) between 3-cores bunched together and the sheath 8 μF,

(ii) between conductor and the other two connected together to the sheath 5 μF. Calculate the capacitance to neutral and the total charging kVA when the cable is connected to a 11 kV 50Hz 3-phase supply.

Solution: Let c_c be the capacitance between any two cores and C_s be the capacitance between any core and sheath as shown in Fig. P. 9.4.

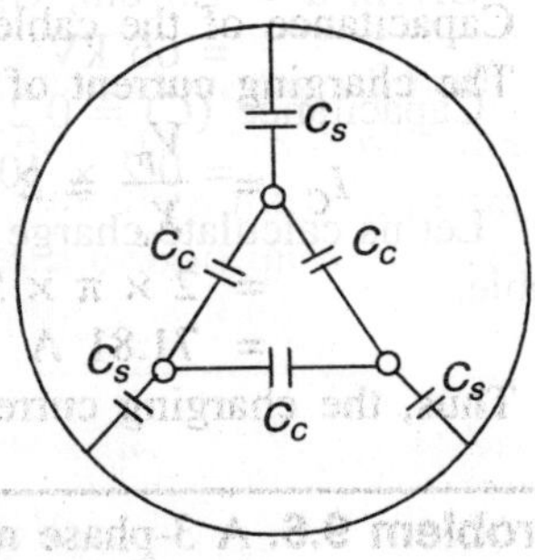

Fig. P. 9.4

When two conductors are bunched with sheath, the equivalent capacitance is given by

$$C_1 = 2C_c + C_s$$

Again, when all the conductors are bunched, the equivalent capacitance is given by $C_2 = 3C_s$

Given, From (i) $C_2 = 8\ \mu F$ or $3C_s = 8\ \mu F$

$$\therefore C_s = \frac{8}{3}\ \mu F$$

Again, From (ii) $C_1 = 5\ \mu F$ or $2C_c + C_s = 5\ \mu F$

$$\text{Or}\quad 2C_c = \left(5 - \frac{8}{3}\right)\ \mu F$$

$$\therefore C_c = \frac{7}{6}\ \mu F$$

The capacitance to neutral in given by

$$C_n = 3C_c + C_s = 3 \times \frac{7}{6} + \frac{8}{3} = 6.17\ \mu F \quad \textbf{Ans.}$$

Given, Operating line voltage = 11 kV

$$\text{Phase voltage} = \frac{11}{\sqrt{3}}\ \text{kV} = 6.351\ \text{kV}$$

The total charging kVA is given by 3 times the per phase kVA

$$\therefore \text{Total kVA} = 3 \times V_p^2\, \omega C_n$$

$$= 3 \times (6.351)^2 \times 10^6 \times 314 \times 6.17 \times 156\ \text{kVA}$$

$$= 234.434\ \text{kVA} \quad \textbf{Ans.}$$

> **Problem 9.5.** An H-type cable, 40 km long has a capacitance per km between any two-conductors of 0.15 μF. The supply voltage is 3-phase 33 kV at 50 Hz. Determine the charging current.

Solution: Given;

Line voltage (V_c) = 33 kV

$$\text{Phase voltage } (V_p) = \frac{33}{\sqrt{3}}\ \text{kV} = 19.05\ \text{kV}$$

Operating frequency (f) = 50 HZ

Capacitance per unit length = 0.15 μF/km

Total capacitance of the cable = $0.15 \times 10^{-6} \times 40$ F = 6 μF

Capacitance of the cable to neutral = $2 \times 6\ \mu F = 12\ \mu F$
The charging current of the cable is given by

$$I_C = \frac{V_p}{X_c} = 2\ \pi f c V_p$$
$$= 2 \times \pi \times 50 \times 12 \times 10^{-6} \times 19.05 \times 10^3 \text{ A}$$
$$= 71.81 \text{ A} \quad \textbf{Ans.}$$

Thus, the charging current through the cable is 71.81 A.

Problem 9.6. A 3-phase metal sheathed cable one km long gave the following results on a test for capacitance :

(i) Capacitance between two conductors bunched with the sheath and the third conductor 0.5 μF.

(ii) Capacitance between bunched conductors and sheath 1 μF.

With the sheath insulated, find the capacitance (a) between any two cores, (b) between any two bunched conductors and the third conductor, and (c) calculate the charging current phase per km when connected to 11 kV 50 Hz supply.

Solution: Let C_s be the capacitance between each core and sheath and C_c be the capacitance between any two cores as shown in Fig. P. 9.6 (a)-I.

When two conductors are bunched with sheath, the equivalent capacitance is given by $C_1 = 2C_c + C_s$ as shown in Fig. P. 9.6 (a)-II.

Fig. P. 9.6 (a)

When all the conductors are bunched, the equivalent capacitance is given by $C_2 = 3C_s$

Given, $C_1 = 0.5\ \mu F/km$ *ie* $2C_c + C_s = 0.5 \times 10^{-6}$ ....(1)

and $C_2 = 1 \times 10^{-6}$ F/km, $C_2 = 3C_s = 1 \times 10^{-6}$ ....(2)

Now, to calculate capacitance to neutral, apply the relation

$$C_o = \frac{3}{2}C_1 - \frac{C_2}{6} = \left(\frac{3}{2} \times 0.5 - \frac{1}{6}\right) \mu F/km$$
$$= 0.5833\ \mu F/km$$

(a) The capacitance between any two cores is given by

$$C_c = \frac{C_0}{2} = \frac{0.5833}{2}\ \mu F = 0.2917\ \mu F/km \quad \textbf{Ans.}$$

(b) From equation (2) $C_s = 0.333\ \mu F/km$

Again, from equation (1),

$$2C_c + C_s = 0.5 \times 10^{-6} \text{ or } 2C_c = 0.5 \times 10^{-6} - 0.333 \times 10^{-6}$$

$$\therefore \quad C_c = 0.0835\ \mu F/km$$

The equivalent circuit for measuring capacitance between two bunched conductors and the third conductor can be drawn as shown in Fig. P. 9.6 (b).

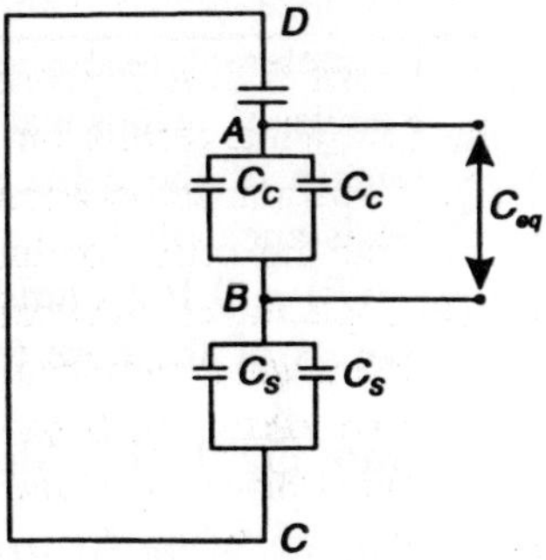

Fig. P. 9.6 (b)

The equivalent capacitance between B and C is given by

$$C_{BC} = C_s + C_s = 2C_s$$

Now, The equivalent capacitance between A and B is given by

$$C_{eq} = C_{AB} = 2C_c + \frac{C_{BC} \times C_{AD}}{C_{BC} + C_{AB}} = 2C_c + \frac{2C_S \times C_S}{3C_S}$$

$$= 2C_c + \frac{2}{3}C_s = (2 \times 0.0835 + \frac{2}{3} \times 0.333)\ \mu F/\ km$$

$$= 0.389\ \mu F/km$$

Thus, Equivalent capacitance between any two bunched conductors and third conductor = 0.389 μF/km **Ans.**

(c) Given, Line voltage (V_L) = 11 kV

$$\therefore \text{ Phase voltage } (V_L) = \frac{11}{\sqrt{3}}\ kV = 6.351\ kV$$

Frequency (f) = 50 Hz.

The charging current per phase per km is given by

$$= V_p \times \omega C_o = V_p \times 2\pi f \times C_o$$

$$= (6.351) \times 10^3 \times 2 \times \pi \times 50 \times 0.5833 \times 10^{-6}$$

$$= 1.163\ A/km \quad \textbf{Ans.}$$

Problem 9.7. A 3-phase underground cable consists of 3 single core cables each of radius 0.75 cm and spaced 5 cm apart in equilateral formation. The diameter of the lead sheath is 2.3 cm and the sheath thickness 0.15 cm. The specific resistance of lead is

22.0 × 10^{-6} Ω-cm at the working temperature and the conductor resistance 0.162 ohm per km at 65° C, for a cable length of 1.6 km and a load of 200 amps determine (i) the ratio of sheath loss, to core loss and (ii) the induced voltage without bonding.

Solution: Given, $\rho_s = 22.0 \times 10^{-6}$ Ω-cm.

Sheath thickness $(t) = 0.15$ cm

Distance between the centers of two covers $(D) = 5$ cm

Diameter of lead sheath $(d) = 2.3$ cm

Radius of lead sheath $(r) = 1.15$ cm

Length of the cable $(l) = 1.6$ km

Resistance of the conductor

$(R) = 0.162$ ohm/Km

Radius of the core $(r_c) = 0.75$ cm

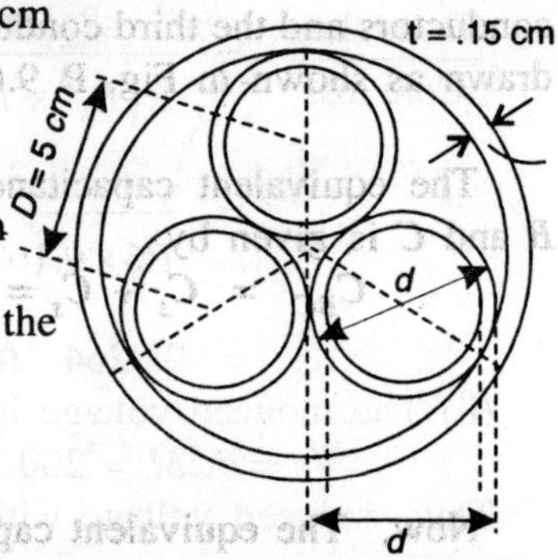

Fig. P. 9.7

From Fig. P. 9.7, Mean radius of the

$$\text{sheath } (r_m) = r - \frac{t}{2}$$

$$= 1.15 - \frac{0.15}{2}$$

$$= 1.075 \text{ cm}$$

The mutual inductance between the core is given by

$$M = 2 \times 10^{-7} \ln \frac{D}{r_m} \text{ H/m}$$

$$= 2 \times 10^{-7} \ln \frac{5}{1.075} \text{ H/m}$$

$$= 3.074 \times 10^{-7} \text{ H/m}$$

Mutual inductance of 1.6 km long cable $= M \times l$

$$= 3.074 \times 10^{-7} \times 1.6 \times 1000 \text{ H} = 4.92 \times 10^{-4} \text{ H}$$

Now, let us calculate the resistance of the sheath as we know that

$$R = \rho \frac{l}{A}$$

$$\text{Now, Area of the sheath } (A) = \frac{\pi\left[d^2 - (d-2t)^2\right]}{4}$$

$$= \frac{\pi}{4}[(2.3)^2 - (2.3 - 2 \times 0.15)^2] = 1.013 \text{ sq.cm}$$

Let us take $l = 1$ km $= 1 \times 10^5$ cm

$$\rho_s = 22.0 \times 10^{-6} \ \Omega\text{–cm}$$

Resistance of the sheath $(R_s) = \rho_s \dfrac{l}{A}$

$$= \frac{22.0 \times 10^{-6} \times 10^{5}}{1.013} \ \Omega/\text{km} = 2.172 \ \Omega/\text{km}$$

$\therefore$ Resistance of the sheath of 1.6 km

$$= R_s \times l = 1.6 \times 2.172 \ \Omega = 3.48 \ \Omega$$

(i) To calculate ratio of the sheath loss to core loss of the cable, apply the relation,

$$\frac{\text{Sheath loss}}{\text{Core loss}} = \frac{R_S M^2 \omega^2}{R\left(R_S{}^2 + M^2\omega^2\right)}$$

$$= \frac{3.48 \times (4.92)^2 \times 10^{-8} \times (314)^2}{1.6 \times 0.162\left[(3.48)^2 + (4.92)^2 \times 10^{-8} \times (314)^2\right]}$$

$$= 0.0264 \ \textbf{Ans.}$$

(ii) The induced voltage is given by

$$V = I\omega M = 200 \times 314 \times 4.90 \times 10^{-4} \text{ V} = 30.89 \text{ V}$$

Thus, induced voltage without bonding = 30.89 kV **Ans.**

Problem 9.8. A single core metal-sheathed cable operating at 66 kV is to be graded by means of a metallic intersheath: (a) Determine the diameter of the intersheath and the voltage at which it must be maintained in order to obtain the minimum overall cable diameter D. The dielectric strength of the material is 50 kV/cm. (b) Compare the conductor and outside diameters with those of an ungraded cable of the same material under the same condition.

Solution: Given,

Operating voltage (V) = 66 kV

Dielectric strength (g_{max}) = 50 kV/cm

(a) Let, d, d_1 and D be the diameters of the conductor, intersheath and *C* overall cable respectively. Let the intersheath be maintained at a potential V_1

From Fig. P. 9.8,

$d_1 = 2.718\ d$ ....(1)

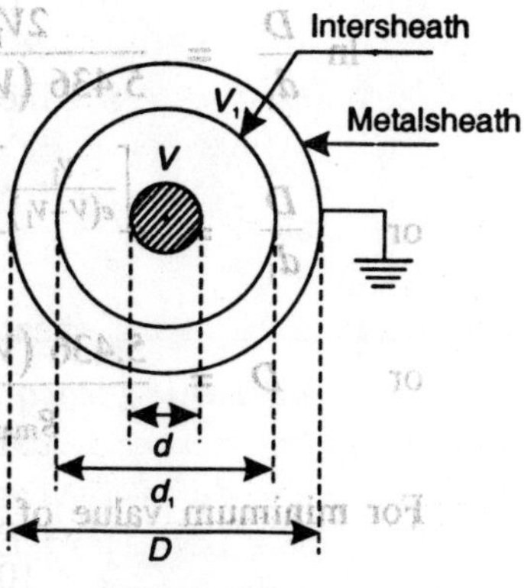

Fig. P. 9.8

For conductor and intersheath, the maximum stress occurs at a distance $d/2$ from the centre of the conductor and its value is given by

$$g_{max} = \frac{V - V_1}{\left(\frac{d}{2}\right)\ln\frac{d_1}{d}} \quad \text{..................} (2)$$

Now, the maximum stress between the intersheath and metal sheath occurs at the surface of the intersheath and in given by

$$g_{max} = \frac{V_1}{\left(\frac{d_1}{2}\right)\ln\frac{d}{d_1}} \quad \text{..................} (3)$$

From, Eqs. (1) and (2) $\dfrac{(V - V_1)\times 2.718}{\left(\frac{d_1}{2}\right)\ln\frac{d_1}{d}} = g_{max}$

$$\therefore \quad d_1 = \frac{5.436(V - V_1)}{g_{max}\ln\frac{d_1}{d}} = \frac{5.436(V - V_1)}{g_{max}\ln\frac{e\times d}{d}}$$

$$= \frac{5.436(V - V_1)}{g_{max}} \quad \text{.......}(4)$$

Substituting the value of d_1 in Eq. (3), we get

$$g_{max} = \frac{2V_1 \times g_{max}}{\ln\frac{D}{d_1}\times 5.436(V - V_1)}$$

$$\ln\frac{D}{d_1} = \frac{2V_1}{5.436\ (V - V_1)} \text{ or } \ln\frac{D}{d_1} = \frac{2V_1}{2\times e(V - V_1)}$$

$$\text{or} \quad \frac{D}{d_1} = e^{\left[\frac{V_1}{e(V-V_1)}\right]} \text{ or } D = d_1\ e^{\frac{V_1}{2.718\ (V-V_1)}}$$

$$\text{or} \quad D = \frac{5.436\ (V - V_1)}{g_{max}}\ e^{\frac{V_1}{2.718\ (V-V_1)}} \quad \text{.........} (5)$$

For minimum value of D, $\dfrac{dD}{dV_1} = 0$

$$\text{or } \frac{5.436\left(V-V_1\right)}{g_{\max}} \times \frac{V_1}{2.718\left(V-V_1\right)}\, e^{\frac{V_1}{2.718(V-V_1)}} - \frac{5.436}{g_{\max}}\, e^{\frac{V_1}{2.718(V-V_1)}} = 0$$

on simplifying, this yeilds $\frac{V}{2.718\left(V-V_1\right)} = 1$

or $V = 2.718\ V - 2.718\ V_1$ or, $V_1 = \frac{1.718}{2.718}\ V$

or $V_1 = 0.632\ V$ ....(6)

Now, from equation (4),

$$d_1 = \frac{5.436\left(V-V_1\right)}{g_{\max}} = \frac{5.436(1-0.632)\times 66}{50}$$

$$= 2.64 \text{ cm}$$

Then, Diameter of intersheath (d_1) = 2.64 cm **Ans.**

From Eq. (6), Voltage at the interseath = 0.632 × V

= 0.632 × 66 = 41.712 kV **Ans.**

(b) From equation (1)

$$d = \frac{d_1}{2.718} = \frac{2.64}{2.718} = 0.971 \text{ cm}$$

∴ diameter of the conductor (d) = 0.971 cm. **Ans.**

Now, from equation (5),

$$D = \frac{5.436\left(V-0.632\ V\right)}{g_{\max}}\, e^{\frac{0.632\ V}{2.718(1-0.632)V}}$$

$$= \frac{3.76\ V}{g_{\max}} = \frac{3.76\times 66}{50} = 4.96 \text{ cm}$$

Overall diameter of the cables = 4.96 cm **Ans.**

Again, When the cable is ungraded,

$D = ed = 2.718$ d

The maximum electric stress is given by

$$g_{max} = \frac{V}{(d/2)\ln\frac{D}{d}}$$

$$\text{or} \quad d = \frac{2V}{g_{\max}\ \ln\frac{e\times d}{d}} = \frac{2V}{g_{\max}}$$

In this case, Core diameter $= \dfrac{2 \times 66}{50} = 2.64$ cm **Ans.**

Overall diameter of the cable (D)

$= 2.718 \times d = 2.718 \times 2.64 = 7.18$ cm **Ans.**

Thus, When the cable is ungraded, the overall diameter of the cable is increased to 7.18 cm.

Problem 9.9. Determine the maximum working voltage of a single core lead sheathed cable having a conductor 1 cm dia and sheath of 5 cm dia inside. Two insulating materials with permitivities and maximum stresses 4, 2.5 and 60 kV/cm and 50 kV/cm respectively are used.

Solution: Given,

Dia of the conductor (d) = 1 cm

Inner dia of the sheath (D) = 5 cm

Radius of the conductor (r) = 0.5 cm

Inner radius of the sheath (r) = 2.5 cm

As we know that the insulating material having highest permitivity is placed nearer to the conductor as shown in Fig. P. 9.9.

Let $\varepsilon_1 = 4, \varepsilon_2 = 2.5$

$g_{1max} = 60$ kV/cm

$g_{2max} = 50$ kV/cm

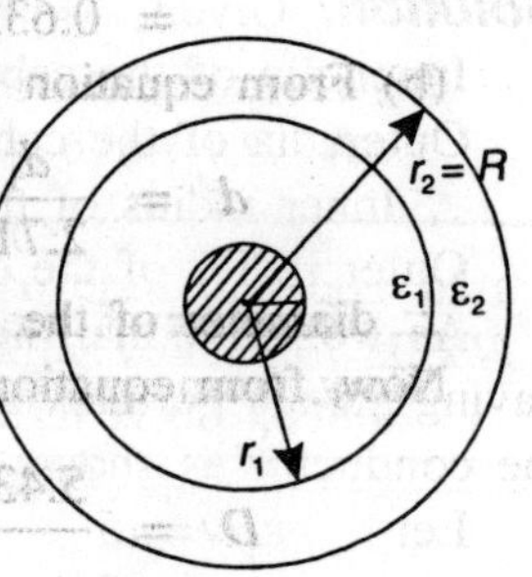

Fig. P. 9.9

Let λ be the charge permit length of the cable

The maximum electric stress (g_{1max}) at the conductor surface is given by

$$g_{1max} = \frac{\lambda}{2\pi\varepsilon_0\varepsilon_1 r} \quad \text{.........(1)}$$

Again, the maximum electric stress (g_{2max}) at the surface of inner insulating material is given by

$$g_{2max} = \frac{\lambda}{2\pi\varepsilon_0\varepsilon_2 r_1} \quad \text{.........(2)}$$

From equations (1) and (2)

$$\frac{\lambda}{2\pi\varepsilon_0} = r\varepsilon_1\, g_{1max} = r_1\varepsilon_2\, g_{2max}$$

or $1.5 \times 5 \times 38 = r_1 \times 3 \times 28$

This yeilds, $r_1 = 0.96$ cm

Now, the operating voltage of the cable is given by

$$V = rg_{1max} \ln \frac{r_1}{r} + r_1 g_{2max} \ln \frac{R}{r_1} \text{ kV}$$

$$= 0.50 \times 60 \ln \frac{0.96}{0.50} + 0.96 \times 50 \ln \frac{2.5}{0.96} \text{ kV}$$

$$= 65.51 \text{ kV}$$

Thus, maximum working voltage of the cable (V)

$= 65.51$ kV **Ans.**

Problem 9.10. The inner and outer dia of a cable are 3 cm and 8.5 cm. The cable is insulated with two materials having permitivies of 5 and 3 respectively with corresponding stresses of 38 kV/cm and 28 kV/cm. Calculate the radial thickness of each insulating layer and the safe working voltage of the cable.

Solution: Given,

Inner dia of the cable (d) = 3 cm

Outer dia of the cable (D) = 8.5 cm

∴ Inner radius of the cable (r) = 1.5 cm

Outer radius of the cable (R) = 4.25 cm

As we know that the insulating material having greater permitivity is placed nearer to the conductor as shown in Fig. P. 9.10.

Let $\varepsilon_1 = 5$

$\varepsilon_2 = 3$

$g_{1max} = 38$ kV/cm

$g_{2max} = 28$ kV/cm

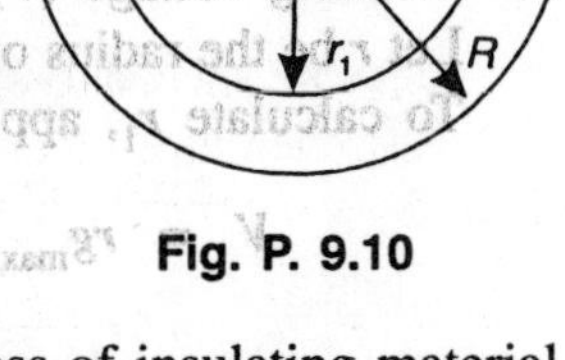

Fig. P. 9.10

Let r_1 be the radius and t_1 be the thickness of insulating material nearer to the core and t_2 be the thickness of the outer insulating material.

Again, let α be the charge per unit length of the cable.

The maximum electric stress (g_{1max}) at the surface of the conductor is given by

$$g_{1max} = \frac{\lambda}{2\pi\varepsilon_0\varepsilon_1 r} \quad \text{.............(1)}$$

Again, the maximum electric stress at the outer surface of the inner insulating material is given by

$$g_{2max} = \frac{\lambda}{2\pi\varepsilon_0\varepsilon_2 r_l} \quad \text{.............(2)}$$

From equations (1) and (2)

$$\frac{\lambda}{2\pi\varepsilon_0} = r\varepsilon_1 g_{1max} = r_1\varepsilon_2\, g_{2max}$$

$1.5 \times 5 \times 38 = r_1 \times 3 \times 28$

This yeilds $r_1 = 3.39$ cm

Now, Thickness of the inner insulating material (t_1)

$= r_1 - r = 3.39 - 1.5 = 1.89$ cm Ans.

Thickness of the outer insulating material (t_2)

$= R - r_1 = 4.25 - 3.39 = 0.86$ cm $= 8.6$ mm **Ans.**

The maximum safe working voltage of the cable is given by

$$V = rg_{1max} \ln \frac{r_1}{r} + r_1 g_{2max} \ln \frac{R}{r_1}$$

$$= 1.5 \times 38 \times \ln \frac{3.39}{1.5} + 3.39 \times 28 \times \ln \frac{4.25}{3.39} \text{ kV}$$

$$= 67.94 \text{ kV}$$

Thus, maximum safe operating voltage = 67.94 kV **Ans.**

Problem 9.11. Determine the economic overall diameter of a single core cable metal sheathed for a working voltage of 80 kV, if dielectric strength of the insulating material is 60 kV/cm.

Solution: Given

Working voltage (*V*) = 80 kV, g_{max}= 60 kV/cm

Let *r* be the radius of the conductor and *R* be the dia of the sheath

To calculate r_1, apply the relation.

$$V = rg_{max} \ln \frac{R}{r}$$

For economic size, $\frac{R}{r} = e$

$$\therefore \quad V = rg_{max} \times \ln e \text{ or } V = rg_{max}$$

$$\therefore \quad r = \frac{V}{g_{max}} = \frac{80}{60} = 1.33 \text{ cm}$$

Diameter of the conductor (*d*) = $2 \times r$

$= 2 \times 1.33 = 2.66$ cm

The overall diameter of the single core cable metal sheathed in given by

$$D = de = 2.66 \times e = 7.23 \text{ cm} \quad \textbf{Ans.}$$

Problem 9.12. A single core lead covered cable is to be designed for 60 kV to earth. Its conductor radius is 0.4 cm and its here insulating materials *A*, *B* and *C* have relative permitivities of 3.5, 4 and 5 with maximum permissible stress 40, 30 and 50 kV/cm respectively. Determine the minimum internal diameter of diameter of the sheath. Discuss the arrangement of insulating materials.

Solution: Given

Radius of the conductor (r) = 0.4 cm

$\varepsilon_A = 3.5, \varepsilon_B = 4, \varepsilon_C = 5$

$g_{Amax} = 40$ kV/cm, $g_{Bmax} = 30$ kV/cm, $g_{Cmax} = 50$ kV/cm

Let r_A, r_B and r_C be the radius of the insulating materials A, B. and C respectively. Let R be the radius of the lead sheath.

Now, to have minimum internal diameter of the lead sheath and consistency of electric stresses. The material having maximum value of product of electric stresses and relative permitivity should be placed nearer to the conductor, i.e. the insulating materials are arranged in the decreasing order of their respective product.

Now, $\varepsilon_A g_{Amax} = 3.5 \times 40 = 140,$

$\varepsilon_B g_{Bmax} = 4 \times 30 = 120$

$\varepsilon_C g_{Cmax} = 5 \times 50 = 250$

Thus, the insulating materials can be arranged as ε_C, ε_A and ε_B as shown in Fig. P. 9.12.

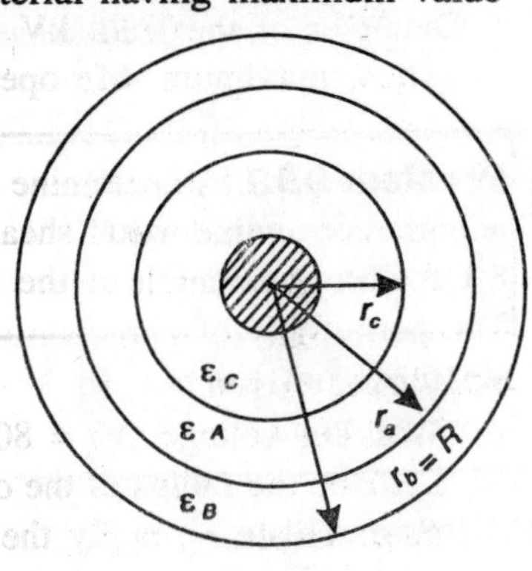

Fig. P. 9.12

Let q be the charge per unit length of the cable then.

$$g_{cmax} = \frac{q}{2\pi\varepsilon_o\varepsilon_c r}$$

$$g_{amax} = \frac{q}{2\pi\varepsilon_o\varepsilon_A r_c}$$

$$g_{bmax} = \frac{q}{2\pi\varepsilon_o\varepsilon_B r_a}$$

From these three equations.

$$\frac{q}{2\pi\varepsilon_o} = r\varepsilon_C \, g_{cmax} = r_a\varepsilon_B \, g_{bmax} = r_c\varepsilon_A \, g_{amax}$$

Now, $r\varepsilon_c\, g_{cmax} = r_c\varepsilon_A\, g_{amax}$ or $0.4 \times 5 \times 50 = r_c \times 3.5 \times 40$

This yields $r_c = \dfrac{0.4\times5\times50}{3.5\times40} = 0.714$ cm

Again, $r\varepsilon_c\, g_{cmax} = r_a\varepsilon_B\, g_{bmax}$, ε $0.4 \times 5 \times 50 = r_a \times 4 \times 30$

This yeilds $r_a = \dfrac{0.4\times5\times50}{4\times30} = 0.833$ cm

The maximum operating voltage is given by

$$V = rg_{cmax} \ln\frac{r_C}{r} + r_c\, g_{amax} \ln\frac{r_a}{r_C} + r_a g_{bmax} \ln\frac{R}{r_a}$$

or $$60 = 0.4 \times 50 \ln\frac{0.714}{0.4} + 0.714 \times 40 \ln\frac{0.833}{0.714} + 0.833 \times 30 \ln\frac{R}{r_a}$$

or $\ln\dfrac{R}{r_a} = 2.001$

$\therefore \dfrac{R}{r_a} = e^{2.001}$

or $R = r_a \times e^{2.001} = 7.396 \times 0.833 = 6.16$ cm.

Diameter of the lead sheath $(D) = 2 \times R = 2 \times 6.16$

$= 12.32$ **Ans.**

Problem 9.13. A conductor of 2 cm dia passes centrally through a porcelain cylinder of internal diameter 4 cm and external dia 8 cm. The cylinder is surrounded by tightly fitting metal sheath. The permitivity of porcelain is 6 and the peak voltage gradient in air must not exceed 36 kV/cm Determine the maximum safe working voltage.

Solution: Given,

Diameter of the conductor $(d) = 2$ cm

$\therefore$ Radius of the conductor $(r) = 1$ cm

$g_{1max} = 36$ kV/cm, or $d_1 = 4$ cm

$\therefore r_1 = 2$ cm and $d_2 = 8$ cm

$\therefore r_2 = 4$ cm

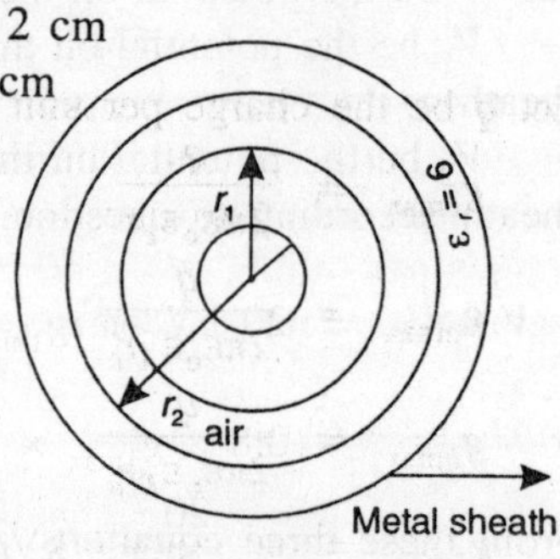

Fig. P. 9.13

Let g_{1max} be the maximum electric stress on the surface of the conductor and g_{2max} be the maximum electric stress on the inner surface of the procelain.

Again, Let q be the charge per unit length of the conductor then

$$g_{1max} = \frac{q}{2\pi\varepsilon_o r} \qquad ....(1)$$

and $$g_{2max} = \frac{q}{2\pi\varepsilon_o\varepsilon_r r_1} \qquad ....(2)$$

From equations (1) and (2) $\frac{q}{2\pi\varepsilon_o} = rg_{1max} = \varepsilon_r\ r_1\ g_{2max}$

This yields

$$g_{2max} = \frac{r \times g_{1max}}{\varepsilon_r \times r_1} = \frac{1 \times 36}{6 \times 2} = 3 \text{ kV/cm}$$

The maximum working voltage is given by

$$V = rg_{1max} \ln \frac{r_1}{r} + r_1 g_{2max} \ln \frac{r_2}{r} \text{ kV}$$

$$= 1 \times 36 \ln \frac{2}{1} + 2 \times 3 \times \ln \frac{4}{2} \text{ kV}$$

$$= 29.112 \text{ kV}$$

Then, maximum safe working voltage = 29.112 kV **Ans.**

Problem 9.14. A 60 kV concentric cable with two intersheaths has a core diameter 1.6 cm. Dielectric material 3 mm thick constitutes the three cover of insulation. Determine the maximum stress in each of the three layers if 20 kV is maintained across each of inner two.

Solution: Given,

Diameter of the core (d) = 1.6 cm

∴ Radius of the core (r) = 0.8 cm

The dielectric material can be arranged as shown in Fig. P. 9.14.

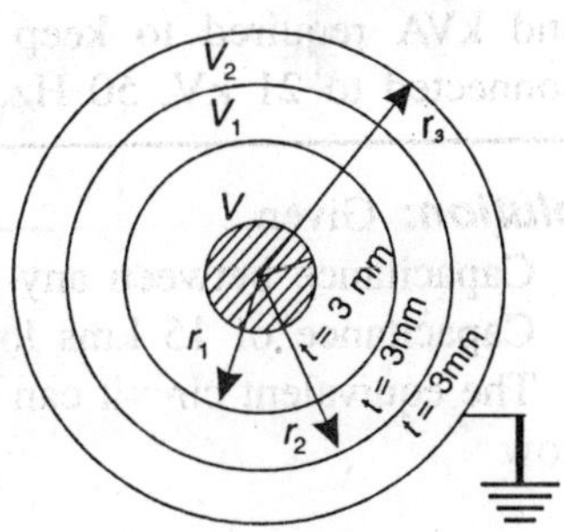

Fig. P. 9.14

Let, V be the potential on the core.

V_1 be the potential on the first intersheath

V_2 be the potential on the 2nd intersheath according to question.

$$V - V_1 = 20 \text{ kV or } rg_{1max} \ln \frac{r_1}{r} = 20$$

or $$0.8 \times g_{1max} \times \ln \frac{(0.8 + .3)}{0.8} = 20$$

$$\therefore \quad g_{1max} = \frac{20}{0.8 \ln \frac{1.1}{0.8}} \text{ kV/cm} = 78.50 \text{ kV/cm} \quad \textbf{Ans.}$$

Again, we have

$$V_1 - V_2 = 20 \text{ kV or } r_1 g_{2max} \ln \frac{r_2}{r_1} = 20$$

or $$1.1 \times g_{2max} \ln \frac{1.4}{1.1} = 20, \text{ this yields}$$

$$g_{2max} = \frac{20}{1.1 \ln \frac{1.4}{1.1}} \text{ kV} = 75.39 \text{ kV/cm} \quad \textbf{Ans.}$$

Let us calculate potential V_2

$$V_2 = V - (V - V_1) - (V_1 - V_2) = 60 - 20 - 20 = 20 \text{ kV}$$

or $$r_2 g_{3max} \ln \frac{r_3}{r_2} = 20 \text{ or } 1.4\, g_{3max} \ln \frac{1.7}{1.4} = 20$$

Thus yeilds, $$g_{3max} = \frac{20}{1.4 \ln \frac{1.7}{1.4}} \text{ kV/cm} = 73.58 \text{ kV/cm}$$

Thus, the maximum stress at second intersheath is $= 73.58$ kV/cm **Ans.**

> **Problem 9.15.** The capacitance of the 3-core lead-sheathed cable measured between any two of the conductors with sheathed earthed is 0.14 μF/km. Determine the equivalent star connected capacity and kVA required to keep 15 km of the cable charged when connected to 21 kV, 50 Hz.

Solution: Given,

Capacitance between any two of the conductors = 0.14 μF/km.

Capacitance of 15 kms long cable = 0.14 × 15 μF = 2.1 μF

The equivalent circuit can be drawn as below

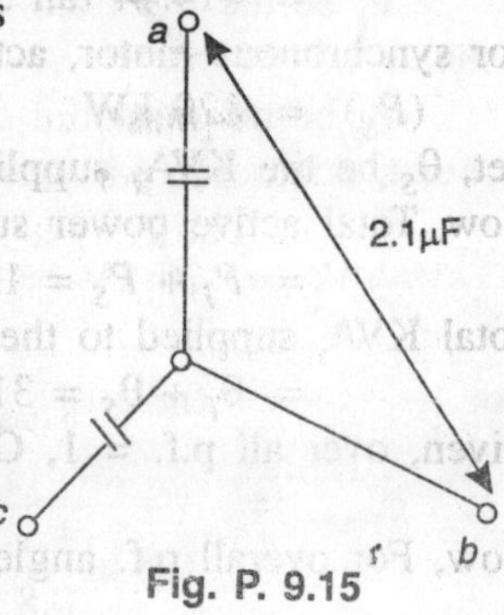

Fig. P. 9.15

The capacitance per phase will be

$$2 \times C_{ab} = 2 \times 2.1 \text{ μF} = 4.2 \text{ μF}$$

Now, 3-phase MVA required

$$= V^2 \omega c = 2\pi f C \times V^2$$
$$= 2 \times \pi \times 50 \times 4.2 \times 10^{-6} \times (21 \times 10^3)^2$$
$$= 581.89 \text{ kVA} \quad \textbf{Ans.}$$

Chapter 10

Voltage Control

Problem 10.1. A 3-phase induction motor delivers 500 HP at an efficiency of 90% when the operating p.f. is 0.8 lag. A loaded synchronous motor with a power consumption of 120 kW is connected in parallel with the induction motor. Calculate the necessary kVA and the operating p.f. of the synchronous motor if the overall p.f. is to be unity.

Solution: Given,

Power output of I.M. (P_o) = 500 HP = 500 × 746 W = 373 kW

Efficiency of the I.M. = 0.90

As we know that efficiency of induction motor = $\dfrac{\text{Output}}{\text{Input}}$

$\therefore$ Input power (P_I) = $\dfrac{373}{0.9}$ = 414.44 kW

Operating power factor of I.M. = 0.8 (log)

$\therefore$ p.f angle (θ_I) = 36.87°

KVAR supplied to the induction motor (θ_I) = P_I tan (36.87°)

= 414.44 tan 36.87° = 310.83 kVA_r

For synchronous motor, active power consumption in S.M.

(P_S) = 120 kW

Let, θ_S be the KVA_r supplied by the S.M.

Now, Total active power supplied to the system (P)

= $P_I + P_S$ = 120 + 414.44 = 534.44 kW

Total KVA_r supplied to the system (θ)

= $\theta_I + \theta_S$ = 310.83 + θ_S in kVA_r

Given, over all p.f. = 1, Overall p.f. angle = θ

Now, For overall p.f. angle $\tan\theta = \dfrac{Q}{P}$ or $\tan 0° = \dfrac{Q}{P}$

$\therefore \quad Q = p \times \tan 0° \text{ or } Q_S = 0 \text{ or } Q_S + 310.83 = 0$

$\therefore \quad Q_S = -310.83 \text{ kVA}$

The –ve sign shows that the synchronous motor is operating at leading p.f.

The power phasor for the system can be drawn as shown in Fig. P. 10.1.

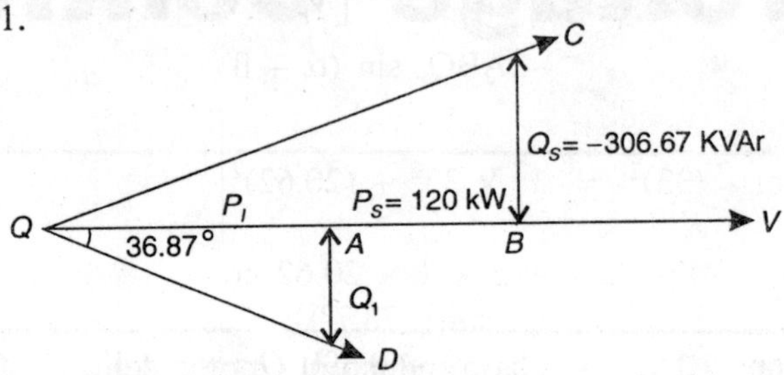

Fig. P. 10.1

From Fig. P. 10.1,

KVA supplied to the synchronous motor

$$= \sqrt{QB^2 + Bc^2} = \sqrt{(120)^2 + (-310.83)^2}$$

$$= 333.18 \text{ kVA}$$

$\therefore$ KVA supplied to the synchronous motor = 333.18 kVA **Ans.**

Operating P.F. or the synchronous motor

$$= \cos\left[\tan^{-1}\left(\frac{Q_S}{P_S}\right)\right] = \cos\left[\tan^{-1}\left(\frac{-310.83}{120}\right)\right]$$

$$= 0.3601 \text{ (lead)} \quad \textbf{Ans.}$$

Problem 10.2. A 3-phase line having an impedance of (5 + *j* 20) ohm per phase delivers a load of 30 MW at a p.f. of 0.8 lag and voltage 33 kV. Determine the capacity of the phase modifier required to be installed at the receiving end if the voltage at the sending end is to be maintained at 33 kV.

Solution: Given,

$$Z = 5 + j\,20\ \Omega/\text{phase} = 20.62\ \angle 78.7°\ \Omega/\text{phase}$$

$$= B\ \angle\beta\ \Omega/\text{phase}$$

$V_S = V_r = 33$ kV, $\cos\phi = 0.8$, This yields $\sin\phi = 0.6$

$P_r = 30$ MW

The sending end voltage is given by

$$V_s = AV_r + BI_r = V_r + ZI_r$$

So_1 we have A = $1\angle 0°$, B = Z = $20.62 \angle 78.7°$

i.e. $\alpha = 0°$, $\beta = 78.7°$

To calculate kVA_r (Q_r), apply the relation

$$V_S^2 = A^2V_r^2 + B^2\left[\frac{P_r^2}{V_r^2}+\frac{Q_r^2}{V_r^2}\right] + 2AB\,P_r \cos(\alpha - \beta)$$

$$-2ABQ_r \sin(\alpha - \beta)$$

or $$(33)^2 = 1^2 \times 33^2 + (20.62)^2\left[\left(\frac{30}{33}\right)^2 + \frac{Q_r^2}{(33)^2}\right]$$

$$+ 2 \times 1 \times 20.62 \cos(-78.7°) - 2 \times 1 \times 20.62 \times Q_r \sin(-78.7°)$$

or $0 = 351.39 + 0.390\, Q_r^2 + 8.08 + 40.44\, Q_r$

or $0.390\, Q_r^2 + 40.44\, Q_r + 359.47 = 0$

or $Q_r^2 + 103.69\, Q_r + 921.72 = 0$

This fields $$Q_r = \frac{-103.69 \pm \sqrt{(103.69)^2 - 4\times 1\times 921.72}}{2}$$

$$= \frac{-103.69 \pm 84.05}{2}$$

Neglecting lower –ve value

$$Q_r = \frac{-103.69 + 84.05}{2} = -9.82 \text{ kVA}_r$$

Since Q_r is negative, the phase modifier supplies 9.82 kVAr more to the requirement of the load 30 MW

the KVA_r phase modifies capacity is

$$Q_r' = Q_r + \frac{P_r}{\cos\varphi} \times \sin\theta$$

$$= 9.82 + \frac{30}{0.8} \times 0.6 = 32.32 \text{ KVAr}$$

∴ Phase modifier capacity = $\sqrt{P_r^2 + Q_r^2}$

$$= \sqrt{(30)^2 + (32.32)^2} = 44.1 \text{ MVA} \quad \textbf{Ans.}$$

Problem 10.3. Determine the tap ratio in problem 10.2 if the receiving end voltage is to be maintained at 0.92 p.u. of the sending end voltage. The line is fed through a 33/220 V transformer. Assume $t_s t_r = 1$.

Solution: Given, $t_s t_r = 1$

Sending end voltage $(V_1) = 220$ kV

Sending end voltage per phase $(V_{1P}) = \dfrac{220}{\sqrt{3}}$ kV = 127.02 kV

Receiving end voltage of the line $(V_{2P}) = 0.92 \times 127.02$ kV

Load active power $(P) = 30$ MW

Load active power per phase $(P) = 10$ MW

Load reactive power $(Q) = 30 \tan \cos^{-1} (0.8) = 22.5$ MVAr

Load reactive power per phase $(q) = \dfrac{22.5}{3}$ MVAr = 7.5 MVAr

Resistance of the line per phase $(r) = 5\ \Omega$

Reactance of the line per phase $(x) = 20\ \Omega$

Let t_s be the tap ratio. To calculate t_s, apply the relation

$$t_s^2 \left(1 - \frac{rP + xq}{V_{1p} V_{2p}}\right) = \frac{V_{2P}}{V_{1p}}$$

or
$$t_s^2 \left(1 - \frac{5 \times 10 \times 10^6 + 7.5 \times 10^6 \times 20}{0.92 \times 127.02 \times 127.02 \times 106}\right) = \frac{0.92 \times 127.02}{127.02}$$

This yeilds $t_s = 0.9657$

Thus, ∴ Tap ratio = 0.9657 **Ans.**

Problem 10.4. Determine the transformer tap ratios when the receiving end voltage is equal to the sending end voltage, the high voltage line operators at 230 kV and transmits 80 MW at 0.8 p.f. and the impedance of the line is $(40 + j150)$ ohm. Assume $t_a t_r = 1$.

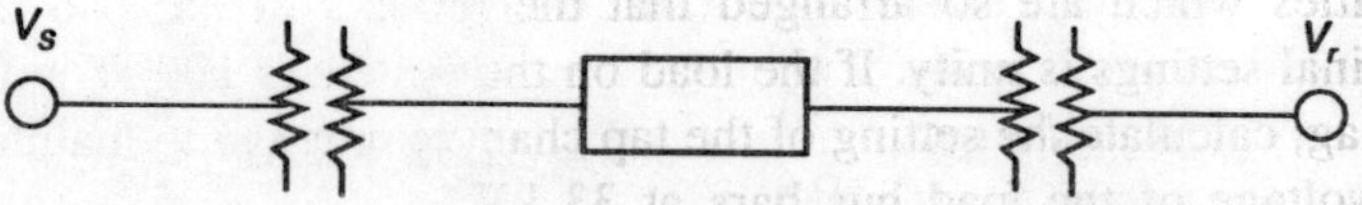

Fig. P. 10.4

Solution: Let the tap ratio be $t_s : 1$

Given, operating live voltage at the sending end $V_2 = 230$ kV

operating line voltage at the sending end per phase

$$(V_{2P}) = \frac{230}{\sqrt{3}} \text{ kV} = 132.79 \text{ kV}$$

Load active power (P) = 80 MW

Load active power per phase $(P) = \dfrac{80}{3} = 26.67$ MW

Load reactive power per phase (q) = 26.67 tan $\cos^{-1}$ 0.8
= 20.003 KVAr

resistance of the line $(r) = 40\ \Omega$

Reactance of the line $(x) = 150\ \Omega$

According to question $t_s t_r = 1$ and operating line voltage at the sending end per phase $V_{1P} = V_{2P} = 132.79$ kV

To calculate t_s, apply the relation,

$$t_s^2\left(1-\frac{r_p + x_q}{V_{1P}\times V_{2P}}\right) = \frac{V_{2P}}{V_{1P}}$$

or $$t_s^2\left(1-\frac{40\times 26.67\times 10^6 + 150\times 20.003\times 10^6}{132.79\times 132.79\times 10^6}\right)$$

$$= \frac{132.79}{132.79}$$

or $t_s^2\ (1 - 0.2307) = 1$

This yeilds $t_s = 1.14$

Thus, the required top ratio = 1.14 **Ans.**

Problem 10.5. A 132 kV line is fed through 33 kV/132 kV from a constant 33 kV supply as shown in Fig. P. 10.5. The total impedance of the line and transformers at 132 kV is (25 + *j* 60) ohms. Both the transformers are equipped with tap changing facilities which are so arranged that the product of the two off-nominal settings is unity. If the load on the system is 100 W at 0.6 p.f. lag, calculate the setting of the tap changer required to maintain the voltage of the load bus bars at 33 kV.

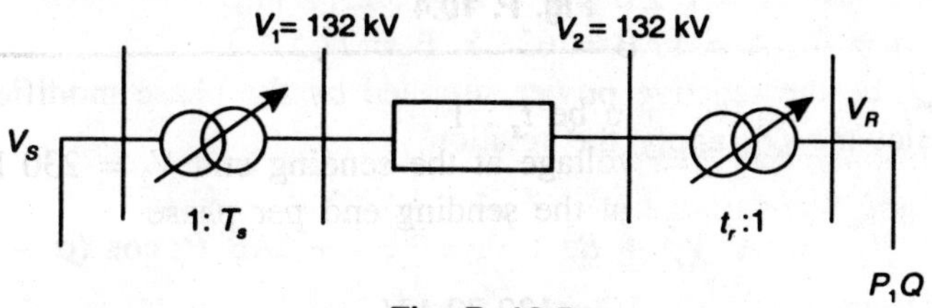

Fig. P. 10.5

Solution: Given, $V_S = V_r$ and $t_s t_r = 1$

Sending end voltage referred to transmission line $V_S = 132$ kV

Receiving end no voltage referred to transmission line

V_r = 132 kV

Resistance of the line and transformer (R) = 25 Ω

Reactance of the line and transformer (x) = 60 Ω

Load active power (P) = 100 MW

Load reactive power (Q) = 100 tan $\cos^{-1}$ 0.6 = 133.33 KVAr

Let t_s be the setting of the tap. To calculate t_s, apply the relation,

$$t_s^2 = \left(1 - \frac{RP + xQ}{V_s \times V_r}\right) = \frac{V_r}{V_s}$$

$$t_s^2 \left(1 - \frac{100 \times 25 \times 10^6 + 133.33 \times 60 \times 10^6}{\left(132 \times 10^3\right)^2}\right) = 1$$

or $\quad t_s^2$ (1 – 0.6026) = 1

This yeilds, t_s = 1.586

Thus, the required top setting (t_s) = 1.586 **Ans.**

Problem 10.6. A 3-phase line has an impedance of (20 + j60) ohm per phase. The sending end voltage is 142 kV while the receiving end voltage is maintained at 132 kV for all loads by an automatic phase modifier. If the kVA r of the modifier has the same value for zero load as for a load of 50 MW, determine the rating of the modifier and the p.f. of this load.

Solution: Given Impedance of the line (z)

= 20 + j 60 Ω/phase = 63.25 ∠71.57° Ω/phase

Sending end voltage (V_s) = 142 kV

Receiving end voltage (V_r) = 132 kV

The sending end voltage of the transmission line is given by

$$V_s = AV_r + BI_r = V_r + ZI_r$$

So, we have A = 1 ∠0°, B = Z = 6.25 ∠71.57°

Here α = 0°, A = 1, B = 63.25, B = 71.57°

Let Q_r be the reactive power supplied by the phase modifier

To calculate Q_r, apply the relation,

$$V_s^2 = A^2 V_r^2 + B^2 \left(\frac{P_r^2}{V_r^2} + \frac{Q_r^2}{V_r^2}\right) + 2AB\ P_r \cos(\alpha - \beta)$$

$$- 2\text{AB}\ Q_r \sin(\alpha - \beta)$$

At No-load, $P_r = 0$

$$\therefore \quad V_s^2 = A^2V_r^2 + \frac{B^2Q_r^2}{V_r^2} - 2\text{ AB } Q_r \sin(\alpha - \beta)$$

$$\text{or } (142)^2 = (1)^2 \times (132)^2 + \frac{(63.25)^2 \times Q_r^2}{(132)^2} - 2 \times 1$$

$$\times 63.25\, Q_r \sin(0 - 71.57)$$

or $\quad 0.22\ 96\ Q_r^2 + 120.01\ Q_r - 2740 = 0$

or $\quad 2 + 522.69\ Q_r - 11933.798 = 0$

This yields

$$Q_r = \frac{-522.69 \pm \sqrt{(522.69)^2 + 4 \times 11933.798}}{2}$$

$$= \frac{-522.69 \pm 566.52}{2}$$

Neglecting lower negative value

$Q_r = 21.92$ MVAr

Rating of the phase modifier = 21.92 MVAr

Since the KVAr or the modifier has the same value for zeroload or for a load of 50 MW

$\therefore \quad P_r = 50$ MW

$$\text{Power factor of this load} = \cos \tan^{-1}\left(\frac{Q_r}{P_r}\right)$$

$$= \cos \tan^{-1}\left(\frac{21.92}{50}\right) = 0.9159$$

Thus, Power factor of the load ($\cos \phi_r$) = 0.9159 **Ans.**

Problem 10.7. A typical line has the following paraneters :

$A = D = 0.96\ \angle 10°,\ B = 100\ \angle 80°$

(i) If the line supplies a load of 30 MW at 0.8 p.f. lag, 110 kV, find the sending end voltage and hence the regulation of the line

(ii) For a load of 50 MW at 0.8 p.f. lag, 110 kV, find the reactive power supplied by the line and by the synchronous capacitor if the sending end voltage is 120 kV. Also, determine the p.f. of the line at the receiving end.

(iii) Find the maximum power that can be transmitted if the sending and receiving end voltages are as in (ii).

Solution: $A = D = 0.96 \angle 1°$, $B = 100 \angle 80°\ \Omega$

(i) Here, load power (P_r) = 30 MW, p.f. ($\cos \phi_r$) = 0.8
Load p.f. angle (Q_r) = 36.9°, Receiving voltage = 110 kV

$$\text{Receiving end voltage/phase} = \frac{110}{\sqrt{3}}\text{ kV} = 63508.53\text{ V}$$

$$\text{Receiving end current } (I_r) = \frac{P_r}{\sqrt{3} \times V_L \times \cos P_r}$$

$$= \frac{30 \times 10^6}{\sqrt{3} \times 110 \times 10^3 \times 0.8} = 196.82\text{ A}$$

Let us take receiving end voltage as reference phasor.

$\therefore \quad \bar{V}_r = 63508.53 \angle 0°\ V$ and $\bar{I}_r = 196.82 \angle -36.9°$

Now, the sending end voltage of the transmission line is given by

$$\bar{V}_S = A\bar{V}_r + B\bar{I}_r$$

or $\quad V_S = 0.96 \angle 1° \times 63508.53 \angle 0° + 100 \angle 80° \times 196.82 \angle -36.9° =$

$= 60968.19 \angle 1° + 19682 \angle 43.1°$

$= 75329.96 + j\ 14512.24 = 76715.11 \angle 10.9°$ V

Sending end line voltage $= \sqrt{3} \times V_S = \sqrt{3} \times 7615.11$

$= 132.87$ kV **Ans.**

Percentage regulation of the line

$$= \frac{V_S/A - V_r}{V_r} \times 100\%$$

$$= \frac{76715.11/0.96 - 63508.53}{63508.53} \times 100\%$$

$= 25.83\%$ **Ans.**

(ii) Here, $V_S = 120$ kV, $V_r = 110$ kV, p.f. = 0.8, $P_r = 50$ MW
Here $X = 100 \sin 80° = 98.48\ \Omega$

Let Q be the reactive power supplied by the line. To calculate Q, apply the relation.

$$V_r = V_s - \frac{Q}{V_S}X \text{ or } 110 = 120 - \frac{Q}{120} \times 98.48$$

This yeilds $Q = 12.19$ MVAr

Thus, reactive power supplied by the line = 12.19 MVAr

Let Q_r be the reactive power supplied by the phase modifier.

To calculate Q_r, apply the relation,

$$V_{s2} = A^2V_r^2 + B^2\left(\frac{P_r^2}{V_r^2}+\frac{Q_r^2}{V_r^2}\right) + 2AB\ P_r \cos(\alpha - \beta)$$

$$- 2AB\ Q_r \sin(\alpha - \beta)$$

or $$(120)^2 = (0.96)^2 \times (110)^2 + (100)^2 \left(\frac{(50)^2}{(110)^2}+\frac{Q_r^2}{(110)^2}\right)$$

$$+ 50 \times 2 \times 0 \times 0.96 \times 100 \cos(1^\circ - 80^\circ) - 2 \times 0.96 \times 100 \times Q_r \sin(1^\circ - 80^\circ)$$

or $$0.826\ Q_r^2 + 188.47\ Q_r^2 + 649.24 = 0$$

or $$Q_r^2 + 288.17\ Q_r + 786.0 = 0$$

This yields

$$Q_r = \frac{-288.17 \pm \sqrt{(288.17)^2 - 4 \times 786}}{2}$$

$$= \frac{-288 \pm 282.66}{2}$$

Neglecting negative lower value,

$$Q_r = -2.67 \text{ MVAr}$$

Since Q_r is negative, the phase modifier supplies 2.67 MVA more than the requirement of the load 50 MW.

The reactive power supplied by the synchronous capacitor is given by

$$Q_{PM} = |Q_r| + P_r \tan Q_r = 2.67 + 50 \tan 36.9^\circ$$

$$= 40.21 \text{ MVAr}$$ **Ans.**

The p.f. of the line at receiving end is given by

$$\text{p.f.} = \cos \tan^{-1} \frac{Q_r}{P_r}$$

$$= \cos \tan^{-1}\left(\frac{2.67}{50}\right)$$

$$= 0.9985$$

Since Q_r is negative the p.f. will be leading.

Receiving end power factor = 0.9985 (leading) **Ans.**

(iii) Here, $V_s = 120$ kV, $V_r = 110$ kV, A = 0.96,

$\alpha = 1^\circ$, $B = 100$, $\beta = 80^\circ$

Now, the maximum power transmitted is given by

$$P_{max} = \frac{V_S V_r}{B} - \frac{A V_r^2}{B} \cos(\beta - \alpha)$$

$$= \frac{110 \times 120}{100} - \frac{0.96 \times (10)^2}{100} \cos(80° - 1) \text{ MW}$$

$$= 109.84 \text{ MW}$$

∴ Maximum power transmitted through the line

$= 109.84$ MW **Ans.**

Chapter 11

Neutral Grounding

Problem 11.1. A transmission line has a capacitance of 0.1 F per phase. Determine the inductance of Peterson coil to neutralize the effect of capacitance of (i) complete length of line, (ii) 97% of the line, (iii) 90% length of the line. The supply frequency is 50 Hz.

Solution: Given,

Capacitance of the line per phase (C) $= 1 \times 10^{-6}$ F

Operating frequency (f) = 50 Hz

$$\text{Capacitive reactance of the line } (X_c) = \frac{1}{\omega c} = \frac{1}{2\pi f C}$$

$$= \frac{1}{2 \times 3.14 \times 50 \times 1 \times 10^{-6}} \Omega = 3.185 \text{ k}\Omega$$

Let X_L be the reactance of the coil for necessary neutralization of the capacitance of the line is given by

The inductance of the coil to neutralise the capacitance of the line is given by

$$L = \frac{X_c}{3\omega\alpha}$$

Where α is the percentage length of the line the capacitance of which is to be neutralized.

(i) For complete length of the line, α = 100% of the line = 1

Required inductance of the coel

$$= \frac{X_c}{3\omega\alpha} = \frac{3.185 \times 10^3}{3 \times 2 \times 3.14 \times 50 \times 1} = 3.38 \text{ H} \quad \textbf{Ans.}$$

(ii) For 97% of the line, α = 0.97

Required inductance of the coil

$$(L) = \frac{X_c}{3\omega\alpha} = \frac{3.185 \times 10^3}{3 \times 314 \times 0.97} = 3.49 \text{ H} \quad \textbf{Ans.}$$

(iii) For 90% of the line, $\alpha = 0.90$

Required inductance of the coil

$$= \frac{X_c}{3\omega\alpha} = \frac{3.185 \times 10^3}{3 \times 314 \times 0.90} = 3.76 \text{ H} \quad \textbf{Ans.}$$

> **Problem 11.2.** A 132 kV, 50 Hz, 3-phase, 100 km long transmission line has a capacitance of 0.012 μF per km per phase. Determine the inductive reactance and kVA rating of the arc suppression coil suitable for the line to eliminate arcing ground phenomenon.

Solution: Given,

Length of the line (l) = 100 km

operating frequency (f) = 50 Hz

operating voltage (V) = 132 kV

Capacitance of the line per phase (C_n) = 0.012×10^{-6} F/km/phase

Capacitance of the 100 km long line (C) = $C_n \times 1$

$= 0.012 \times 10^{-6} \times 100$ F/phase = 1.2 μF/phase

To calculate inductive reactance of the arc suppressor coil,

Apply the relation, $V_{ph} \times 3 \times X_L = V_p \times X_C$

or $$X_L = \frac{X_c}{3} = \frac{1}{3\omega C} = \frac{1}{3 \times 314 \times 1.2 \times 10^{-6}}$$

$= 884.64\ \Omega$ **Ans.**

The kVA rating of the suppressor coil is given by

$$\frac{V^2}{3X_L} = \frac{(132)^2 \times 10^6}{3 \times 884.64} = 6565 \text{ kVA}$$

kVA rating of the coil = 6565 kVA **Ans.**

> **Problem 11.3.** A 132 kV, 3-phase, 50 Hz overhead line of 100 km length has a capacitance to earth of each line of 0.01 μF per km. Determine inductance and kVA rating of the arc suppression suitable for this line.

Solution: Given,

Operating frequency (f) = 50 Hz

Operating voltage (V) = 132 kV

Length of the line (l) = 100 km

Capacitance of the line per phase (c_n) = 0.1×10^{-6} F/km

Capacitance of the 100 km long line per phase (C)

$$= C_n \times l = 0.01 \times 10^{-6} \times 100 = 1 \times 10^{-6} \text{ F}$$

$$= 1 \ \mu\text{F}$$

To calculate inductance of the suppression coil, apply the relation.

$$V_{pn} \times 3 \times X_L = V_P \times X_C$$

$$\text{or} \quad X_L = \frac{X_c}{3} = \frac{1}{3\omega c} = \frac{1}{3 \times 2 \times 3.14 \times 50 \times 4.5 \times 10^{-6}}$$

$$= 235.9 \ \Omega$$

Reactance of the coil (X_L) = 235.9 Ω **Ans.**

Problem 11.4. A 142 kV, 3-phase, 50 Hz transmission line 200 km long consists of three conductors of effective diameter 24 cm arranged in a vertical plane with 5 m spacing and regularly transposed. Find the inductance and kVA rating of the arc suppressor coil in the system

Solution: Given,

Operating frequency (t) = 50 Hz

Diameter of each conductor = 24 mm

$$\therefore \text{ Radius of each conductor} = \frac{24}{2} = 12 \text{ mm} = 12 \times 10^{-3} \text{ m}$$

Mutual GMD of these lines (D_m) = $\sqrt[3]{5 \times 5 \times 10}$ = 6.3 m

The capacitance of the line per phase is given by

$$C_n = \frac{10^{-9}}{18 \ln \frac{d_m}{r}} \text{ F/m} = \frac{1}{18 \ln \frac{6.3}{12 \times 10^{-3}}} \ \mu\text{F/km}$$

$$= 8.87 \times 10^{-9} \text{ F/km}$$

$\therefore$ Capacitance of 200 km long line = $C_n \times l$

$$C = 8.87 \times 10^{-9} \times 200 \text{ F}$$

$$C = 1.774 \ \mu\text{F}$$

To calculate inductance of the suppressor apply the relation.

$$V_{pn} \times 3\omega C = \frac{V_p}{\omega_L} \text{ or } L = \frac{1}{3\omega^2 C}$$

$$= \frac{1}{3 \times (2 \times 3.14 \times 50)^2 \times 1.774 \times 10^{-6}} = 1.906 \text{ H}$$

∴ Inductance = 1.906 H **Ans.**

The MVA rating of the suppressor coil is given by

$$\frac{V^2}{3\omega l} = \frac{\left(142\times 10^3\right)^2}{3\times 2\times 3.14\times 50\times 1.906} = 11.231 \text{ MVA}$$

∴ Rating of each suppressor coil = 11.231 MVA **Ans.**

Problem 11.5. A 50 Hz overhead line has line to earth capacitance of 2 μf. It is decided to use an earth fault neutralizer. Determine the reactance to neutralize the capacitance of
(i) 100% of the length of the line.
(ii) 80% of the length of the line and
(iii) 50% of the length of the line.

Solution: Given,

Capacitance of neutral (C) = 2×10^{-6} F
Operating frequency (f) = 50 Hz

$$\text{Capacitive reactance of the line } (X_c) = \frac{1}{\omega C}$$

$$= \frac{1}{2\times\pi\times f\times C} = \frac{1}{2\times 3.14\times 50\times 2\times 10^{-6}} = 1.59 \text{ k}\Omega$$

Let X_L be the reactance of the coil.
The reactance to neutralize the capacitance of the line is given by

$$X_L = \frac{X_C}{3\alpha}$$

When α is the percentage length of the line whose capacitance is to be neutralized.

(i) For 100% line, α = 100% = 1

$$\text{Required reactance} = \frac{X_c}{3\times 1},\ X_L = \frac{1.59\times 10^3}{3\times 1} \text{ ohm}$$

X_L = 530 Ω **Ans.**

(ii) For 80% of the length of the line α = 80% = 0.8
Here,

$$\text{Required reactance} = \frac{X_c}{3\alpha},\ X_L = \frac{1.59\times 10^3}{3\times 0.8},\ X_L = 662.5\ \Omega$$ **Ans.**

(iii) For 50% of the length of the line, $\alpha = 50\% = 0.50$
Here,

$$\text{Required reactance } (X_L) = \frac{X_c}{3\alpha} = \frac{1.59 \times 10^3}{3 \times 0.50} = 1060\ \Omega \quad \textbf{Ans.}$$

Problem 11.6. Calculate the reactance of a coil suitable for 33 kV, 3-phase, 50 Hz. Transmission system of which the capacitance to earth of earth conductor is 4.5 μF.

Solution: Given,

Operating voltage (V) = 33 kV
Operating frequency (t) = 50 Hz
Capacitance of the line (C) = 4.5×10^{-6} F/phase
Let X_L be the reactance of the coil.
To calculate reactance of the coil, apply the relation

$$V_{ph} \times 3\omega C = \frac{V_p}{\omega L} \text{ or } L = \frac{1}{8\omega^2 C} = \frac{1}{3 \times (314)^2 \times 1 \times 10^{-6}} \text{ H}$$

$$= 3.381 \text{ H}$$

∴ Inductance of the coil (L) = 3.381 H
Now the kVA rating of this suppressor coil in given by

$$\frac{V^2}{3\omega l} = \frac{(132)^2 \times 10^6}{3 \times 314 \times 3.381} = 5.471 \text{ MVA}$$

∴ MVA rating of this suppression coil = 5.471 MVA **Ans.**

Chapter 12

Transient in Power Systems

Problem 12.1. The effective inductance and capacitance of a faulted system as viewed from the contacts of a breaker are 2.5 mH and 600 pF respectively. Determine the respectively voltage across the breaker contacts when a fault current of 150 amps is chopped

Solution: Given,

Inductance of the system (L) = 2.5 MH = 2.5×10^{-3} H

Capacitance of the system (c) = 600 PF = 600×10^{-12} F

Fault current (i_f) = 150 A

The restricting voltage across the circuit breaker contacts for any fault current is given by

$$V = i_f \sqrt{\frac{L}{C}} = 150 \times \sqrt{\frac{2.5 \times 10^{-3}}{600 \times 10^{-12}}} = 306.2 \text{ kV}$$

∴ Restricting voltage across the breaker contacts

= 306.2 kV **Ans.**

Problem 12.2. A rectangular surge of 2 μsec duration and magnitude 100 kV travels along a line of surge impedance 500 ohms. The latter is connected to another line of equal impedance through an inductor of 500 H. Calculate the maximum value of surge transmitted to the second line.

Solution: Given,

Magnitude of the surge voltage (V) = 100 kV

Time-period of the surge voltage (τ) = 2 μ sec = 2×10^{-6} sec.

Surge impedance of each line (Z_N) = 500

Inductance of the inductor (L) = 500 μH = 500×10^{-6} H

The circuit diagram of the given situation is as follows

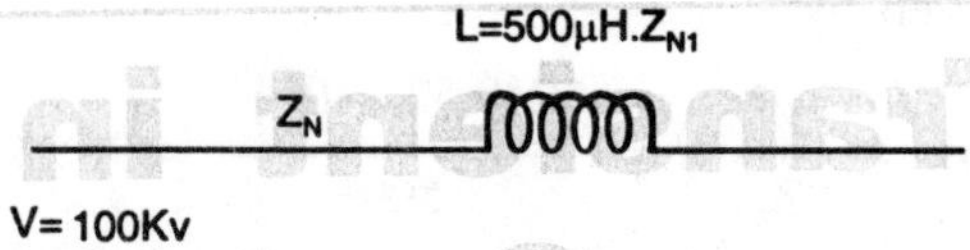

Fig. P. 12.1

$V = 100$ kV

The maximum value if surge transmitted to the second live is given by

$$V'' = V\left[1-e^{-\frac{2Z_N\tau}{L}}\right] = 100\times10^3\left[1-e^{-\frac{2\times500\times2\times10^{-6}}{500\times10^{-6}}}\right]$$

$$= 100\,[1-e^{-4}]\text{ kV} = 98.17\text{ kV}$$

∴ Maximum value of surge transmitted to the second line $= 98.17$ kV **Ans.**

> **Problem 12.3.** A long overhead line has a surge impedance of 500 ohms and an effective resistance of 6 ohms per km. If a surge of 400 kV enters the line at a certain point, calculate the magnitude of this surge after it has transversed 100 km and calculate the power loss and heat loss of the wave over this distance. Assume velocity of wave as 3×10^8 m/sec.

Solution: Given,

Initial surge voltage $(V_o) = 400$ kV

Surge impedance of the line $(Z_n) = 500\ \Omega$

Resistance of the conductor $(R) = 65\ \Omega$/km

Distance $(x) = 100$ km, where we have to calculate surge voltage.

Velocity of the wave $(V) = 3 \times 10^8$ m/sec.

Conductance of the line $(G) = 0.0$ ℧

As the wave travels along the line, it gets attenuated in magnitude due to the resistance of the conductor. The value of surge voltage at a distance x is given by

$$V = V_o e^{-\frac{1}{2}\left(-\frac{R}{Z_N}+GZ_N\right)x} = 400\times10^3\ e^{-\frac{1}{2}\left(\frac{6}{500}+0\times500\right)\times100}$$

$$= 400 \times e^{-0.6}\text{ kV} = 219.52\text{ kV}$$

∴ Surge voltage at 100/c for the incident point = 219.52 kV **Ans.**

As we know that the power loss is an instantaneous quantity, *i.e.* it depends upon the distance *x*. The power loss during travelling 100 km is given by

$$\text{Power loss} = \frac{V^2}{Z_n}\text{W} = \frac{\left(219.52\times 10^3\right)^2}{500} = 96.38 \text{ MW} \quad \textbf{Ans.}$$

Now, The heat loss of the wave over this distance is given by

$$\text{Heat loss} = \int_0^t vidt$$

Given $v = 3 \times 10^8$ m/sec

Let *t* be the tie taken to travel 100 km.

$$\text{We know that, Speed} = \frac{\text{distance }(X)}{\text{time }(t)}$$

$$\therefore \quad t = \frac{100\times 1000}{3\times 108}\text{sec.} = 3.33 \times 10^{-4} \text{ sec.}$$

For this line, the instantaneous values of voltage and current wave are given by

$$v = V_o e^{-\frac{1}{2}\frac{Rvt}{Z_n}} \text{ and } i = 20\, e^{-\frac{Rvt}{2Z_n}}$$

Now,

$$\text{Heat loss} = \int_0^t V_o e^{-\frac{Rvt}{2Z_n}} \times 20\, e^{-\frac{Rvt}{2Z_n}}\, dt$$

$$= \int_0^t V_o \times \frac{V_o}{Z_n} e^{-\frac{Rvt}{2Z_n}}\, dt = \frac{V_o^2}{Z_n}\int_0^t e^{-\frac{Rvt}{2Z_n}}\, dt$$

$$= \frac{V_o^2}{Z_n}\left[-\frac{Z_N}{Rv}e^{-\frac{Rvt}{Z_N}}\right]_0^t = \frac{V_o^2}{Z_n}\left[-\frac{Z_N}{Rv}e^{-\frac{Rvt}{Z_N}} + \frac{Z_N}{Rv}\right]$$

$$= \frac{V_o^2}{Rv}\left[1-e^{-\frac{Rvt}{Z_N}}\right]$$

$$= \frac{\left(400\times 10^3\right)^2}{6\times 3\times 10^8 \times 10^{-3}}\left[1-e^{-\frac{6\times 3\times 10^8\times 3.33\times 10^{-4}\times 10^{-3}}{500}}\right]$$

$$= \frac{1600}{18}\left[1-e^{-1.199}\right]\text{kJ} = 62.09 \text{ kJ}$$

Thus, Heat loss of the wave over 100 km = 62.09 kJ **Ans.**

Problem 12.4. An overhead line with surge impedance 400 ohms bifurcates into two lines of surge impedance 400 ohms and 40 ohms respectively. If a surge of 20 kV is incident on the overhead line, determine the magnitudes of voltage and current which enter the bifurcate lines.

Solution: Given,

$Z_N = 400\ \Omega$, $V = 20$ kV, $Z_1 = 400\ \Omega$, $Z_2 = 40\ \Omega$

This is the situation of T-Junction

The diagram of the given problem is as follows

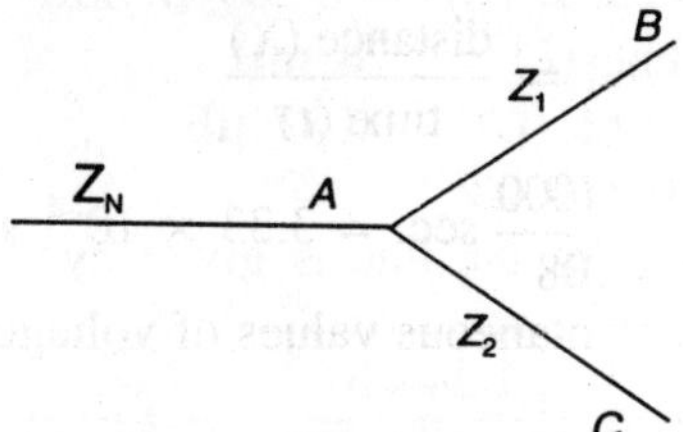

Fig. P. 12.4

As the surge voltage reaches to junction, it reflects and transmitted voltage will be same in both parallel branches Z_1 and Z_2. The transmitted voltage is given by

$$V'' = \frac{2V/Z_N}{\dfrac{1}{Z_N}+\dfrac{1}{Z_1}+\dfrac{1}{Z_2}} = \frac{2V}{1+\dfrac{Z_N}{Z_1}+\dfrac{Z_N}{Z_2}}$$

$$= \frac{2\times 20\times 10^3}{1+\dfrac{400}{400}+\dfrac{400}{40}} = \frac{40\times 10^3}{12} = 3.33 \text{ kV}$$

$\therefore$ Transmitted voltage = 3.33 kV **Ans.**

The surge current entering through line B is given by

$$I'' = \frac{V''}{Z_1} = \frac{3.33\times 10^3}{400} = 8.33 \text{ A} \quad \textbf{Ans.}$$

The surge current entering through line C is given by

$$I_2'' = \frac{V''}{Z_2} = \frac{3.33\times 10^3}{40} = 0.833 \text{ A} \quad \textbf{Ans.}$$

Problem 12.5. An overhead line with inductance and capacitance per km of 1.24 mH and 0.087 μF respectively is connected in series with an underground cable having inductance and capacitance of 0.185 mH/km and 0.285 μF/km. Calculate the values of transmitted and reflected waves of voltage and current at the junction due to a voltage surge of 110 kV travelling to the junction (i) along the line towards the cable, and (ii) along the cable towards the line.

Solution: Given

Inductance of the caste (L_b) = 0.185 mH/km
Capacitance of the cable (C_b) = 0.285 μF/km
Inductance of the line (L) = 1.24 mH
Capacitance of the line (C) = 0.087 μF
Voltage surge (V) = 110 kV
The surge impedance of the line is given by

$$Z_N = \sqrt{\frac{L}{C}} = \sqrt{\frac{1.24 \times 10^{-3}}{0.087 \times 10^{-6}}} = 119.39\ \Omega$$

The surge impedance of the cable is given by

$$Z_b = \sqrt{\frac{L_b}{C_b}} = \sqrt{\frac{0.185 \times 10^{-3}}{0.285 \times 10^{-6}}} = 25.48\ \Omega$$

(i) When voltage surge travels towards the cable
The diagram is as follows

Z_N = 119.39 Ω, Z_N = 25.48 Ω

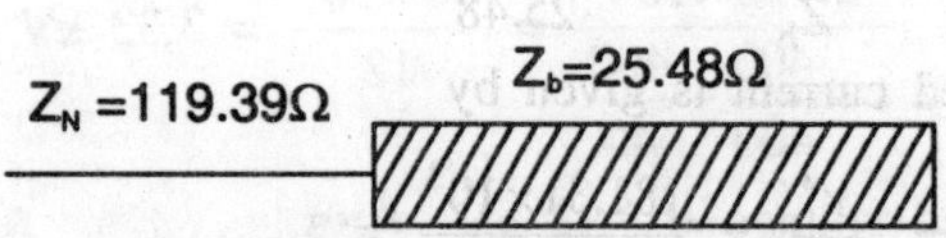

Fig. P. 12.5 (a)

The transmitted voltage is given by

$$V'' = \frac{2Z_b V}{Z_N + Z_b} = \frac{2 \times 25.48 \times 110 \times 10^3}{119.39 + 25.48} = 38.69\ \text{kV}\ \textbf{Ans.}$$

The reflected voltage is given by

$$V' = \frac{Z_b - Z_N}{Z_b + Z_N} \times V = \frac{25.48 - 119.39}{25.48 + 119.39} \times 110 \times 10^3$$

$$= -71.31\ \text{kV}\ \textbf{Ans.}$$

The transmitted current is given by

$$I'' = \frac{2V}{Z_b + Z_N} = \frac{2 \times 110 \times 1000}{25.48 + 119.39} = 1518.6 \text{ A} \quad \textbf{Ans.}$$

The reflected current is given by

$$I' = -\frac{V'}{Z_N} = -\frac{(-71.31 \times 10^3)}{119.39} = 597.29 \text{ A} \quad \textbf{Ans.}$$

(ii) When the voltage surge travels along the cable to overhead line.

The diagram is an follows.

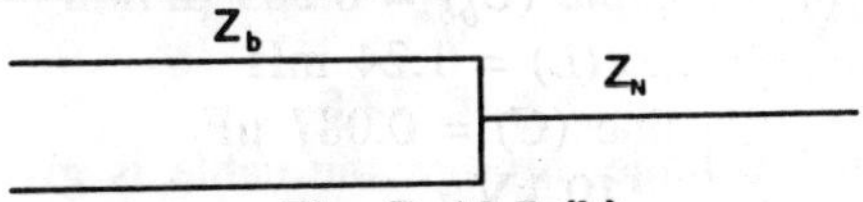

Fig. P. 12.5 (b)

The transmitted voltage is given by

$$V'' = \frac{2VZ_N}{Z_b + Z_N} = \frac{2 \times 110 \times 10^3 \times 119.39}{119.39 + 25.48}$$

$$= 181.31 \text{ kV} \quad \textbf{Ans.}$$

The reflected voltage is given by

$$V' = \frac{Z_N - Z_b}{Z_N + Z_b} \times V = \frac{119.39 - 25.48}{119.39 + 25.48} \times 110 \times 10^3$$

$$= 71.31 \text{ kV} \quad \textbf{Ans.}$$

The reflected current is given by

$$I' = -\frac{V'}{Z_b} = -\frac{71.31 \times 10^3}{25.48} = 2798.67 \text{ A} \quad \textbf{Ans.}$$

The refracted current is given by

$$I'' = \frac{V''}{Z_N} = \frac{181.31 \times 10^3}{119.39} = 1518.64 \text{ A} \quad \textbf{Ans.}$$

Problem 12.6. Explain why a short length of cable is connected between the dead end tower and the terminal apparatus in a station. An overhead line with surge impedance 400 ohms is connected to a terminal apparatus through a short length of cable of surge impedance 40 ohms.

A traveling wave of constant magnitude 100 kV and infinite duration originates in the overhead line and travels towards the junction with the cable. Calculate the energy transmitted into the cable during a period of 5 sec after the arrival of the wave at the junction.

Solution: Given,

Surge impedence of the line (Z_N) = 400 Ω

Surge impedence of the cable (Z_b) = 40 Ω

Surge voltage (V) = 100 kV, t = 5 μsec = 5 × 10^{-6} sec.

The diagram of this problem is as follows

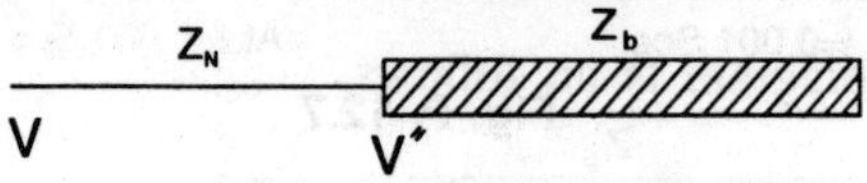

Fig. P. 12.6

The transmitted voltage through the cable is given by

$$V'' = \frac{2Z_b V}{Z_b + Z_N} = \frac{2\times 40\times 100\times 1000}{40+400} = 18.18 \text{ kV} \quad \textbf{Ans.}$$

Now, the energy transmitted into the cable during 5 μ sec is given by, Energy transmitted

$$= \frac{V''^2}{Z_b} \times t \text{ W-sec.} = \frac{\left(18.18\times 10^3\right)^2 \times 5\times 10^{-6}}{40} \text{ W-sec.}$$

$$= 41.3 \text{ W-sec}$$

Thus, energy transmitted through the cable is 5 μsec

$$= 41.3 \text{ W-sec} \quad \textbf{Ans.}$$

Problem 12.7. An overhead transmission line with surge impedance 400 ohms is 300 km long. One end of this line is short-circuited and at the other end a source of 11 kV is suddenly switched in. Calculate the current at the source end 0.005 sec after the voltage is applied.

Solution: Given,

Applied voltage (V) = 11 kV

Surge impedance of the line (Z_N) = 400 Ω

Length of the line (l) = 300 km = 3 × 10^5 m

As we know that velocity of propagation of wave is same as that of light, *i.e.* 3 × 10^8 m/sec.

Time taken by the wave to ravel one end to shorted end

$$(t) = \frac{l}{v} = \frac{3\times 10^5}{3\times 10^8} = 1\times 10^{-3} \text{ sec.} = 0.001 \text{ sec.}$$

At the instant of application of voltage,

$$\text{Current } (I) = \frac{V}{Z_N} = \frac{11 \times 1000}{400} = 27.5 \text{ A}$$

At $t = 0.001$ sec, the reflection takes place. The voltage is reflected with change is sign due to the –ve reflection coefficient, while the current is reflected with positive sign due to positive reflection coefficient as shown in Fig. P. 12.7.

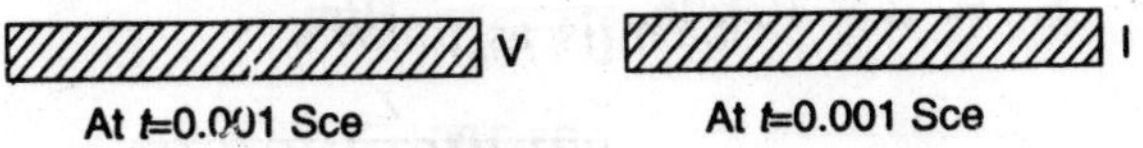

Fig. P. 12.7

At $t = 0.002$ sec, the current is 2 times the first travelling
At $t = 0.003$ sec, $I''' = 3 \times I$ at the receiving end
At $t = 0.004$ sec, $I'''' = 4 \times I$ at the sending end
At $t = 0.005$ sec, $I''''' = 5 \times I$ at the receiving end

The current at the source end just 0.005 sec after voltage is applied, is given by

$$I'''' = 5 \times I = 5 \times 27.5 = 137.5 \text{ A} \quad \textbf{Ans.}$$

Problem 12.8: A three phase transmission line has conductor dia 2 cm spaced 2 m in equilateral formation. The resistance and leakage reactance's are negligible. Calculate:
(i) Natural impedance of the line, (ii) the line current if a voltage wave of 11 kV travels along the line, (iii) rate of energy absorption, rate of reflection and the state and the form of reflection if the line is terminated through a star connected load of 900 Ω per phase, (iv) value of the terminating resistance for no reflection, and (v) amount reflected and transmitted power if the line is connected to a cable extension with inductance and capacitance per phase per cm of 0.4×10^{-8} H and 1×10^{-6} μF respectively

Solution: Given,

Diameter of the conductor = 2 cm
Radius of the conductor (r) = 1 cm = 1×10^{-2} m
Distance between the conductor (D) = 2 m
Inductance of the cable (L_C) = 0.4×10^{-8} H
Capacitance of the cable (C_c) = 1×10^{-6} μF
Operating voltage (V) = 11 kV

(i) To calculate natural impedance of the line, we have to calculate inductance and capacitance of the line

Now, Here, mutual GMD of the line = $\sqrt[3]{2 \times 2 \times 2} = 2$

The capacitance of the line to natural per phase is given by

$$C = \frac{2\pi\varepsilon_o}{r\ln\frac{D}{r}} \text{ F/m} = \frac{1}{18\ln\frac{\sqrt[3]{2\times2\times2}}{1\times10^{-2}}}$$

$$= 1.049 \times 10^{-8} \text{ F/km}$$

The inductance of the line per phase is given by

$$L = 2\times10^{-7}\ln\frac{D}{0.7788r} \text{ H/m}$$

$$= 2\times10^{-4}\ln\frac{2\times100}{0.7788\times1} \text{ H/km}$$

$$= 11.097\times10^{-4} \text{ H/km}$$

∴ Natural impedance (Z_n)

$$= \sqrt{\frac{L}{C}} = \sqrt{\frac{11.097\times10^{-4}}{1.049\times10^{-8}}}\ \Omega = 325.25\ \Omega \quad \textbf{Ans.}$$

(ii) Line current $= \frac{V}{\sqrt{3}\times Z_n} = \frac{11\times1000}{\sqrt{3}\times325.25} = 19.53$ A **Ans.**

(iii) Given, terminating load (L) = 900 Ω per phase since the terminating load is greater than the natural inpedance of the line reflection will be with positive sign.

The voltage across the terminating load is given by

$$V'' = \frac{2Z_L\times\left(V/\sqrt{3}\right)}{Z_L+Z_n} = \frac{2\times900\times11\times1000}{\sqrt{3}(900+325.25)} = 9.33 \text{ kV}$$

∴ The rate of energy consumption

$$= \frac{3V''^2}{Z_L}\text{ W} = \frac{3\times\left(9.33\times10^3\right)^2}{900} = 290.16 \text{ kW} \quad \textbf{Ans.}$$

The reflected voltage is given by

$$V' = \frac{Z_L-Z_N}{Z_L+Z_N}\times\left(\frac{V}{\sqrt{3}}\right) = \frac{900-325.25}{900+325.25}\times\frac{11}{\sqrt{3}} \text{ kV}$$

$$= 2.98 \text{ kV}$$

∴ Reflected voltage = 2.98 kV **Ans.**

The rate of reflected energy is given by $= \frac{3V'^2}{Z_n}$

∴ Rate of reflected energy

$$= \frac{3\times\left(2.98\times10^3\right)^2}{325.25}\text{ W} = 81.91 \text{ kW} \quad \textbf{Ans.}$$

(iv) As we know that whenever any line is terminated through an impedance equal to the surge impedance of the line, there will be no reflection.

∴ Load resistance = single impedance = 325.25 Ω **Ans.**

(v) The surge impedance of the cable is given by

$$Z_{cb} = \sqrt{\frac{L_C}{C_C}} = \sqrt{\frac{0.4\times10^{-8}}{1\times10^{-12}}} = 63.25\ \Omega$$

In this case, the reflected and refracted voltage is given by

$$\frac{Z_{cb}-Z_N}{Z_N+Z_{cb}}\left(\frac{V}{\sqrt{3}}\right) \text{ and } \frac{2\times Z_{cb}}{Z_N+Z_{cb}}\left(\frac{V}{\sqrt{3}}\right) \text{ respectively}$$

$$\therefore \text{ Reflected voltage } (V') = \frac{Z_{cb}-Z_N}{Z_N+Z_{cb}}\times\left(\frac{V}{\sqrt{3}}\right)$$

$$= \frac{63.25-325.25}{63.25+325.25}\times\frac{11\times1000}{\sqrt{3}}\text{ V} = 4.28 \text{ kV}$$

∴ Reflected power

$$= \frac{3V'^2}{Z_{cb}} = \frac{3\times\left(4.28\times10^3\right)^2}{63.25} = 868.86 \text{ kW} \quad \textbf{Ans.}$$

$$\text{Transmitted voltage } (V'') = \frac{2\times Z_{cb}}{Z_N+Z_{cb}}\times\frac{V}{\sqrt{3}}$$

$$= \frac{2\times63.25}{63.25+325.25}\times\frac{11\times1000}{\sqrt{3}}\text{ V} = 2.07 \text{ kV}$$

$$\text{Transmitted power} = \frac{3V''^2}{Z_{cb}} = \frac{3\times\left(2.07\times10^3\right)^2}{63.25}$$

$$= 203.24 \text{ kW} \quad \textbf{Ans.}$$

Problem 12.9. A surge of 11 kV magnitude travels along a cable towards its junction with an overhead line. The inductance and capacitance of the cable and overhead line are respectively 0.2 mH,

0.3 μF and 1 mH and 0.010 μF/km respectively. Find the voltage rise at the junction.

Solution: Given,

Capacitance of the line (C) = 0.010 μF/km

Inductance of the line (L) = 1 mH

Capacitance of the cable (C_b) = 0.3 μF/km

Inductance of the cable (L_b) = 0.2 H

The surge impedance of the line is given by

$$Z_n = \sqrt{\frac{L}{C}} = \sqrt{\frac{1\times10^{-3}}{0.010\times10^{-6}}} = 316.23\ \Omega$$

The surge impedance of the cable is given by

$$Z_{cb} = \sqrt{\frac{L_b}{C_b}} = \sqrt{\frac{0.2\times10^{-3}}{0.3\times10^{-6}}} = 25.82\ \Omega$$

Here, the surge travels from cable towards the overhead line, hence there will be positive voltage reflection at the junction.

The voltage rise at the junction is given by

$$V'' = \frac{2\times Z_N\times V}{Z_{cb}+Z_N} = \frac{2\times316.23\times11\times1000}{(316.23+25.82)} = 20.34\ \text{kV}$$

∴ Voltage rise at the junction = 20.34 kV **Ans.**

Problem 12.10. A surge of 110 kV traveling in a line of natural impedance 500 Ω arrives at junction with two lines of impedances 700 Ω and 100 Ω respectively. Find the surge voltage and currents transmitted into each branch line.

Solution: Given,

Z_1 = 500 Ω, V = 110 kV, Z_2 = 700 Ω, Z_1 = 100 Ω

This is the situation of T–Junction

The given circuit diagram can be drawn as given in Fig. P. 12.10.

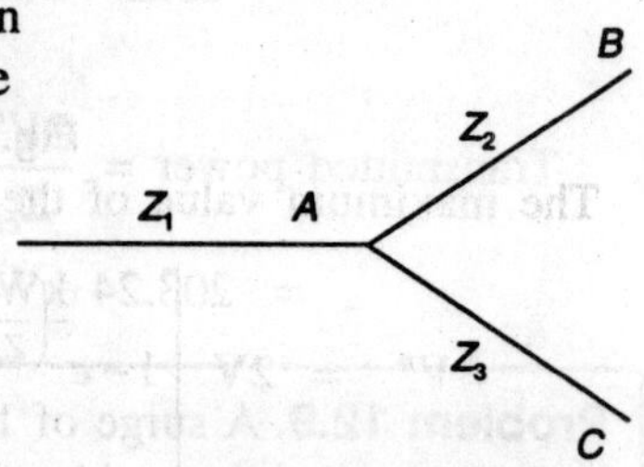

Fig. P. 12.10

As the surge reaches to junction, it reflects and the transmitted voltage will be same in both paralled branches Z_2 and Z_3

The transmitted voltage is given by

$$V'' = \frac{2V/Z_1}{\frac{1}{Z_1}+\frac{1}{Z_2}+\frac{1}{Z_3}} = \frac{2\times 110\times 10^3}{\left(1+\frac{500}{700}+\frac{500}{100}\right)} = 32.97 \text{ kV}$$

$\therefore$ Transmitted voltage $(V'') = 32.79$ kV **Ans.**

The surge current transmitted through line B is given by

$$I_2'' = \frac{V''}{Z_2} = \frac{32.79\times 1000}{700} = 46.84 \text{ A} \quad \textbf{Ans.}$$

Now, the surge current transmitted through line C is given by

$$I_3'' = \frac{V''}{Z_3} = \frac{32.79\times 1000}{100} = 327.9 \text{ A} \quad \textbf{Ans.}$$

Problem 12.11. A 440 kV, 3 μsec rectangular surge on a line having a surge impedance of 360 Ω approaches a station at which concentrated earth capacitance is 2 × 10^{-9} F. Determine the maximum value of the transmitted wave.

Solution: Given,

Surge voltage (V) = 440 kV

Time period of the pulse (τ) = 3 × 10^{-6} sec

Surge impedance (Z_N) = 360 Ω

The diagram of the given situation is as shown in Fig. P. 12.11.

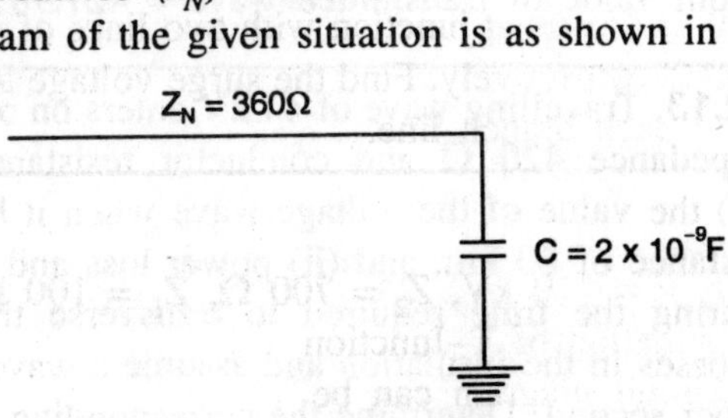

Fig. P. 12.11

The maximum value of the transmitted wave is given by

$$V'' = 2\text{V}\left[1-e^{-\left(\frac{t}{Z_N C}\right)}\right]$$

$$= 2 \times 440 \times 10^3 \left[1 - e^{-\left(\frac{3\times10^{-6}\times10^{9}}{2\times360}\right)} \right]$$

$$= 880\ [1 - e^{-4.167}] = 866.36 \text{ kV}$$

$\therefore$ Maximum transmitted voltage = 866.36 kV **Ans.**

Problem 12.12. An inductance of 600 μH connects two sections of a transmission line, each having a surge impedance of 360 Ω. A 440 kV, 3 μs rectangular surge travels along the line towards the inductance. Determine the maximum value of the transmitted wave.

Solution: Given,

Inductance (L) = 600 μH = 600 × 10^{-6} H
Surge impedance of each line (Z_N) = 360
Surge voltage (V) = 440 kV
Time period of pulse = 3 μS = 3 × 10^{-6} S
The maximum value of the transmitted wave is given by

$$V'' = V\left[1 - e^{-\frac{2Z_N C}{L}}\right] = 440 \times 10^3 \left[1 - e^{-\frac{2\times360\times3\times10^{-6}}{600\times10^{-6}}}\right]$$

$$= 440\ [1 - e^{-3.6}] = 427.98 \text{ kV}$$

$\therefore$ Maximum value of transmitted wave = 427.98 kV **Ans.**

Problem 12.13. Travelling wave of 44 kV enters on overhead line of surge impedance 420 Ω and conductor resistance 5 Ω/km. Determine (i) the value of the voltage wave when it has travelled through a distance of 60 km, and (ii) power loss and heat loss of the wave during the time required to transverse this distance. Neglect the losses in the insulation and assume a wave velocity of 2 × 10^5 km per second. Determine the corresponding values for a cable having surge impedance of 42 Ω and relative permitivity 4.2.

Solution: Given,

Initial wave voltage (V_o) = 44 kV
Surge impedance (Z_n) = 420 Ω
Resistance of conductor (R) = 5 Ω
Distance (x) = 60 km

Conductance of the line $(G) = 0.0$ ℧

(i) As the wave travels, if gets attenuated in magnitude because line has some resistance.

The value of the wave at a distance X is given by

$$V = V_o\, e^{-\frac{1}{2}\left(\frac{R}{Z_N}+GZ_N\right)x}$$

$$= 44 \times 10^3\, e^{-\frac{1}{2}\left(\frac{5}{420}+0\times 420\right)\times 60}$$

$$= 44\, e^{-\frac{1}{2}\left(\frac{5}{420}\times 60\right)}\ \text{kV} = 44\, e^{-0.357}\ \text{kV}$$

$$= 30.79\ \text{kV}$$

∴ Value of the wave at a distance 60 km = 30.79 kV **Ans.**

(ii) In this case, the power loss is instantaneous quantity, i.e. it depends upon the distance x.

The value of the travelling wave at 60 km = 30.79 kV.

The power loss during this travelled distance is given by

$$\text{Power loss} = \frac{V^2}{Z_n}\ \text{W} = \frac{\left(30.79\times 10^3\right)^2}{420} = 2.257\ \text{MW}$$

Thus, required power loss = 2.257 MW **Ans.**

The heat loss of the wave is given by Heat loss = $\int_0^t vi\,dt$

Let us calculate t,

Given, $v = 2 \times 10^5$ km/s

We know that $v = \dfrac{x}{t}$

$$\therefore t = \frac{x}{v} = \frac{60}{2\times 10^5} = 30 \times 10^{-5}\ \text{sec.} = 3 \times 10^{-4}\ \text{sec.}$$

For this line the instantaneous values of voltage and current wave are given by

$$v = V_o e^{-\frac{1R}{2Zn}vt} \quad \text{and } 2 = I_o e^{-\frac{R}{2Z_N}vt}$$

Now,

$$\text{Heat loss} = \int_0^t v_2\,dt = \int_0^t V_0 e^{-\frac{Rvt}{2Z_N}} \times I_o e^{-\frac{Rvt}{2Z_N}}\,dt$$

$$= \int_0^t V_o \times \frac{V_o}{Z_N} e^{-\frac{Rvt}{2\,Z_N}} dt$$

$$= \frac{V_o^2}{Z_N} \int_0^t e^{-\frac{Rvt}{Z_N}} dt = \frac{V_o^2}{Z_N}\left[-\frac{Z_N}{Rv} e^{-\frac{Rvt}{Z_N}}\right]_0^t$$

$$= \frac{V_o^2}{Z_N}\left[-\frac{Z_N}{Rv} e^{-\frac{Rvt}{Z_N}+\frac{Z_N}{Rv}}\right] = \frac{V_o^2}{Z_N} \times \frac{Z_N}{Rv}\left[1-e^{-\frac{Rvt}{Z_N}}\right]$$

$$= \frac{\left(44\times10^3\right)^2}{5\times2\times10^5}\left[1-e^{-\frac{5\times2\times10^5\times30\times10^{-5}}{420}}\right]$$

$$= 1936\ [\ 1 - e^{-0.714}] = 987.977 \text{ J} = 0.988 \text{ kJ}$$

$\therefore$ Heat loss = 0.988 kJ **Ans.**

Chapter 13

Symmetrical Components and Fault Calculation

Problem 13.1. The line currents in a 3-phase supply to an unbalanced load are respectively $I_a = 10 + j\,20$, $I_b = 12 - j10$ and $I_c = -3 - j5$ amperes. The phase sequence is *abc*. Determine the sequence components of currents.

Solution: Given

$I_a = 10 + j\,20$ A, $I_b = 12 - \text{j}\,10$ A, $I_c = -3 - j\,5$ A

The positive sequence component of the current is given by

$$I_{a1} = \frac{1}{3}(I_a + \lambda I_b + \lambda^2 I_c)$$

$$= \frac{1}{3}[10 + j\,20 + (-0.5 + j\,0.866)\,(12 - j\,10) + (-0.5 - j\,866)\,(-3 - j\,5)]$$

$$= \frac{1}{3}[10 + j\,20 + (-6 + 5\,j + j\,12 \times 0.866) + (1.5 + j\,2.5 + j\,0.866 \times 3 - 5 \times 0.866)]$$

$$= 3.277 \times j\,13.50 \text{ A}$$

The negative sequence component of the current is given by

$$I_{a2} = \frac{1}{3}\,(I_a + \lambda^2 I_b + \lambda I_c)$$

$$= \frac{1}{3}(10 + j\,20 + (-0.5 - j\,0.866)\,(12 - j\,10) + (-0.5 + j\,0.866)\,(-3 - j\,5))]$$

$$= \frac{1}{3}[10 + j\,20 - 6 + 5\,j - j\,12 \times 0.866 - 8.66 + 1.5 + 2.5\,j - 3 \times j\,0.866 + 5 \times 0.866]$$

$$= 0.39 + j\,4.837 \text{ A}$$ **Ans.**

And, the two sequence component of the current is given by

$$I_{a0} = \frac{1}{3}(I_a + I_b + I_c)$$

$$= \frac{1}{3}(10 + j\,20 + 12 - j\,10 - 3 - j\,5)$$

$$= \frac{1}{3}(19 + j\,5) = 6.333 + j\,1.667 \text{ A} \quad \textbf{Ans.}$$

Problem 13.2. The voltages across a 3-phase unbalanced load are $V_a = 300$ V, $V_b = 300\angle{-90^\circ}$ V and $V_c = 800\angle 143.1^\circ$ V respectively. Determine the sequence components of voltages. Phase sequence is abc.

Solution: Given,

$V_a = 300\angle 0^\circ$ V, $V_b = 300\angle{-90^\circ}$ V, $V_c = 800\angle 143.1^\circ$V

The positive sequence of the voltage is given by

$$V_a = \frac{1}{3}(V_a + \lambda V_b + \lambda^2 V_c)$$

$$= \frac{1}{3}[300\angle 8^\circ + 1\angle 120 \times 300\angle{-90^\circ} + 1\angle{-120^\circ} \times 800\angle 143.1^\circ]$$

$$= \frac{1}{3}[300\angle 0^\circ + 300\angle 30^\circ + 800\angle 23.1^\circ]$$

$$= 431.888 + j\,154.62 = 458.73\angle 19.7^\circ \text{ V} \quad \textbf{Ans.}$$

The negative sequence component of the voltage is given by

$$V_{a2} = \frac{1}{3}(V_a + \lambda^2 V_b + \lambda V_c)$$

$$= \frac{1}{3}[300\angle 0^\circ + 1\angle{-120^\circ} \times 300\angle{-90^\circ} + 1\angle 120^\circ \times 800\angle 143.1^\circ]$$

$$= \frac{1}{3}[300\angle 0^\circ + 300\angle{-120^\circ} + 800\angle 263.1^\circ]$$

$$= -18.639 - j\,214.735 = 215.54\angle{-95^\circ} \text{ V} \quad \textbf{Ans.}$$

The two sequence component of the voltage is given by

$$V_{a0} = \frac{1}{3}[V_a + V_b + V_c]$$

$$= \frac{1}{3}[300\angle 0^\circ + 300\angle{-90^\circ} + 800\angle 143.1^\circ]$$

$$= -113.25 + j\,60.112 = 128.215\angle 152.04^\circ \text{ V} \quad \textbf{Ans.}$$

Problem 13.3. Three 6.6 kV, 12 MVA, 3-phase alternators are connected to a common set of bus bars. The positive, negative and zero sequence impedances of each alternator are 15%, 12% and 4.5% respectively. If an earth fault occurs on one busbar, determine the fault current:

(i) if all the alternator neutrals are solidly grounded;

(ii) if one only of the alternator neutrals is solidly earthed and the others are isolated;

(iii) if one of the alternator neutrals is earthed through a reactance of 0.5 ohm and the others are isolated.

Solution: Let us choose 12 MVA and 6.6 kV as base quantities

Base kVA = 12 MVA, V_b = 6.6 kV

$$\therefore \text{ Base current } (I_b) = \frac{\text{Base kVA}}{\sqrt{3} \times V_b} = \frac{12 \times 10^6}{\sqrt{3} \times 6.6 \times 10^3} = 1049.72 \text{ A}$$

Given, positive sequence impedance (Z_1) = 0.15 p.u.

Negative sequence impedance (Z_2) = 0.12 p.u.

Zero sequence impedance (Z_0) = 0.045 p.u.

As we know that the positive, negative and zero sequences components are equal, i.e.

$$I_{a1} = I_{a2} = I_{a0}$$

$$\text{and fault current } (I_a) = \frac{3Ea}{Z_1 + Z_2 + Z_0 + 3\,Z_n}$$

(i) For solid grounded system, $Z_n = 0$

Here, E_a = 1.00 + j 0.0 p.u. since

$$E_{a\,pu} = \frac{V}{V_b} = \frac{6.6}{6.6} = 1 + j\,0.0$$

The fault current due to each generator is given by Fig. P. 13.3 (a).

$$I_a = \frac{3Ea}{Z_1 + Z_2 + Z_0} \times I_b$$

$$= \frac{3 \times 1 \times 1049.73}{0.15 + 0.12 + 0.045}$$

$$= 9.997 \text{ kA}$$

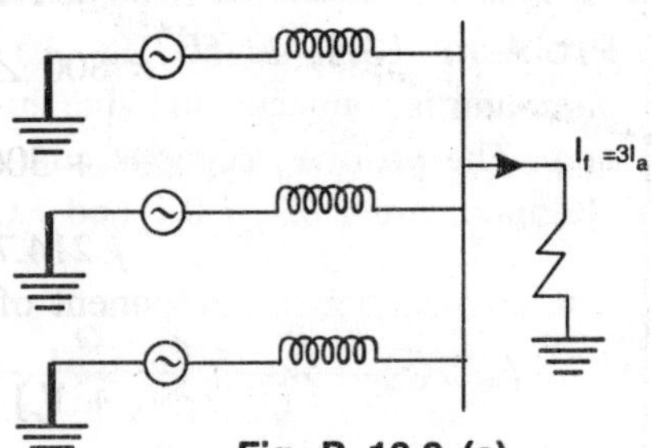

Fig. P. 13.3 (a)

$\therefore$ Fault current occurs at the bus bar = sum of fault current due to each generator

$$= 3 \times I_a = 3 \times 9.997 \text{ kV} = 29.991 \text{ kA} \quad \textbf{Ans.}$$

(ii) When two generators are isolated, The fault current at bus bar is given by Fig. P. 13.3 (b).

$$I_f = 2 \times \frac{3 \times E_g}{Z_1 + Z_2} \times I_b$$

$$= 2 \times \frac{3 \times 1}{0.15 + 0.12} \times 1049.73$$

$$= 23.327 \text{ kA}$$

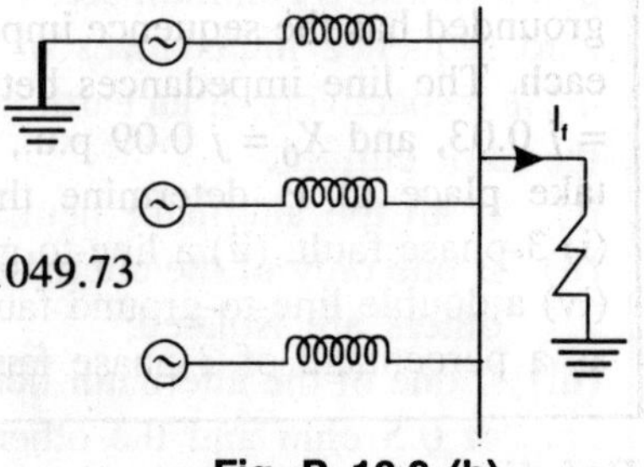

Fig. P. 13.3 (b)

Thus, fault current at the bus bar = 23.327 kA

(iii) When only one generated is earthed through a reactance of 0.5 ohm,

Here, $Z_n = 0.5\ \Omega$

In this case, the fault current at the bus is given by Fig. P. 13.3 (c).

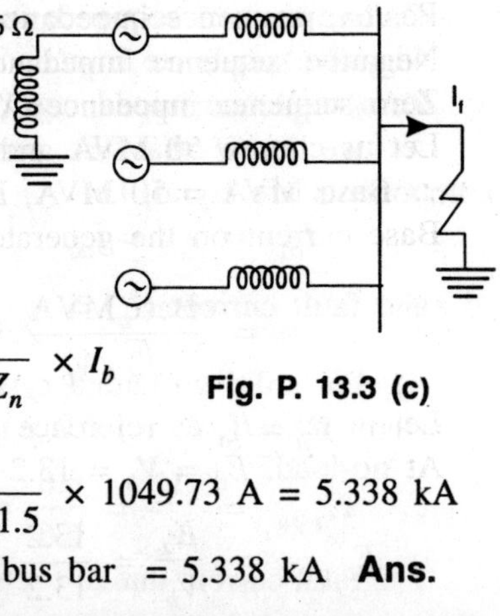

$$I_f = 3 \times I_a \times I_b$$

$$I_f = 3 \times \frac{3 \times Ea}{Z_1 + 2_f + 3 \times Z_n} \times I_b$$

Fig. P. 13.3 (c)

$$= 3 \times \frac{3 \times 1}{0.15 + 0.12 + 1.5} \times 1049.73 \text{ A} = 5.338 \text{ kA}$$

In this case, fault current at the bus bar = 5.338 kA **Ans.**

Problem 13.4. A 50 Hz, 50 MVA, 13.2 kV star grounded alternator is connected to a line through a Δ/Y transformer as shown here. The positive, negative and zero sequence impedances of the alternator are *j* 0.1, *j* 0.1 and *j* 0.05 p.u. respectively.

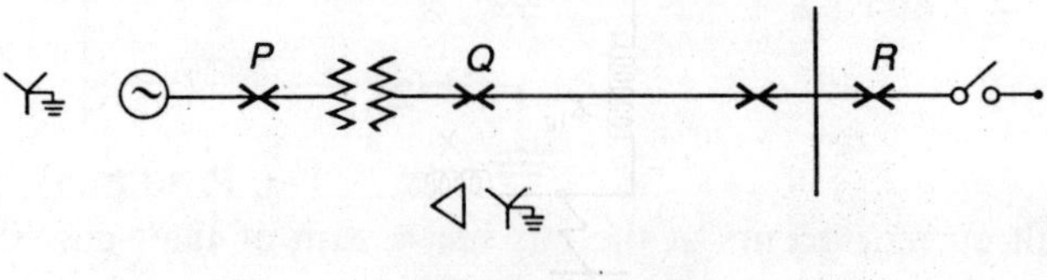

Fig. P. 13.4 (a)

The transformer rated at 13.2 kV Δ/120 MVA with star solidly grounded has the sequence impedances of $X'' = X_2 = X_0 = j\,0.1$ p.u. each. The line impedances between Q and R are $X'' = j\,0.03$, $X_2 = j\,0.03$, and $X_0 = j\,0.09$ p.u., respectively. Assuming the fault to take place at P, determine the subtransient fault current for a (i) 3-phase fault, (ii) a line-to-ground fault, (iii) a line-to-line fault, (iv) a double line-to-ground fault, Also express these fault currents as a percentage of 3-phase fault current as calculated in (i).

Solution: Given

For 50 Hz, 50 MVA, 13.2 kV generator
positive sequence impedance $(X_{1g}) = j\,0.1$ p.u.
Negative sequence impedance $(X_{2g}) = j\,0.1$ p.u.
Zero sequence impedance $(X_{0g}) = j\,0.05$ p.u.
For 13.2 kVΔ/120 kVY. 50 MVA Transformers.
Positive sequence impedance $(X'') = j\,0.1$ p.u.
Negative sequence impedance $(X_2) = j\,0.1$ p.u.
Zero sequence inpedance $(X_0) = j\,0.1$ p.u.
Let us choose 50 MVA and 13.2 kV as base quantities
∴ Base MVA = 50 MVA, Base voltage (V_b) = 13.2 kV
Base current on the generator side (I_b)

$$= \frac{\text{Base MVA}}{\sqrt{3} \times V_b} = \frac{50 \times 10^6}{\sqrt{3} \times 13.2 \times 10^3} = 2186.93 \text{ A}$$

Let in take E_a as reference phasor.
At no-load, $E_a = V_b = 13.2$ kV

$$E_{a\ \text{p.u.}} = \frac{E_a}{V_b} = \frac{13.2}{13.2} = 1\angle 0° \text{ p.u.}$$

Now, Let us calculate positive, Negative and zero sequence impedance

The positive sequence network of the given problems is shown in Fig. P. 13.4 (b).

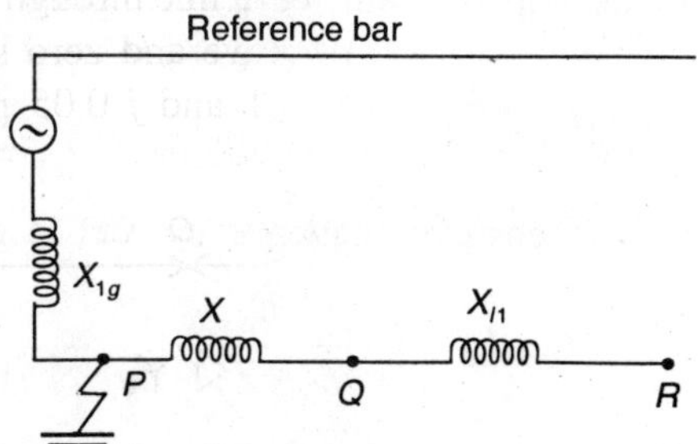

Fig. P. 13.4 (b)

From Fig. P. 13 (b),

Positive sequence impedance (Z_1) $= X_{1g} = j\ 0.1$ p.u.

The negative sequence network of the given problem can be drawn as in Fig. P. 13.4 (c).

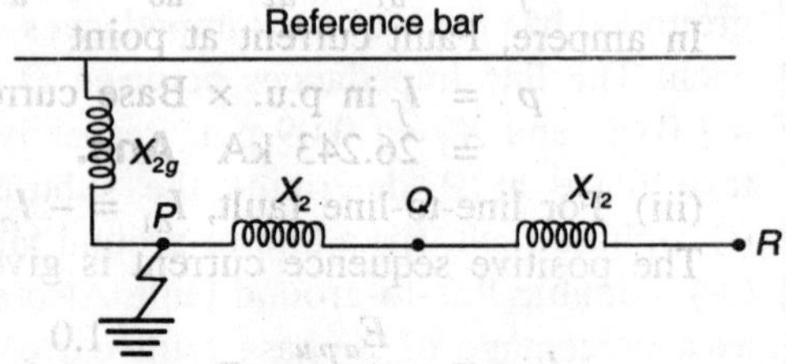

Fig. P. 13.4 (c)

From Fig. P. 13.4 (c),
Negative sequence impedance (Z_2) $= X_{2g} = j\ 0.1$ p.u.

The Zero sequence network is as shown in Fig. P. 13.4 (d)

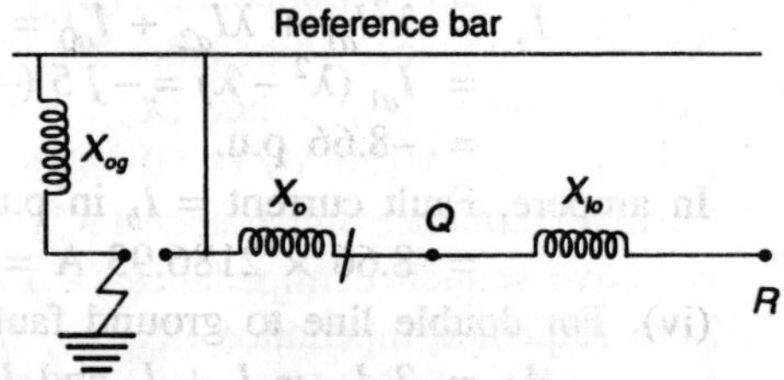

Fig. P. 13.4 (d)

From Fig. P. 13.4 (d),
Zero sequence impedance (Z_0) $= X_{go} = j\ 0.05$ p.u.

(i) For three phase fault, $|I_a| = |I_b| = |I_c| = |I_{a1}|$

$$I_{a2} = I_{a0} = 0 \text{ and } I_{a1} = \frac{E_a}{Z_1}$$

The positive sequence current (I_{a1}) is given by

$$I_{a1} = \frac{E_{a\,p.u}}{Z_1} = \frac{1.0}{j0.1} = -j\ 10 \text{ p.u.}$$

The fault current at point p is given by

$$I_f = |I_a| = |I_{a1} + I_{a2} + I_{a0}| = |I_{a1}| = 10 \text{ p.u.}$$

In ampere, Fault current at point p = I_f in p.u. × base current

$$= 10 \times 2186.93 = 21.869 \text{ kA} \quad \textbf{Ans.}$$

(ii) For line-to-ground fault, $I_{a1} = I_{a2} = I_{a0} = \dfrac{I_a}{3}$

The positive sequence current is given by

$$I_{a1} = \frac{E_{a\,p.u}}{Z_1 + Z_2 + Z_3} = \frac{1.0}{j(0.1 + 0.1 + 0.05)}$$

$$= \frac{1.0}{j.25} = -j\ 4 \text{ p.u.}$$

The fault current is given by

$$I_f = I_{a1} + I_{a2} + I_{a0} = 3\ I_a = 3X - j\ 4 = -\ j\ 12 \text{ p.u.}$$

In ampere, Fault current at point

$$p = I_f \text{ in p.u.} \times \text{Base current} = 12 \times 2186.93$$
$$= 26.243 \text{ kA} \quad \textbf{Ans.}$$

(iii) For line-to-line fault, $I_{a1} = -\ I_{a2}$ and I_{a0}

The positive sequence current is given by

$$I_{a1} = \frac{E_{a\,p.u}}{Z_1 + Z_2} = \frac{1.0}{j(0.1+0.1)} = -\ j\ 5 \text{ p.u.}$$

$\therefore$ Negative sequence current $(I_{a2}) = -\ I_{a1} = j\ 5$ p.u.

Now, The fault current at point p is given by

$$I_b = \lambda^2 I_{a1} + \lambda I_{a2} + I_{a0} = \lambda^2 I_{a1} - \lambda I_{a1} + 0$$
$$= I_{a1}(\lambda^2 - \lambda) = -\ j\ 5\ (-0.5 - j\ 0.866 + 0.5 - j\ 0.866)$$
$$= -8.66 \text{ p.u.}$$

In ampere, Fault current $= I_b$ in p.u. $\times$ base current

$$= 8.66 \times 2186.93 \text{ A} = 18.94 \text{ kA} \quad \textbf{Ans.}$$

(iv) For double line to ground fault,

$$I_f = 3\ I_{a0} = I_b + I_c \text{ and } I_b = -I_c \text{ i.e. } |I_b| = |I_c|$$

The positive sequence current is given by

$$I_{a1} = \frac{E_a\ p.u(Z_2 + Z_0)}{Z_1(Z_0 + Z_2) + Z_0 Z_2}$$

$$= \frac{1.0\ (j\,0.1 + j\,0.05)}{j\,0.1(j\,0.1 + j\,0.05) + (j\,0.1 \times j\,0.05)} = -\ j\ 7.5 \text{ p.u.}$$

The zero sequence current is given by

$$I_{a0} = \frac{E_a\ p.u - Ia_1 Z_1}{Z_0} = -\frac{1.0 - (-j\,7.5 \times j\,0.1)}{j\,0.05}$$
$$= j\ 5 \text{ p.u.}$$

$\therefore$ Fault current at point $p = 3 \times |I_{a0}| \times$ base current

$$= 3 \times 5 \times 2186.93 = 32.80 \text{ kA} \quad \textbf{Ans.}$$

% Fault current in terms of three phase fault is given by

% Fault current in double line to ground fault

$$= \frac{32.80}{21.869} \times 100 = 150\% \quad \textbf{Ans.}$$

Problem 13.5. Solve problem 13.4 when fault: is at point *Q*.

Solution: Given,

For alternator, positive sequence impedance of the alternator

$(X_{1g}) = j\ 0.1$ p.u.

Negative sequence impedance of the alternator $(X_{2g}) = j\ 0.1$ p.u.

Zero sequence impedance of the alternator $(X_{0g}) = j\ 0.05$ p.u.

For 13.2 kV Δ/120 kVY, 50 MVA,

Positive sequence impedance $(X'') = j\ 0.1$ p.u.

Negative sequence impedance $(X_2) = j\ 0.1$ p.u.

Zero sequence impedance $(X_0) = j\ 0.1$ p.u.

Let us calculate positive sequence impedance, Negative sequence impedance and zero sequence impedance upto fault point Q.

Let us choose 50 MVA, 13.2 kV as base quantities.

$MVA_{new} = 50$ MVA , $V_{new} = 13.2$ kV

Now, let us calculate the impedances of the transformer in the new base.

For transformer, $MVA_{old} = 50$ MVA, $V_{old} = 13.2$ kV

To calculate impedances in the new base quantities, apply the relation $Z_{p.u.\ new} = Z_{p.u.\ old} \times \dfrac{KVA_{new}}{KVA_{old}} \times \left(\dfrac{V_{old}}{V_{new}}\right)^2$

Now, positive sequence impedance of the transformer

$$(X_n'') = j\ 0.1 \times \frac{50}{50} \times \left(\frac{13.2}{13.2}\right)^2 = j\ 0.1 \text{ p.u.}$$

Similarly,

$X_{2n} = j\,0.1 \text{ p.u.} = X_{0n}$

The positive sequence network of this problem is as shown in Fig. P. 13.5 (a).

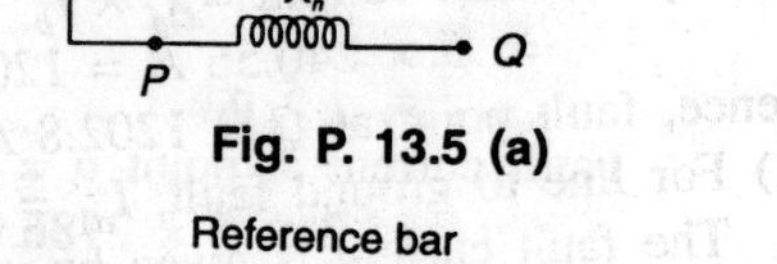

Fig. P. 13.5 (a)

Here, from Fig. P. 13.5 (a),

Positive sequence impedance

$(Z_1) = X_{1g} + X_n''$

$= j\ 0.1 + j\ 0.1$

$= j\ 0.2$ p.u.

The negative sequence network can be drawn as shown in Fig. P. 13.5 (b).

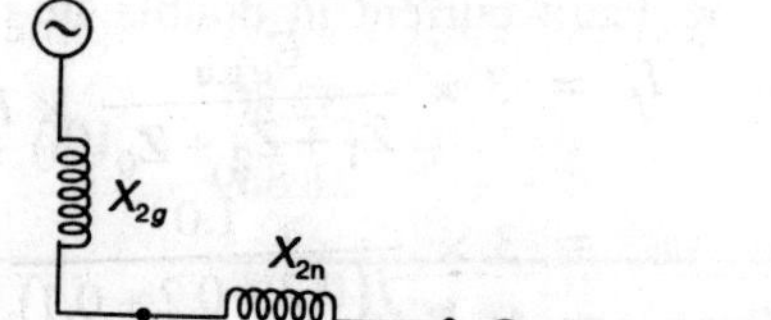

Fig. P. 13.5 (b)

From Fig. 13.5 (b)

$\therefore$ Negative sequence impedance $(Z_2) = X_{2g} + X_{2n}$

$= j\,0.1 + j\,0.1 = j\,0.2$ p.u.

The zero sequence network can be drawn as shown in Fig. P. 13.5 (c).

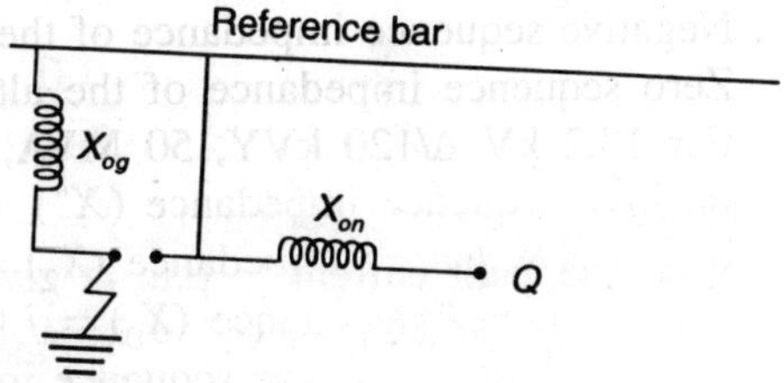

Fig. P. 13.5 (c)

From Fig. P. 13.5 (c)

Negative sequence impedance $(Z_0) = X_{0n} = j\,0.1$ p.u.

Base voltage referred to fault side (V_{bf})

$$= \text{Transformation ratio} \times V_b = \frac{120}{13.2} \times 13.2$$

$$= 120 \text{ kV}$$

Base current to the fault side $(I_B) = \dfrac{MVA_b}{\sqrt{3} \times V_{of}}$

$$= \frac{50 \times 10^6}{\sqrt{3} \times 120} = 240.56 \text{ A}$$

Let us take E_a as reference phasor.

At no load, $V_b = E_a = 13.2$ kV

$$\therefore \quad E_{a\,\text{p.u.}} = \frac{E_{a\,\text{p.u}}}{V_b} = \frac{13.2}{13.2} = 1\angle 0° \text{ p.u.}$$

(i) For three phase fault, the fault current is given by

$$I_f = \frac{E_{a\,\text{p.u}}}{Z_1} = \frac{1.0}{j0.5} = -j\,5 \text{ p.u.}$$

In ampere, fault current $= I_f \times I_b$

$= 5 \times 240.56$ A $= 1202.8$ A

Hence, fault point at $Q = 1202.8$ A **Ans.**

(ii) For line to ground fault, $I_{a1} = I_{a2} = I_{a0}$, $I_f = I_a = 3\,I_{a1}$

The fault current is given by

$$I_f = 3\,I_{a1} \text{ in p.u.}$$

$$I_f = 3 \times \frac{E_{a\,\text{p.u}}}{Z_1 + Z_2 + Z_0} \times I_B \text{ A}$$

$$= 3 \times \frac{1.0}{j(0.2 + 0.2 + 0.1)} \times 240.56 \text{ A}$$

$$= -j\,1443.36 \text{ A}$$

Thus, fault current of point (Q) = 1443.36 A **Ans.**

(iii) For line to line fault, $I_{a1} = -I_{a2}$, $I_{a0} = 0$

The positive sequence current is given by

$$I_{a1} = \frac{E_{a\,p.u}}{Z_1 + Z_2} = \frac{1.0}{j\,0.2 + j\,0.2} = -j\,2.5 \text{ p.u.}$$

Negative sequence current $(I_{a2}) = -I_{a1} = j\,2.5$ p.u.

Now, the fault current in p.u. is given by

$$I_f = \lambda^2 I_{a1} + \lambda I_{a2} = \lambda^2 I_{a1} - \lambda I_{a1} = I_{a1}(\lambda^2 - \lambda)$$
$$= I_{a1}(-0.5 - j\,0.866 + 0.5 - j\,0.866)$$
$$= (-j\,2.5) \times (2 \times -j\,.866) = -4.33 \text{ p.u.}$$

In ampere, fault current = I_f in p.u. × Base current (I_B)

$$= 4.33 \times 240.56 = 1041.62 \text{ A}$$ **Ans.**

(iii) For double line to ground fault,

The positive sequence (I_{a1}) is given by

$$I_{a1} = \frac{E_{a\,p.u}(Z_2 + Z_0)}{Z_1(Z_0 + Z_2) + Z_0 Z_2}$$

$$= \frac{1.0(j\,0.2 + j\,0.1)}{j\,0.2(j\,0.2 + j\,0.1) + j\,0.1 \times j\,0.2} \text{ p.u.} = -j\,3.75 \text{ p.u.}$$

The zero sequence current is given by

$$I_{a0} = -\frac{E_{a\,p.u} - Ia_1 Z_1}{Z_0} = \frac{1.0 - j\,0.2 \times (-13.75)}{j\,0.1}$$

$$= j\,2.5 \text{ p.u.}$$

For double line to ground fault, Fault current

$$= 3\,I_{a0} = 3 \times j\,2.5 \text{ p.u.} = j\,7.5 \text{ p.u.}$$

In ampere, fault current (I_f) = Fault current in p.u. × base current

$$= 7.5 \times 240.56 \text{ A} = 1804.2 \text{ A}$$ **Ans.**

Problem 13.6. Solve Problem 13.5 when fault is at point *R*.

Solution: Given

For 50 Hz, 50 MVA, 13.2 kV, generator,

Positive sequence impedance $(X_{1g}) = j\,0.1$ p.u.

Negative sequence impedance $(X_{2g}) = j\,0.1$ p.u.

Zero sequence impedance $(X_{0g}) = j\,0.05$ p.u.

For 13.2 kV Δ/ 120 VY, 50 MVA transformer.

Positive sequence impedance $(X'') = j\,0.1$ p.u.

Negative sequence impedance $(X_2) = j\,0.1$ p.u.

Zero sequence impedance $(X_0) = j\ 0.1$ p.u.
For transmission line,
Positive sequence impedance $(X_l'') = j\ 0.03$ p.u.
Negative sequence impedance $(X_{l2}) = j\ 0.03$ p.u.
Zero sequence impedance $(X_{l0}) = j\ 0.09$ p.u.
Let us choose 50 MVA, 13.2 kV as base quantities
$\therefore\ MVA_b = 50$ MVA, $V_b = 13.2$ kV
Since the base quantities and ratings of the transformer are same, impedances remain same on the base of base quantities.

Now, let us calculate positive, negative and zero sequence impedances.

The positive sequence impedance of the given problem is as shown in Fig. P. 13.6 (a).

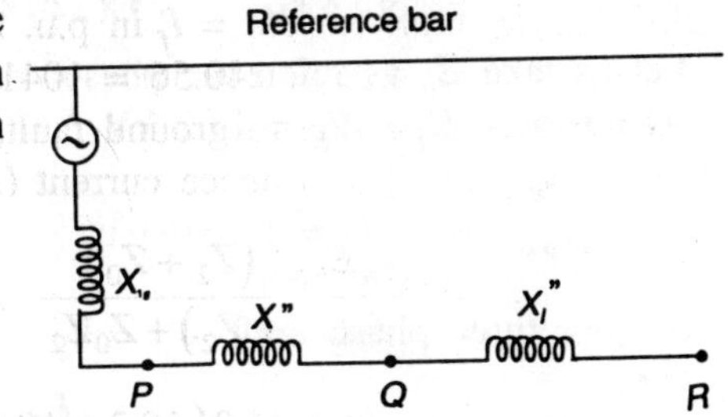

Fig. P. 13.6 (a)

From Fig. P. 13.6 (a)

Positive sequence impedance $(Z_1) = X_{1g} + X'' + X_l''$
$= j\ 0.1 + j\ 0.1 + j\ 0.03 = j\ 0.23$ p.u.

The negative sequence impedance is as follows

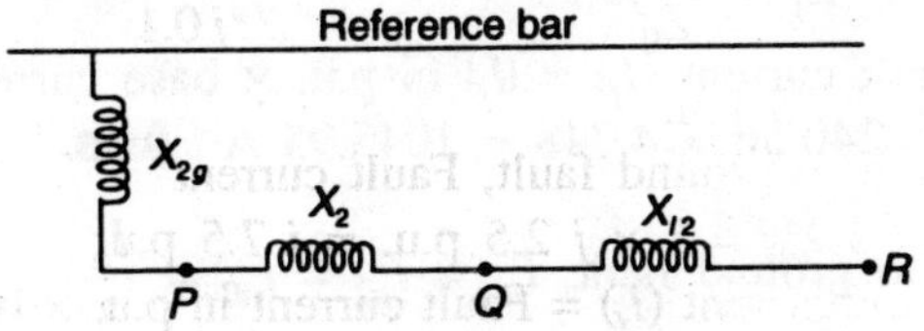

Fig. P. 13.6 (b)

From Fig. P. 13.6 (b),
Negative sequence impedance $(Z_2) = X_{2g} + X_2 + X_{l2}$
$= j\ 0.1 + j\ 0.1 + j\ 0.03 = j\ 0.23$ p.u.

The zero sequence impedance of the given problem upto the fault point R can drawn shown in Fig. P. 13.6 (c).

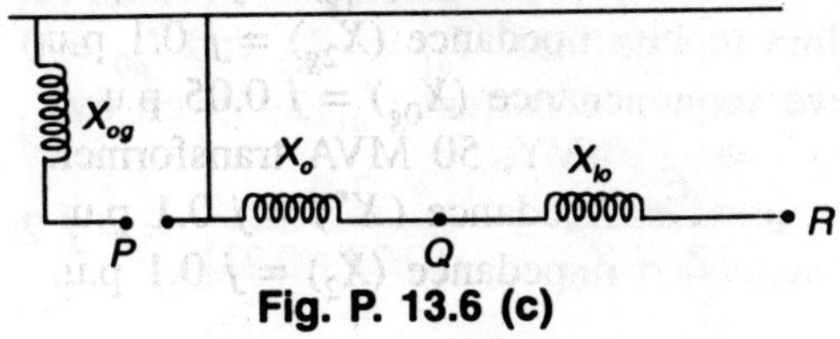

Fig. P. 13.6 (c)

From Fig. P. 13.6 (c),
Zero sequence impedance (Z_0)

$$= X_0 + X_{l0} = j\,0.1 + j\,0.09 = j\,0.19 \text{ p.u.}$$

The base voltage referred to line side is given by

$$V_{bl} = \text{Transformation ratio} \times V_b$$

$$= \left(\frac{120}{13.2}\right) \times 13.2 = 120 \text{ kV}$$

$\therefore$ Base current referred to line side

$$I_{bl} = \frac{MVA_b}{\sqrt{3} \times V_{bl}} = \frac{50 \times 10^6}{\sqrt{3} \times 120 \times 10^3} = 240.56 \text{ A}$$

Let us take E_a as reference phasor.
At no load, $E_a = V_b = 13.2$ kV

$$\therefore \quad E_{a\text{ p.u.}} = \frac{E_a}{V_b} = 1 \angle 0° \text{ p.u.}$$

(i) For three phase fault, $|I_a| = |I_b| = |I_c| = |I_{a1}|$

$$I_{a2} = I_{a0} = 0, \text{ and } I_{a1} = \frac{E_a}{Z_1}$$

For three phase fault, fault current $(I_f) = I_{a1}$

$$= \frac{E_{a\text{ p.u}}}{Z_1} = \frac{1.0}{j0.23} = -j\,4.348 \text{ p.u.}$$

In ampere, fault current $(I_f) = |I_f|$ in p.u. × base current

$$= 240.56 \times 4.348 = 1045.95 \text{ A} \quad \textbf{Ans.}$$

(ii) For line to ground fault, $I_{a1} = I_{a2} = I_{a0} = \dfrac{I_a}{3}$

The positive sequence current is given by

$$I_{a1} = \frac{E_{a\text{ p.u}}}{Z_1 + Z_2 + Z_0} = \frac{1.0}{j(0.23 + 0.23 + 0.19)}$$

$$= -j\,1.538 \text{ p.u.}$$

Fault current at point $R = 3\,|I_{a1}| \times$ Base current, A

$$= 3 \times 1.538 \times 240.56 = 1110.28 \text{ A} \quad \textbf{Ans.}$$

(iii) For line to line fault, $I_{a1} = -I_{a2}$ and $I_{a0} = 0$
The positive sequence current (I_{a1}) is given by

$$= \frac{E_{a\text{ p.u}}}{Z_1 + Z_2} = \frac{1.0}{j(0.23 + 0.23)} = -j\,2.174 \text{ p.u.}$$

$\therefore$ Negative sequence current $(I_{a2}) = -I_{a1} = j\ 2.174$ p.u.

In this case, the fault current is given by

$$I_f = \lambda^2 I_{a1} + \lambda I_{a2} + I_{a0} = \lambda^2 I_{a1} - I_{a1}$$
$$= (\lambda^2 - \lambda) I_{a1}$$
$$= (-0.5 - j\ 0.866 + 0.5 - j\ 0.866) \times (-j\ 2.174)$$
$$= -3.432 \text{ p.u.}$$

$\therefore$ Fault current is ampere = $|I_f|$ in p.u. × base current

= 3.765 × 240.56 A = 905.80 A **Ans.**

(iv) For double line to ground fault, fault current $(I_f) = 3\ I_{a0}$

The positive sequence current is given by

$$I_{a1} = \frac{E_{a\,\text{p.u}}(Z_0 + Z_2)}{Z_1(Z_0 + Z_2) + Z_0 Z_2}$$

$$= \frac{1.0(j0.19 + j0.23)}{j0.23(j0.23 + j0.19) + (j0.19 \times j0.23)}$$

$$= -j\ 2.993 \text{ p.u.}$$

The zero sequence current is given by

$$I_{a0} = -\frac{E_{a\,\text{p.u}} - Ia_1 Z_1}{Z_0} = -\frac{1.0 + (j2.993) \times (j0.23)}{j0.19}$$

$$= j\ 1.64 \text{ p.u.}$$

The fault current in ampere is given by

$$I_f = 3 \times |I_{a0}| \times \text{base current}$$
$$= 3 \times 1.64 \times 240.65 \text{ A} = 1183.59 \text{ A}$$

Thus, fault current point (*R*) = 1183.59 A **Ans.**

Problem 13.7. Solve Problem 13.4 when the neutral of the alternator is grounded through an impedance of j 0.2 ohm.

Solution: Given

For 50 Hz, 50 MVA, 13.2 kV generator,

Positive sequence impedance $(X_{1g}) = j\ 0.1$ p.u.

Negative sequence impedance (X_{2g}) = j 0.1 p.u.

Zero sequence impedance $(X_{0g}) = j\ 0.1$ p.u.

Neutral impedance $(Z_n) = j\ 0.2\ \Omega$

Let us choose 50 MVA and 13.2 kV as base quantities

$\therefore MVA_b = 50$ MVA, $V_b = 13.2$ kV

$$\therefore \text{Neutral impedance in p.u.} = \frac{Z_n \times MVA_b}{(KV_b)2}$$

$$= \frac{j0.2 \times 50}{(13.2)^2} = 10.057 \text{ p.u.}$$

For 13.2 kV Δ/120 kVY, 50 MVA transformer,
Positive sequence impedance $(X'') = j\ 0.1$ p.u.
Negative sequence impedance $(X_2) = j\ 0.1$ p.u.
Zero sequence impedance $(X_0) = j\ 0.1$ p.u.
Since the base quantities and ratings of the transformer are same, the impedances of the transformer on the base quantities remain same.

Now, Let us calculate the positive, negative and zero–sequence impedance of the network, the positive sequence network is as follows

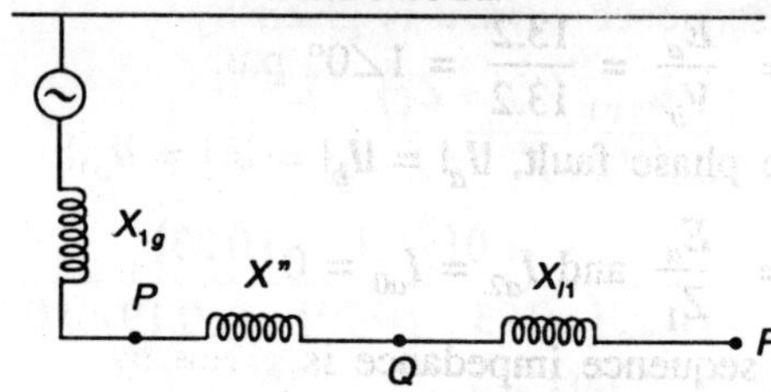

Fig. P. 13.7 (a)

From Fig. P. 13.7 (a).
Positive sequence impedance $(Z_1) = X_{1g} = j\ 0.1$ p.u.
The negative sequence network is an follows

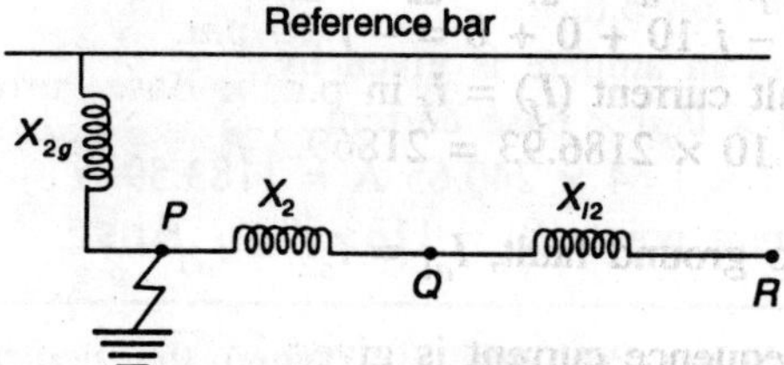

Fig. P. 13.7 (b)

From Fig. P. 13.7 (b),
Negative sequence impedance $(Z_2) = X_{2g} = j\ 0.1$ p.u.
The zero sequence network is as follows

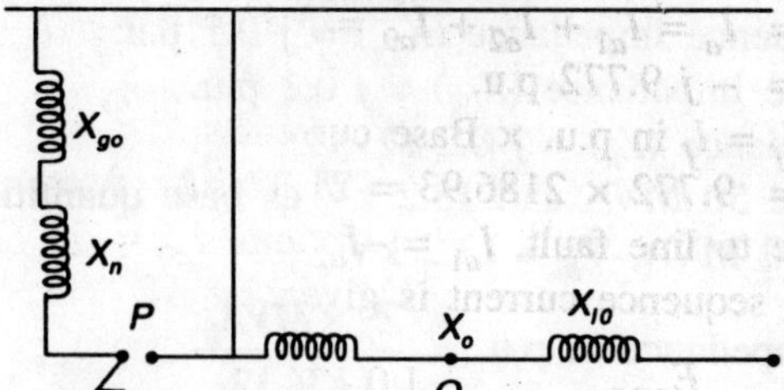

Fig. P. 13.7 (c)

From Fig. P. 13.7 (c),
Zero sequence impedance $(Z_0) = X_n + X_{g0}$
$= j\,0.05 + j\,0.057 = j\,0.107$ p.u.
Base current to the generator side (I_b)

$$= \frac{MVA_b}{\sqrt{3} \times V_b} = \frac{50 \times 10^6}{\sqrt{3} \times (13.2) \times 10^3} = 2186.93 \text{ A}$$

Let us take E_a as reference phasor.
At No load, $E_a = V_b = 13.2$ kV

$$\therefore \quad E_{a\,\text{p.u}} = \frac{E_a}{V_b} = \frac{13.2}{13.2} = 1\angle 0° \text{ p.u.}$$

(i) For three phase fault, $|I_a| = |I_b| = |I_c| = |I_{a1}|$

$$I_{a1} = \frac{E_a}{Z_1} \text{ and } I_{a2} = I_{a0} = 0$$

The positive sequence impedance is given by

$$I_{a1} = \frac{E_{a\,\text{p.u}}}{Z_1} = \frac{1.0}{j0.1}$$
$$= -j\,10 \text{ p.u.}$$

Fault current $(I_f) = I_a = I_{a1} + I_{a2} + I_{a0}$
$= -j\,10 + 0 + 0 = -j\,10$ p.u.
In ampere, fault current $(I_f) = I_f$ in p.u. × Base current
$= 10 \times 2186.93 = 21869.3$ A $= 21.869$ kA **Ans.**

(ii) For line to ground fault, $I_{a1} = I_{a2} = I_{a0} = \dfrac{I_a}{3}$

The positive sequence current is given by

$$I_{a1} = \frac{E_{a\,\text{p.u}}}{Z_1 + Z_2 + Z_0} = \frac{1.0}{j(0.1 + 0.1 + 0.107)}$$
$$= -j\,3.257 \text{ p.u.}$$

The fault current at point p is given by

$$I_f = I_a = I_{a1} + I_{a2} + I_{a0} = 3\,I_a = 3 \times -j\,3.257$$
$$= -j\,9.772 \text{ p.u.}$$

In ampere, $I_f = I_f$ in p.u. × Base current
$= 9.772 \times 2186.93 = 21.37$ kA **Ans.**

(iii) For line to line fault, $I_{a1} = -I_{a2}$ and $I_{a0} = 0$
The positive sequence current is given by

$$I_{a1} = \frac{E_{a\,\text{p.u}}}{Z_1 + Z_2} = \frac{1.0}{j(0.1 + 0.1)} = -j\,5 \text{ p.u.}$$

Negative sequence current $(I_{a2}) = -I_{a1} = j\,5$ p.u.

In this case, the fault current at point p is given by

$$I_f = I_b = \lambda^2 I_{a1} + \lambda I_{a2} + I_{a0}$$
$$= \lambda^2 I_{a1} - \lambda I_{a1} + 0 = I_{a1}(\lambda^2 - \lambda)$$
$$= -j\,5\,(-0.5 - j\,0.866 + 0.5 - j\,0.866)$$
$$= -\,8.66 \text{ p.u.}$$

In ampere, fault current at point p = If in p.u. × base current

$$= 8.66 \times 2186.93 \text{ A} = 18.94 \text{ kA}$$ **Ans.**

(iv) For double line to ground fault, fault current

$$(I_f) = 3\,I_{a0} = I_b + I_c$$

The positive sequence current is given by

$$I_{a1} = \frac{E_{a\,p.u}\left(Z_0 + Z_2\right)}{Z_1\left(Z_0 + Z_2\right) + Z_0 Z_2}$$

$$= \frac{1.0\left(j\,0.107 + j\,0.1\right)}{j\,0.1\left(j\,0.1 + j\,0.107\right) + \left(j\,0.107 \times j\,0.1\right)}$$

$$= -\,j\,6.59 \text{ p.u.}$$

The zero sequence current is given by

$$I_{a0} = \frac{E_{a\,\text{p.u}} - I_{a1} Z_1}{Z_0}$$

$$= -\,\frac{1.0 - \left(-j\,6.59 + j\,0.1\right)}{j\,0.107} = j\,3.187 \text{ p.u.}$$

Fault current at point p = 3 × $|I_{a0}|$ × base current

$$= 3 \times 3.187 \times 2186.93 = 20.91 \text{ KA}$$ **Ans.**

Problem 13.8. A power plant has two generators of 10 MVA, 15% reactance each and two 5 MVA generator's of 10% reactance paralleled at a common bus bar from which load is taken through a number of 4 MVA step up transformer each having a reactance of 5%. Determine the short circuit capacity of the breakers on the (i) low voltage, and (ii) high voltage side of the transformer.

Solution: Given,

Reactance of each 10 MVA generator = 0.15 p.u.

Reactance of each 5 MVA generator = 0.10 p.u.

Reactance of each 4 MVA transformer = 0.05 p.u.

Let us choose 10 MVA as base MVA

$\therefore\ X_g = j\,0.15$ p.u.

The circuit diagram of the power system is as follows

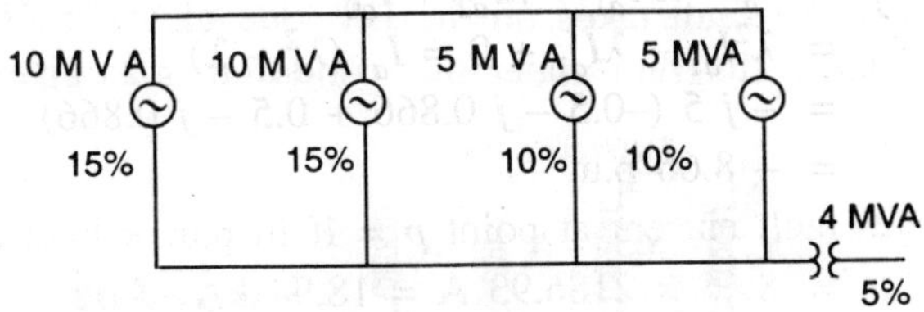

Fig. P. 13.8 (a)

Reactance of each 5 MVA generator on the base of 10 MVA (X_g)

$$= j\,0.10 \times \frac{MVA_b}{\text{Rating of the generator}}$$

$$= j\,0.10 \times \frac{10}{5} = j\,0.20 \text{ p.u.}$$

Reactance of each 4 MVA transformer on the base of 10 MVA

$$= j\,0.05 \times \frac{MVA_b}{\text{Rating of the generator}}$$

$$X_T = j\,0.05 \times \frac{10}{4} = j\,0.125 \text{ p.u.}$$

(i) When the fault takes place or the L.V. side of the transformer, the reactance diagram is as follows:

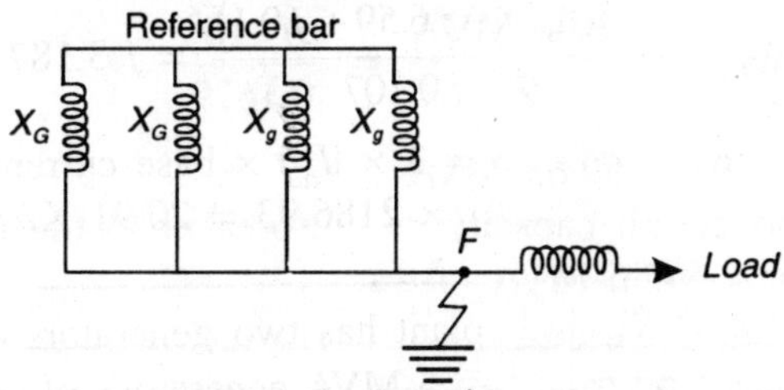

Fig. P. 13.8 (b)

The fault impedance is given by

$$Z_f = X_G \| X_G \| X_g \| X_g = \frac{X_G}{2} \| \frac{X_g}{2}$$

$$= j\,0.075 \| j\,0.10 = \frac{j\,0.075 \times j\,0.10}{j\,0.175} = j\,0.0429 \text{ p.u.}$$

We know that, short circuit capacity $= \dfrac{\text{Base MVA}}{Z_f}$

$$= \frac{10 \times 10^6}{0.0429} = 233.1 \text{ MVA}$$

∴ Short circuit capacity of the breaker or LV side of the transformer = 233.1 MVA **Ans.**

(ii) When the fault takes on the HV side of the transformer. The reactance diagram under this condition is as given below

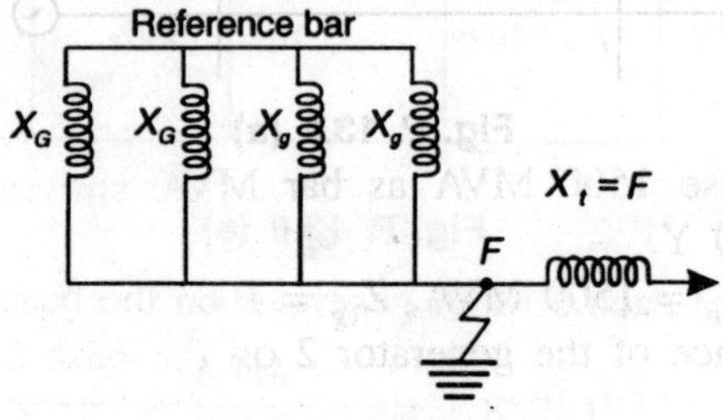

Fig. P. 13.8 (c)

The fault impedance in this case is given by

$$Z_f = X_G \| X_G \| X_g \| X_g + X_T$$

$$= \frac{X_G}{2} \| \frac{X_g}{2} + X_T$$

$$= \frac{j0.15}{2} \| \frac{j0.20}{2} + j\,0.125 = j\,0.0429 + j\,0.125$$

$$= j\,0.1679 \text{ p.u.}$$

We know that short circuit capacity is given by

$$\text{S.C. capacity} = \frac{\text{Base MVA}}{Z_f} = \frac{10 \times 10^6}{0.1679}$$

$$= 59.65 \text{ MVA}$$

∴ Short circuit capacity of the breaker at the HV side of the transformer = 59.56 MVA **Ans.**

Problem 13.9. Two generating stations having short circuit capacities of 1500 MVA and 100 MVA respectively and operating at 11 kV are linked by an interconnected cable having a reactance of 0.6 ohm per phase. Determine the short circuit capacity of each station.

Solution: Given

Impedance of the cable (Z_c) = j 0.6 Ω per phase

Terminal voltage (V_t) = 11 kV

Let us use number 1 and 2 for generators having ratings 1500 MVA and 1000 MVA respectively. The single line diagram of the power system is as shown in Fig. P. 13.9 (a).

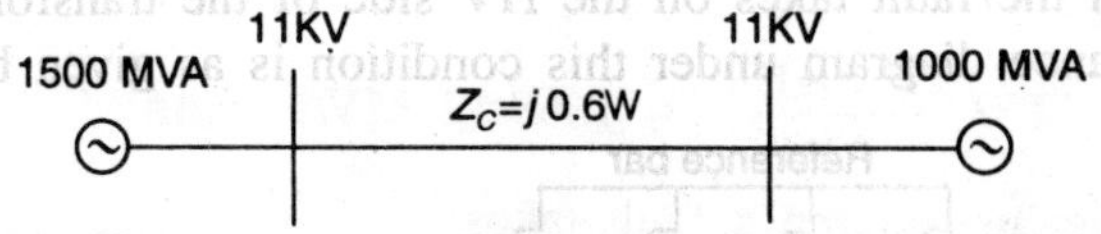

Fig. P. 13.9 (a)

Let us choose 1500 MVA as bar MVA and reactance of this generator be 100 Y.

Hence, MVA_b = 1500 MVA, Z_{1g} = 100 % = j 1.0 p.u.

The impedance of the generator 2 on the base of 1500 MVA is given by

$$Z_{2g} = j\ 1.00 \times \frac{MVA_b}{\text{Rating of the generator}}$$

$$= j\ 1.00 \times \frac{1500}{1000} = j\ 1.5 \text{ p.u.}$$

The percentage impedance of the cable on the base of 1500 MVA a 11 kV is given by

$$Z_{c\ \text{p.u}} = \frac{Z_C \times MVA_b}{(KV)^2} = \frac{j0.6 \times 1500 \times 10^6}{\left(11 \times 10^3\right)^2}$$

$$= j\ 7.438 \text{ p.u.}$$

Let us assume a fault takes place at 1500 MVA plant capacity.

The reactance diagram under this condition is as given in Fig. P. 13.9 (b).

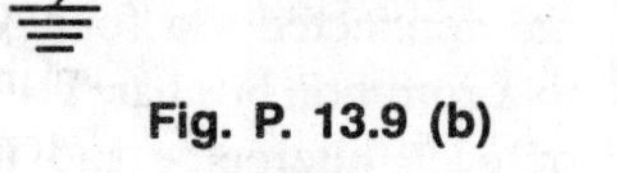

Fig. P. 13.9 (b)

The fault impedance is given by

$$Z_{f\ \text{p.u.}} = Z_{1g} \| (Z_{c\ \text{p.u}} + Z_{2g})$$

$$= \frac{Z_{1g}\left(Z_{c\,\text{p.u}} + Z_{2g}\right)}{Z_{1g} + Z_{2g} + Z_{c\,\text{p.u}}}$$

$$= \frac{j1.0(j7.438 + j1.5)}{j(1.0 + 1.5 + 7.438)} = j\ 0.899 \text{ p.u.}$$

$\therefore$ Short circuit capacity of this station $= \dfrac{MVA_b}{Z_{1\,p.u}}$

$$= \frac{150 \times 10^6}{0.899} = 1668.52 \text{ MVA} \quad \textbf{Ans.}$$

Again, let us assume a fault takes place at 1000 MVA plant capacity.

The reactance diagram under this condition is as given in Fig. P. 13.9 (c).

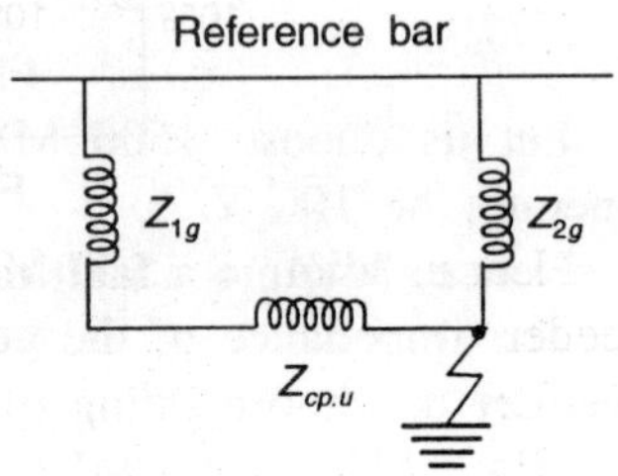

Fig. P. 13.9 (c)

In this condition.

Fault impedance $(Z_{2\text{ p.u.}}) = Z_{2g} \| (Z_{c\text{ p.u.}} + Z_{1g})$

$$= \frac{Z_{2g}\left(Z_{c\,p.u} + Z_{1g}\right)}{Z_{1g} + Z_{2g} + Z_{c\,p.u}}$$

$$= \frac{j1.5(j7.438 + j1.0)}{j(1.0 + 1.5 + 7.438)} = j\,1.274 \text{ p.u.}$$

$\therefore$ Short circuit capacity of this plant $= \dfrac{MVA_b}{Z_{1p.u}}$

$$= \frac{1150 \times 10^6}{1.274} = 1177.4 \text{ MVA} \quad \textbf{Ans.}$$

Problem 13.10. Four 50 MVA generators of 15% reactance each are connected via four 35 MVA reactors, each of 10% reactance to a common bus bar. The feeders are each connected to the junction of each alternator and its reactor. Determine the rating of each feeder circuit breaker.

Solution: Given,

Rating of each generator = 50 MVA

Reactance of each generator = 15%

Rating of each reactor = 35 MVA

Reactance of each reactor = 10%

The power system is as given in Fig. P. 13.10 (a).

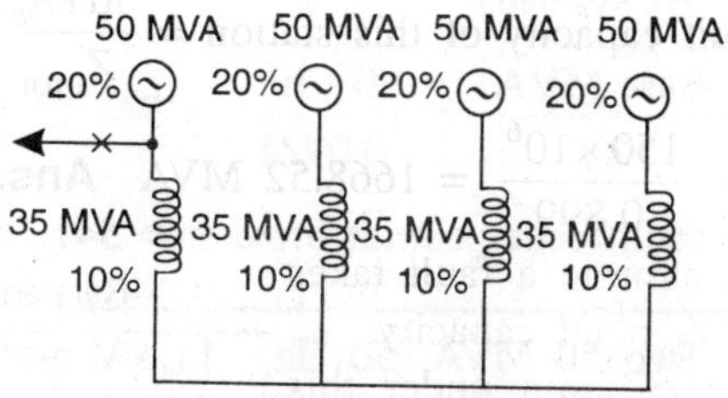

Fig. P. 13.10 (a)

Let us assume a fault takes place at point F on the extreme left feeder

Let us choose rating of the generator a base MVA

Base MVA = 50 MVA, $X_g = j\ 0.15$ p.u.

Reactance of each reactor on the base of

$$50 \text{ MVA} = 10\% \times \frac{MVA_b}{\text{reactor MVA}}$$

$$X_r = 10\% \times \frac{50}{35}$$

$$X_r = 14.3\% = j\ 0.143 \text{ p.u.}$$

For fault F, the reactance diagram is as given in Fig. P. 13.10 (b).

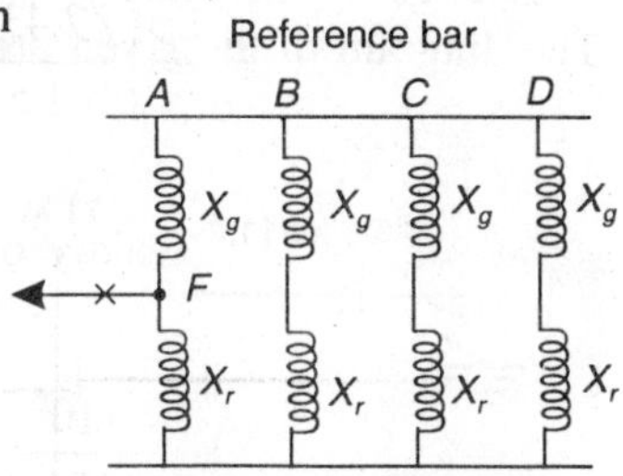

Fig. P. 13.10 (b)

The fault impedance between the reference bus and the fault point F is given by

$$Z_f = X_g \parallel [(X_g + X_r) \parallel (X_g + X_r) \parallel (X_g + X_r)]$$

$$= X_g \parallel \left[X_r + \frac{X_g + K_r}{3} \right]$$

$$= j\ 0.15 \parallel \left[j0.143 + \frac{j0.143 + j0.15}{3} \right]$$

$$= j\ 0.15 \parallel [j\ 0.143 + j\ 0.0977]$$

$$= j\ 0.15 \parallel j\ 0.2407 = \frac{j0.15 + j0.2407}{j(0.15 + 0.2407)}$$

$$= j\ 0.0924 \text{ p.u.}$$

We know that, SC capacity

$$= \frac{\text{Base MVA}}{Z_f} = \frac{50 \times 10^6}{0.0924} = 541.13 \text{ MVA}$$

Thus, rating of each feeder circuit breaker = 541.13 MVA **Ans.**

Problem 13.11. Two 50 MVA, 50 Hz, 11 kV alternators with sub-transient reactance $X'' = j\,0.1$ p.u. and a transformer of 40 MVA 11 kV/66kV and reactance of 0.08 p.u. are connected to a bus *A*. Another generator 60 MVA, 11 kV alternator with reactance of 0.12 p.u. is connected to bus *B*. Bus *A* and *B* are interconnected through a reactor of 80 MVA 20 percent reactance. If a 3-phase fault occurs on the high voltage side of the transformer, calculate the current fed into the fault.

Solution: Given,

Reactance of generators rating 50 MVA = j 0.1p.u.
Reactance of the transformer = j 0.08 p.u.
Reactance of 60 MVA generator = j 0.12 p.u.
Reactance of the reactor = j 0.20 p.u.
The diagram of the given problem is as given in Fig. P. 13.11 (a).

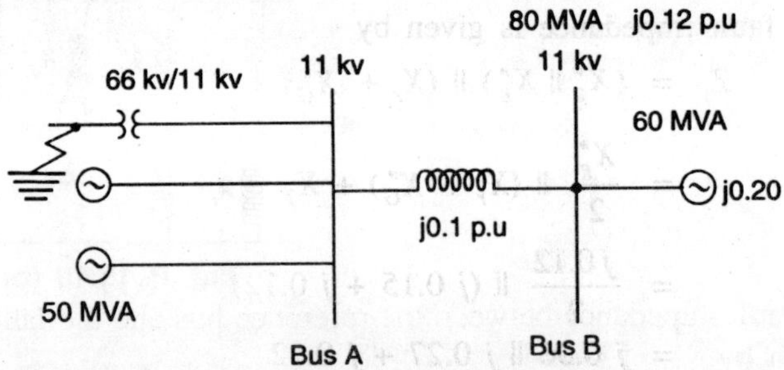

Fig. P. 13.11 (a)

Let us choose 60 MVA and 11 kV as base quantities

$\therefore$ Base MVA = 60 MVA, V_b = 11 kV

$\therefore$ $X''_G = j\,0.12$ p.u.

Reactance of the generator having rating 50 MVA or the base of 60 MVA

$$= j\,0.1 \times \frac{MVA_b}{\text{Rating}}$$

$$X''_g = j\,0.1 \times \frac{60}{50} = j\,0.12 \text{ p.u.}$$

Reactance of the transformer on the base of 50 MVA

$$X_T = j\,0.08 \times \frac{MVA_b}{\text{Rating}} = j\,0.08 \times \frac{60}{40}$$

$$= j\,0.12 \text{ p.u.}$$

The reactance of the reactance on the base of 60 MVA

$$= j\,0.20 \times \frac{MVA_b}{\text{Rating}},\ X_r = j\,0.20 \times \frac{60}{80}$$

$$= j\,0.15 \text{ p.u.}$$

The reactance diagram can be drawn as

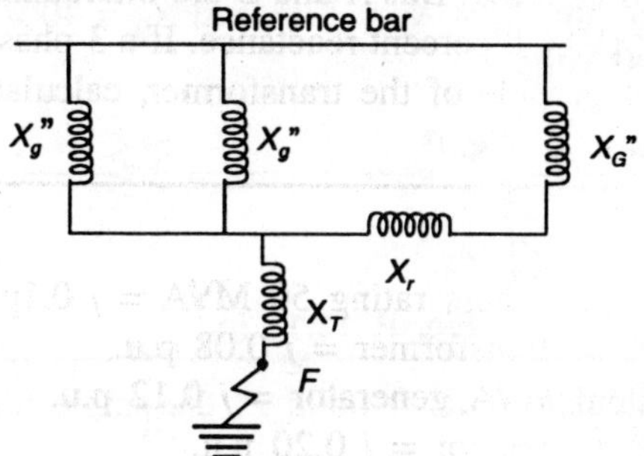

Fig. P. 13.11 (b)

One fault impedance is given by

$$Z_f = (X_g'' \parallel X_g'') \parallel (X_r + X_G'') + X_T$$

$$= \frac{X_g''}{2} \parallel (X_r + X_G'') + X_T$$

$$= \frac{j0.12}{2} \parallel (j\,0.15 + j\,0.12) + j\,0.12$$

$$= j\,0.06 \parallel j\,0.27 + j\,0.12$$

$$= \frac{j0.06 + j0.27}{j(0.06+0.27)} + j\,0.12$$

$$= j\,0.169 \text{ p.u. and } V_f = 1\angle 0° \text{ p.u.}$$

$$\text{Fault current } (I_f) = \frac{V_f}{Z_g} = \frac{1}{j0.169} = -\,j\,5.92 \text{ p.u.}$$

Now, base current on H.V. side of the transformer (I_{bH})

$$= \frac{60\times 10^6}{\sqrt{3}\times 66\times 10^3} = 524.864 \text{ A}$$

Fault current on H.V. side = I_f in p.u. × base current

= 5.92 × 524.864 = 3107.19 A **Ans.**

Problem 13.12. A single line-to-ground fault occurs on phase a at point *P* in the circuit whose single line diagram is shown here. Determine the subtransient current in phase a of machine 1 and in the fault at *P*. Neglect per fault current. Assume that machine 2 is synchronous motor operating at rated voltage. Both machines are rated 1.5 MVA, 600 volts with reactance of $X'' = X_2$ = 8% and X_0 =4%. Each 3-phase transformers rated 1.25 MVA, 60 volts delta/ 4160 volts star with leakage reactance of 5%. The reactance of transmission line are $X_1'' = X_2$ = 12% and X_0 = 40% on a base of 1.25 MVA, 4160 volts.

Fig. P. 13.12 (a)

Solution: To calculate fault at point *P*, the single line diagram can be drawn as below

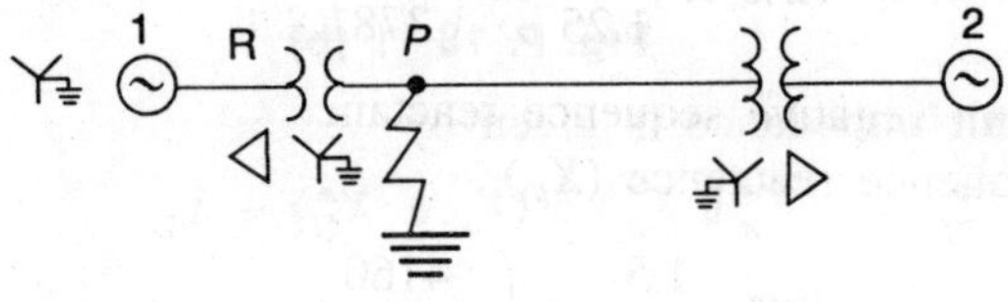

Fig. P. 13.12 (b)

Given,

For 1.5 MVA, 600 V generator

Positive sequence reactance (X_g'') = 8%

Negative sequence reactance (X_{2g}) = 8%

Negative sequence reactance (X_{0g}) = 4%

Let us choose 1.5 MVA and 600 V base quantities

∴ MVA_b = 1.5 MVA and V_b = 600 V

Base voltage referred to line side (V_{b1})

$$= V_b \times \frac{V_H}{V_L} = 600 \times \frac{4160}{660} = 3781.82 \text{ V}$$

For 1.5 MVA, 600 V motor

Positive sequence reactance (X_m'') = 8%

Negative sequence reactance (X_{2m}) = 8%

Zero sequence reactance (X_{0m}) = 4%

For 1.25 MVA, 660 V Δ/4160 Y volts transformer.

Reactance of the transformer = 5%

The reactance of the transformer on the base of 1.5 MVA and 600 V

$$(X_T) = 50\% \times \frac{MVA_b}{MVA_T} \times \left(\frac{V_T}{V_b}\right)^2$$

$$= 5\% \times \frac{1.5}{1.75} \times \left(\frac{660}{600}\right)^2 = 7.26\%$$

For 1.25 MVA, 4160 V transmission lines

Positive sequence reactance = 12%

Negative sequence reactance = 12%

Zero sequence reactance = 40%

The reactance of the transmission lines on the base of 1.5 MVA, 600 V are given.

Positive sequence reactance (X'_{l1})

$$= 12\% \times \frac{MVA_b}{MVA_l} \times \left(\frac{V}{V_{b1}}\right)^2$$

$$= 12\% \times \frac{1.5}{1.25} \times \left(\frac{4160}{3781.82}\right)^2 = 17.42\%$$

Similarly, Negative sequence reactance (X_{l2}) = 17.42%

Two sequence reactance (X_{l0})

$$= 40\% \; \frac{1.5}{1.25} \times \left(\frac{4160}{3781.82}\right)^2 = 58.08\%$$

Now, the positive sequence network of the given problem under fault conditions is as shown in Fig. P. 13.12 (c).

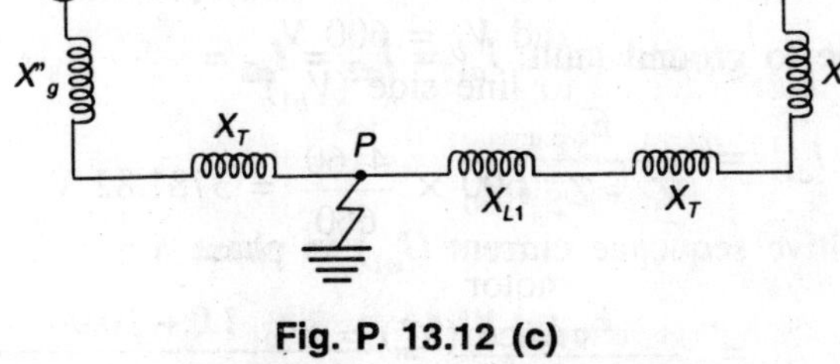

Fig. P. 13.12 (c)

Positive sequence impedance between fault and the reference bus

$$Z_1 = (X''_g + X_T) \parallel (X_{l1} + X_T + X''_m)$$

$$= (j\,0.08 + j\,0.0726) \parallel (j\,0.1742 + j\,0.0726 + j\,0.08)$$
$$= j\,0.104 \text{ p.u.}$$

Again, the negative sequence network of the given problem is as shown in Fig. P. 13.12 (d).

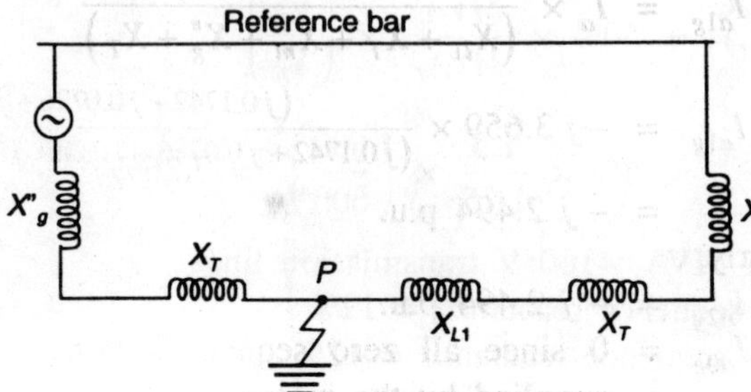

Fig. P. 13.12 (d)

∴ Negative sequence impedance between the fault and reference bus $(Z_2) = (X'_{2g} + X_T) \parallel (X'_{l2} + X_T + X_{2m}) = j\,0.104$ p.u.

The zero sequence network is as shown in Fig. P. 13.12 (e).

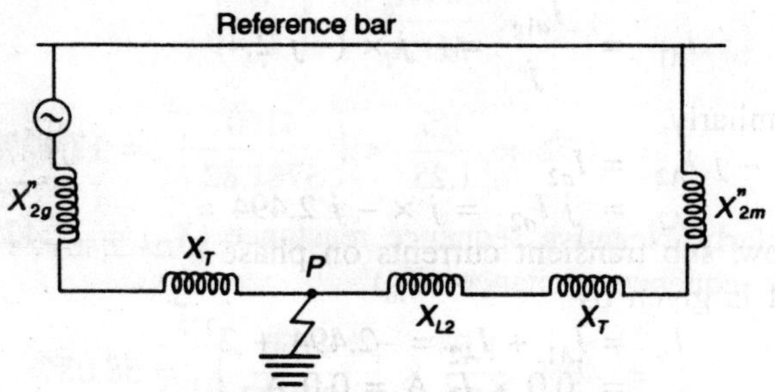

Fig. P. 13.12 (e)

Under this condition

Zero sequence impedance between the fault and reference bus

$$(Z_0) = X_T \parallel (X_{l0} + X_T) = j\,0.0726 \parallel (j\,0.5808 + j\,0.0726)$$
$$= j\,0.065. \text{ p.u.}$$

Let us take E_a as reference phasor ∴ $E_{a\text{ p.u.}} = 1\angle 0°$ p.u.

For line to ground fault. $I_{a1} = I_{a2} = I_{a0} = \dfrac{I_a}{3}$

$$I_{a1} = \frac{E_{a\text{ p.u}}}{Z_1 + Z_2 + Z_0}$$

∴ Positive sequence current (I_{a1}) in phase a

$$= \frac{E_{a\text{ p.u}}}{Z_1 + Z_2 + Z_0} = \frac{1.0 + j\,0.0}{j\,(0.104 + 0.104 + 0.0653)}$$
$$= -j\,3.659 \text{ p.u.}$$

From Fig. P. 13.12 (c), on L.V. side

Now, positive sequence current supplied by the machine 1 at point *P*,

$$I_{a1g} = I_a \times \frac{(X_{l1} + X_T + X''_m)}{(X_{l1} + X_T + X''_m + X''_g + X_T)}$$

$$I_{a1g} = -j\,3.659 \times \frac{(j\,0.1742 + j\,0.0726 + j\,0.08)}{(j\,0.1742 + j\,0.0726 + j\,0.08 + j\,0.08 + j\,0.0726)}$$

$$= -j\,2.494 \text{ p.u.}$$

Similarly,

$$I_{a2g} = -j\,2.494 \text{ p.u.}$$

I_{a0g} = 0 since all zero sequences current at fault are supplied by the motor.

To calculate sub transient currents in machine 1,

Let I_{A1} be the positive sequence current supplied by the machines on H.V. side be the transformer

$$I_{a1g} = j\,I_{A1}$$

$$\therefore \quad I_{A1} = \frac{I_{a1g}}{j} = -j \times (-j\,2.494) \text{ p.u.} = -2.494 \text{ p.u.}$$

Similarly,

$$-j\,I_{A2} = I_{a2}$$

$$\therefore \quad I_{A2} = j\,I_{a2} = j \times -j\,2.494 = 2.494 \text{ p.u.}$$

Now, sub transient currents on phase a on the H.V. side, of the current is given by

$$I_A = I_{A1} + I_{A2} = -2.494 + 2.494 = 0.0 \text{ p.u.}$$

$$= 0.0 \times I_B \text{ A} = 0.0 \text{ A} \quad \textbf{Ans.}$$

Base current on the line side $= \dfrac{MVA_b}{\sqrt{3} \times V_{bl}}$

$$I_{Bl} = \frac{1.5 \times 10^6}{\sqrt{3} \times 3781.82} = 228.997 \text{ A}$$

Similarly,

$$I_B = \lambda^2 I_{A1} + \lambda I_{A2} = -\lambda^2 I_{A2} + \lambda I_{A2}$$

$$= I_{A2}(\lambda - \lambda^2) = I_{A2}(-0.5 + j\,0.866 + 0.5 + j\,0.866)$$

$$= 2.494 \times 2 \times j\,0.866 \text{ p.u.} = j\,4.32 \text{ p.u.}$$

In ampere I_B = I_B in p.u. × Base line current

$$= j\,4.32 \times 288.997 = j\,989.18 \text{ A} \quad \textbf{Ans.}$$

and

$$I_C = \lambda I_{A1} + \lambda^2 I_{A2} = +\lambda^2 I_{A2} - \lambda I_{A2}$$

$$= I_{A2}(\lambda^2 - \lambda)$$

$$= 2.494\,(-0.5 - j\,0.866 + 0.5 - 0.860) \text{ p.u.}$$

$$= -j\,989.18 \text{ A} \quad \textbf{Ans.}$$

Now, fault current at point P is given by

$$I_{fp} = 3 \times I_{a1} \times I_{BC} \text{ A}$$

$$= 3 \times \frac{E_{a\,p.u}}{Z_1 + Z_2 + Z_0} \times I_{B1} \text{ A}$$

$$= 3 \times \frac{1.0 + j\,0.0}{j\,(0.104 + 104 + 0.0653)} \times 228.994 \text{ A}$$

$$= 3 \times -j\,3.659 \times 228.997 = -j\,2513.7 \text{ A} \quad \textbf{Ans.}$$

Problem 13.13. Solve Problem 13.12 when fault is at *R*.

Solution: To calculate fault current at point R, the single-line diagram can be drawn as below:

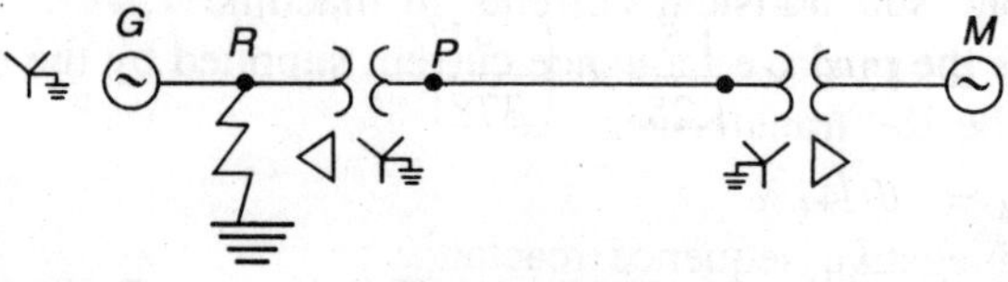

Fig. P. 13.13 (a)

Given,

For 1.5 MVA , 600 V generator,

Positive sequence reactance (X_g'') = 8%

Negative sequence reactance (X_{1g}) = 8%

Zero sequence reactance (X_{0g}) = 4%

Let us choose 1.5 MVA and 600 V as base quantities

$\therefore\ MVA_b$ = 1.5 MVA, V_b = 600 V

Base voltage referred to line side $(V_{b1}) = V_b \times \frac{V_H}{V_L}$

$$= 600 \times \frac{4160}{660} = 3781.82 \text{ V}$$

For 1.5 MVA, 600 V motor,

Positive sequence reactance (X_m'') = 8%

Negative sequence reactance (X_{1m}) = 8%

Zero sequence reactance (X_{0m}) = 4%

For 1.25 MVA, 660 V Δ/ 4160 Y volts transformer, reactance of the transformer = 5%

The reactance of the transformer on the base of 1.5 MVA and 600 V

$$(X_T) = 5\% \times \frac{MVA_b}{MVA_T} \times \left(\frac{V_T}{V_{bL}}\right)^2$$

$$= 5\% \times \frac{1.5}{1.25}\left(\frac{4160}{3781.82}\right)^2$$

$$= 7.26\%$$

For 1.25 MVA, 4160 V transmission line positive reactance = 12%
Negative sequence reactance = 12%
Zero sequence reactance = 40%

The reactance of the transmission lines on the base of 1.5 MVA 600 V are given by

Positive sequence reactance (X_{l1})

$$= 12\% \times \frac{MVA_b}{MVA_l} \times \left(\frac{V_l}{V_{bl}}\right)^2$$

$$= 12\% \times \frac{1.5}{1.25} \times \left(\frac{4160}{3781.82}\right)^2$$

$$= 17.42\%$$

Similarly, Negative sequence reactance (X_{l2}) = 17.42%
Zero sequence reactance of the line (X_{l0})

$$= 40\% \times \frac{MVA_b}{MVA_L} \times \left(\frac{V_l}{V_{bl}}\right)^2$$

$$= 40\% \times \frac{1.5}{1.25} \times \left(\frac{4160}{3781.82}\right)^2$$

$$= 58.08\%$$

Now, the positive sequence network under this condition is as shown in Fig. P. 13.13 (b).

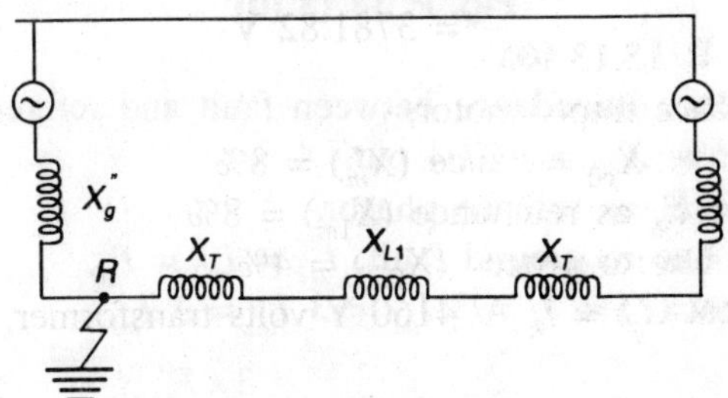

Fig. P. 13.13 (b)

From Fig. P. 13.13 (b),
Positive sequence impedance (Z_1)

$$= X_g'' \parallel (X_T + X_{l1} + X_T + X_m'')$$

$= j\,0.08 \parallel (j\,0.0726 + j\,0.1742 + j\,0.0726 + j\,0.08)$

$$= \frac{j\,0.08 \times (j\,0.0726 + j\,0.1742 + j\,0.0726 + j\,0.08)}{j(0.08 + 0.0726 + 0.1742 + 0.0726 + 0.08)}$$

$= j\,0.0666$ p.u.

The negative sequence network is an below:

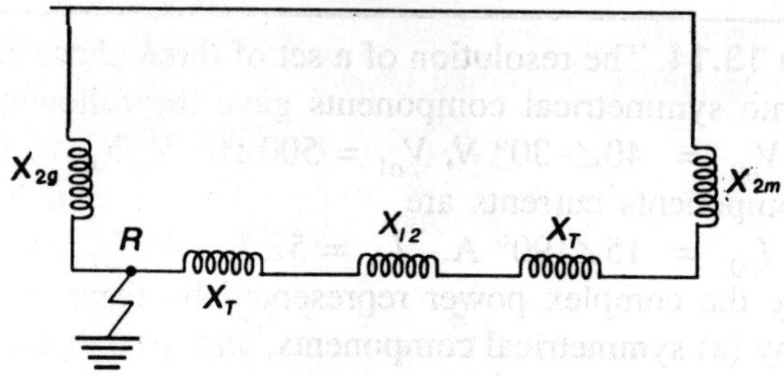

Fig. P. 13.13 (c)

In the same way, from Fig. P. 13.13 (c),

Negative sequence impedance (Z_2)

$= X_{2g} \parallel (X_T + X_{l1} + X_T + X_{2m}) = j\,0.0666$ p.u.

Again, the zero sequence reactance can be drawn as

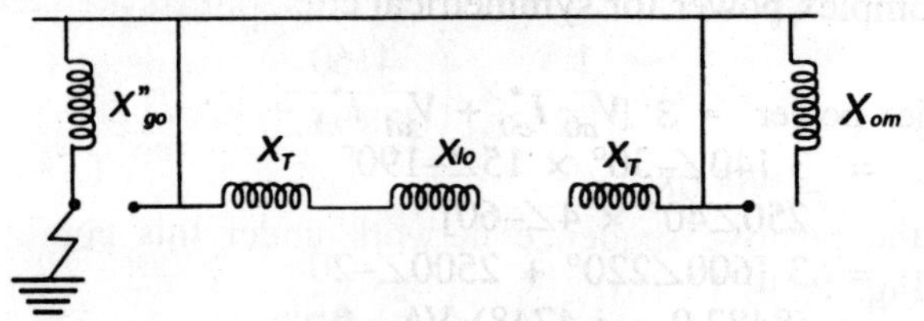

Fig. P. 13.13 (d)

From Fig. P. 13.13 (d),

Zero sequence impedance between fault and reference bus

$(Z_0) = X_{g0} = j\,0.04$ p.u.

Let us take E_a as refence phasor.

For single line to ground fault, $I_{a1} = I_{a2} = I_{a0}$

Fault current $(I_f) = I_a = 3\,I_{a1} = 3\,I_{a2} = 3\,I_{a0}$

Fault current at point R in p.u. $= \dfrac{3 \times E_{a\,\text{p.u}}}{Z_1 + Z_2 + Z_0}$ p.u.

$$I_{f\,\text{p.u.}} = \frac{3 \times (1.0 + j\,0.0)}{j(0.04 + 0.0666 + 0.066)} = -j\,17.32 \text{ p.u.}$$

In ampere, Fault current at point $R = I_{f\,p.u.} \times$ Base current

$$= -j\,17.32 \times \frac{MVA_b}{\sqrt{3} \times V_b}$$

$$= -j\,17.32 \times \frac{1.5 \times 10^6}{\sqrt{3} \times 600}$$

$$= j\,24999.27 \text{ A} \quad \textbf{Ans.}$$

> **Problem 13.14.** The resolution of a set of three-phase unbalanced voltage into symmetrical components gave the following results.
>
> $V_{a0} = 40\angle{-30°}$ V, $V_{a1} = 500\angle 0°$ V, $V_{a2} = 250\angle 40°$ V
>
> The components currents are
>
> $I_{a0} = 15\angle 190°$ A, $I_{a1} = 5\angle 20°$ A, $I_{a2} = 4\angle 60°$ A
>
> Determine the complex power represented by these voltages and currents by (a) symmetrical components, and (b) unbalanced phase components

Solution: Given,

$V_{a0} = 40\angle{-30°}$ V, $V_{a1} = 500\angle 0°$ V, $V_{a2} = 250\angle 40°$ V

The currents are

$I_{a0} = 15\angle 190°$ A, $I_{a1} = 5\angle 20°$ A, $I_{a2} = 4\angle 60°$ A

(a) For symmetrical components

The complex power for symmetrical components are calculated as follows:

$$\text{Complex power} = 3\,[V_{a0}\,I^*_{a0} + V_{a1}\,I^*_{a1} + V_{a2}\,I^*_{a2}]$$

$$= 3\,[40\angle{-30°} \times 15\angle{-190°} + 500\angle 0° \times 5\angle{-20°} + 250\angle 40° \times 4\angle{-60}]$$

$$= 3\,[600\angle 220° + 2500\angle{-20°} + 1000\angle{-20°}] \text{ VA}$$

$$= (8487.9 - j\,4748) \text{ VA} \quad \textbf{Ans.}$$

(b) For unbalanced phase components,

Let us calculate phase voltages V_a, V_b and V_c

Now,

$$V_a = V_{a0} + V_{a1} + V_{a2} = 40\angle{-30°} + 500\angle 0° + 250\angle 40° \text{ V}$$

$$= 726.15 + j\,140.7 \text{ V} = 739.66\angle 10.97° \text{ V}$$

$$V_b = V_{a0} + \lambda^2 V_{a1} + \lambda V_{a2}$$

$$= 40\angle{-30°} + 1\angle{-120°} \times 500\angle 0° + 1\angle 120° \times 250\angle 40°$$

$$= 40\angle{-30°} + 500\angle{-120°} + 250\angle 160°$$

$$= -450.28 - j\,367.51 \text{ V} = 581.22\angle 140.78° \text{ V}$$

and

$$V_c = V_{a0} + \lambda V_{a1} + \lambda^2 V_{a2}$$

$$= 40\angle{-30°} + 1\angle 120° \times 500\angle 0° + 1\angle{-120°} \times 250\angle 40°$$

$$= 40\angle{-30^\circ} + 500\angle{120^\circ} + 25\angle{-80^\circ}$$
$$= -172.0 + j\,116.81 \text{ V} = 239.60\angle{134.12^\circ} \text{ V}$$

Again, Let us calculate phase units I_a, I_b and I_c

Now, The phase current in line a is given by

$$I_a = I_{a0} + I_{a1} + I_{a2}$$
$$= 15\angle{190^\circ} + 5\angle{20^\circ} + 4\angle{60^\circ} = -8.07 + j2.57$$
$$= 8.47\angle{107.66^\circ} \text{ A}$$
$$I_b = I_{a0} + \lambda^2 I_{a1} + \lambda I_{a2}$$
$$= 15\angle{190^\circ} + 1\angle{-120^\circ} \times 5\angle{20^\circ} + 1\angle{120^\circ} \times 4\angle{60^\circ}$$
$$= 15\angle{190^\circ} + 5\angle{-100^\circ} + 4\angle{180^\circ}$$
$$= -19.64 - j\,7.53 \text{ A} = 21.03\angle{159^\circ} \text{ A}$$

and the phase current in line C is given by

$$I_c = I_{a0} + \lambda I_{a1} + \lambda^2 I_{a2}$$
$$= 15\angle{190^\circ} + 1\angle{120^\circ} \times 5\angle{20^\circ} + 1\angle{-120^\circ} \times 4\angle{60^\circ}$$
$$= 15\angle{190^\circ} + 5\angle{140^\circ} + 4\angle{-60^\circ}$$
$$= -16.60 - j\,2.85 = 16.84\angle{170.26^\circ} \text{ A}$$

The three phase complex power in this case in given by

$$S = V_a I_a^* + V_b I_b^* + V_c I_c^*$$
$$= 739.66\angle{10.97^\circ} \times 8.47\angle{-107.66} + 581.22\angle{140.78} \times 21.03\angle{-159^\circ} + 239.60\angle{134.1^\circ} \times 16.84\angle{-170.26}$$
$$= (14138.58 - j\,1243.61) \text{ VA}$$ **Ans.**

Problem 13.15. The line to ground voltages or low voltage side of a step down transformer are 60 kV, 40 kV and 50 kV on phases a, b and c respectively. The voltages of phase a leads that of phase b by 90 and lags that of phase c by 160°. Determine analytically the symmetrical components of voltage.

Solution: Let us take voltage of phase a (V_a) as reference phasor. The phase diagram of the given problem is as follows:

$$|V_a| = 60 \text{ kV}$$
$$|V_b| = 40 \text{ kV}$$
$$|V_c| = 50 \text{ kV}$$

Hence

$$V_a = 60\angle{0^\circ}$$
$$V_b = 40\angle{-90^\circ}$$
$$V_c = 50\angle{160^\circ}$$

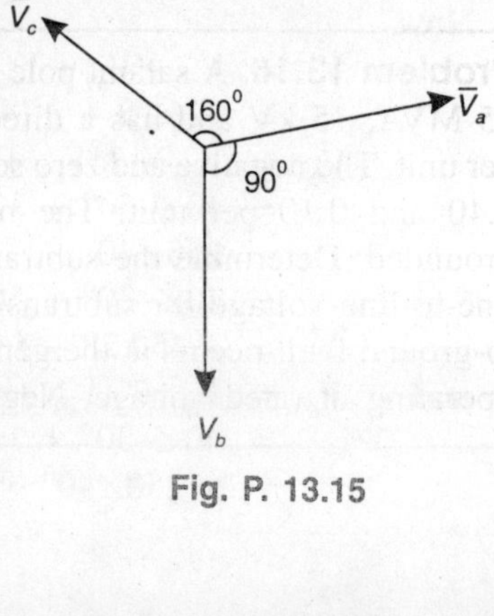

Fig. P. 13.15

The symmetrical components of voltage are positive sequence component, negative sequence component, and zero sequence component of voltage. The positive sequence component of the voltage is given by

$$V_{a1} = \frac{1}{3}(V_a + \lambda V_b + \lambda^2 V_c)$$

$$= \frac{1}{3}[60\angle 0^\circ + 1\angle 120^\circ \times 40\angle -90^\circ + 1\angle -120^\circ \times 50\angle 160^\circ]$$

$$= \frac{1}{3}[60\angle 0^\circ + 40\angle 30^\circ + 50\angle 40^\circ] \quad \textbf{Ans.}$$

$$= (44.31 + j\,17.38)\text{ kV} \quad \textbf{Ans.}$$

The negative sequence component of voltage is given by

$$V_{a2} = \frac{1}{3}(V_a + \lambda^2 V_b + \lambda V_c)$$

$$= \frac{1}{3}[60\angle 0^\circ + 1\angle -120^\circ \times 40\angle -90^\circ + 1\angle 120^\circ \times 50\angle 160^\circ]$$

$$= \frac{1}{3}[60\angle 0^\circ + 40\angle -210 + 50\angle 280^\circ]$$

$$= (11.35 - j\,29.24)\text{ kV} \quad \textbf{Ans.}$$

The zero sequence component of the voltage is given by

$$V_{a0} = \frac{1}{3}[V_a + V_b + V_c]$$

$$= \frac{1}{3}[60\angle 0^\circ + 40\angle -90^\circ + 50\angle 160^\circ]$$

$$= (4.34 - j\,22.90)\text{ kV} \quad \textbf{Ans.}$$

Problem 13.16. A salient pole generator without dampers is rated 25 MVA, 15 kV and has a direct-axis subtransient reactance 0.30 per unit. The negative and zero sequences reactance are respectively, 0.40 and 0.10 per unit. The neutral of the generator is solidly grounded. Determine the subtransient current in the generator and line-to-line voltage for subtransient conditions when a single line-to-ground fault occurs at the generator terminals with the generator operating at rated voltage. Neglect resistance.

Solution: Let us choose 25 MVA, 5 kV as base quantities

The circuit diagram of the given problem is as shown in Fig. P. 13.16 (a).

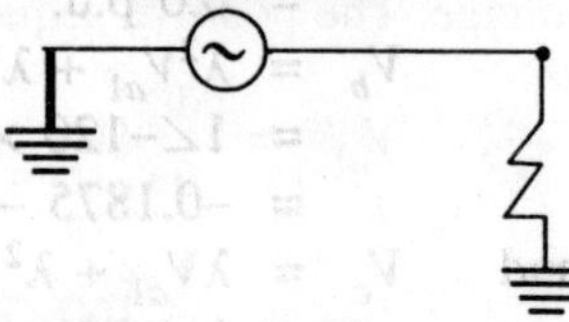

Fig. P. 13.16 (a)

Given,

$Z_1 = 0.30$ p.u.

$Z_2 = 0.040$ p.u.

$Z_0 = 0.10$ p.u.

$V = 15$ kV

At no-load, terminal voltage is equal to internal voltage of the generator.

Hence, $V = E_a = 15$ kV

In p.u.,

$$E_{a\text{ p.u.}} = \frac{E_a}{V_b} = \frac{15}{13} = 1.00\angle 0° \text{ p.u.}$$

$$= 1 + j\,0.0 \text{ p.u}$$

∴ Base current (I_B)

$$= \frac{MVA_b}{\sqrt{3} \times V_b} = \frac{25 \times 10^6}{\sqrt{3} \times 15 \times 10^3} = 962.25 \text{ A}$$

The positive sequence current is given by

$$I_{a1} = \frac{E_{a\text{ p.u}}}{Z_1 + Z_2 + Z_0} \text{ p.u.}$$

$$= \frac{1 + j0.0}{j0.40 + j0.30 + j0.10} = -j\,1.25 \text{ p.u.}$$

For line-to-ground fault, $I_{a1} = I_{a2} = I_{a0}$ and $I_a = 3\,I_{a1}$ the line current in phase a is given by

$$I_a = 3\,I_{a1} = -\,j\,3.75 \text{ p.u.}$$

The actual current in line a is given by

$$I_a = j\,3.75 \times I_b = -\,j\,3.75 \times 962.25 \text{ A}$$

$$= -\,j\,3608.44 \text{ A} \quad \textbf{Ans.}$$

The symmetrical components of the voltages from point a to ground are given by

$$V_{a1} = E_a - I_{a1}\,Z_1 = 1.0 - (-\,j\,1.25) \times (j\,0.30)$$

$$= 0.625 \text{ p.u.}$$

The negative sequence voltage in line a is given by

$$V_{a2} = -\,I_{a2}\,Z_2 = -\,(-j\,1.25) \times j\,0.40 = -\,0.50 \text{ p.u.}$$

and $V_{a0} = -\,I_{a0}\,Z_0 = -\,(-j\,1.25) \times (j\,0.10) = -\,0.125$ p.u.

Now, the line-to-ground voltages are given by

$$V_a = V_{a1} + V_{a2} + V_{a0} = 0.625 - 0.50 - 0.125$$
$$= 0.0 \text{ p.u.}$$
$$V_b = \lambda^2 V_{a1} + \lambda V_{a2} + V_{a0}$$
$$= 1\angle{-120} \times 0.625 + 1\angle 120° \times (-0.50) - 0.125$$
$$= -0.1875 - j\ 0.9743 \text{ p.u.}$$

and
$$V_c = \lambda V_{a1} + \lambda^2 V_{a2} + V_{a0}$$
$$= 1\angle 120° \times 0.625 - 0.50 \times 1\angle{-120°} - 0.125$$
$$= -0.1875 + j\ 0.9743 \text{ p.u.}$$

Now, the line to line voltage are given by

$$V_{ab} = V_a - V_b = 0.0 - (-0.1875 - j\ 0.9743)$$
$$= 0.1875 + j\ 0.9743 = 0.9922\ \angle 97.11° \text{ p.u.}$$
$$= 0.9922\ \angle 79.11° \times V_b \text{ kV}$$
$$= 0.9922\ \angle 79.11° \times 15 \text{ kV}$$
$$= 14.88\ \angle 79.11° \text{ kV} \quad \textbf{Ans.}$$
$$V_{bc} = V_b - V_c$$
$$= -0.1875 - j\ 0.09743 - (-0.1875 + j\ 0.9743)$$
$$= -j\ 1.9486 \text{ p.u.}$$
$$= 1.9486\ \angle 270° \times V_b \text{ kV} = 1.9486\ \angle 270° \times 15 \text{ kV}$$
$$= 29.23 \text{ kV}$$
$$V_{ca} = V_c - V_a = (-0.1875 + j\ 0.9743) - 0.0$$
$$= 0.9922\ \angle 100.89 \text{ p.u.} = 0.9922\ \angle 100.89 \times 15 \text{ kV}$$
$$= 14.88\ \angle 100.89° \text{ kV} \quad \textbf{Ans.}$$

The phasor diagrams before the fault and post fault are as follows:

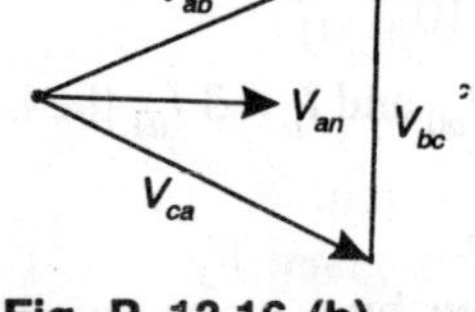

Fig. P. 13.16 (b)

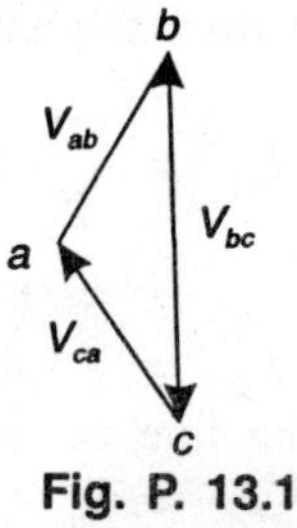

Fig. P. 13.16 (c)

Problem 13.17. A salient pole generator without dampers is rated 25 MVA, 15 kV and has a direct a axis subtransient reactance 0.30 per unit. The negative and zero-sequence reactances are respectively 0.40 and 0.10 per unit. The neutral of the generator is solidly grounded. Find the subtransient currents and the line to line voltages at the fault under subtransient conditions when a line-to-line fault occurs at the terminals of the generator. Assume that the generator is unloaded and operating of rated terminal voltage when the fault occurs. Neglect resistance.

Solution: Let us choose 25 MVA, 15 kV as base quantities

The circuit diagram of the given problem is as follows:

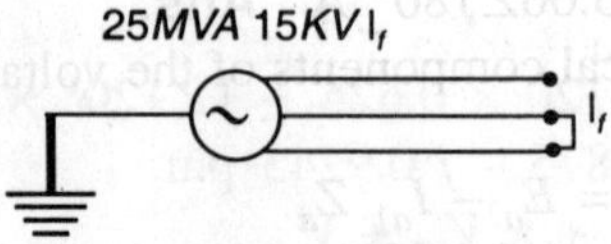

Fig. P. 13.17

Given,

$$Z_1 = 0.30 \text{ p.u.}, Z_2 = 0.40 \text{ p.u.}, Z_0 = 0.10 \text{ p.u.}$$
$$V = 15 \text{ kV}, V_b = 15 \text{ kV}, MVA_b = 25 \text{ MVA}$$

At no-load, terminal voltage is equal to internal voltage of the generator. Hence

$$E_a = V = 15 \text{ kV}$$

$$\therefore \quad E_{a \text{ p.u.}} = \frac{E_a}{V_b} = 1\angle 0° \text{ p.u.}$$

$$\text{Base current } (I_B) = \frac{MVA_b}{\sqrt{3} \times V_b}$$

$$= \frac{25 \times 10^6}{\sqrt{3} \times 15 \times 10^3} = 962.25 \text{ A}$$

For line-to-line fault, $I_{a1} = -I_{a2}, I_{a0} = 0$

$$V_{a1} = V_{a2}, V_{a0} = 0$$

The positive sequence current (I_{a1}) is given by

$$I_{a1} = \frac{E_{a \text{ p.u}}}{Z_1 + Z_2} = \frac{1.0 + j0.0}{j0.40 + j0.30}$$
$$= -j\,1.429 \text{ p.u.}$$
$$\therefore \quad I_{a2} = j\,1.429 \text{ p.u.}, I_{a0} = 0 \text{ p.u.}$$

Now, the line current in phase a is given by

$$I_a = I_{a1} + I_{a2} + I_{a0} = -j\,1.429 + j\,1429 + 0$$
$$= 0.0 \text{ p.u.} = 0.0 \times I_B = 0 \text{ A} \quad \textbf{Ans.}$$

Similarly, the line current in phase b is given by

$$I_b = \lambda^2 I_{a1} + \lambda I_{a2} + I_{a0}$$
$$= 1\angle -120° \times (-j\,1.429) + 1\angle -120° \times j\,1.429 + 0$$
$$= -1.429 + j\,0.0 \text{ p.u.} = 1.429\angle 180° \text{ p.u.}$$
$$= 1.429\angle 180° \times I_B \text{ A} = 1.429\angle 180° \times 962.25 \text{ A}$$
$$= 1375.06\angle 180° \text{ A} \quad \textbf{Ans.}$$

The line current in phase c is given by

$$I_c = \lambda I_{a1} + \lambda^2 I_{a2} + I_{a0}$$
$$= 1\angle 120° \times (-j\,1.429) + 1\angle -120° \times (j\,1.429) + 0.0$$

$= 1.429 + j\,0.0$ p.u. $= 1.429\angle 0°$ p.u.
$= 1.429\angle 0° \times I_B$ A $= 1.429\angle 0° \times 962.25$ A
$= 1375.06\angle 180°$ A **Ans.**

Now, the symmetrical components of the voltage from a to ground are given by

$$V_{a1} = V_{a2} = E_a - I_{a1}\, Z_a$$
$$= 1.0 - (-j\,1.429) \times (j\,0.30) = 0.5713 \text{ p.u.}$$
$$V_{a0} = 0$$

The line-to-ground voltages are given by

$$V_a = V_{a1} + V_{a2} + V_{a0} = 0.5713 + 0.5713 + 0.0$$
$$= 1.1426\angle 0° \text{ p.u.}$$
$$V_b = \lambda^2 V_{a1} + \lambda V_{a2} + V_{a0}$$
$$= 1\angle -120° \times 0.5713 + 1\angle 120° \times 0.5713$$
$$= -0.5712 + j\,.0 \text{ p.u.} = V_c$$

The line-to-line voltages are given by

$$V_{ab} = V_a - V_b = 1.1426 - (-0.5712)$$
$$= 1.7138\angle 0° \text{ p.u.} = 1.7138 \times V_b\angle 0° \text{ kV}$$
$$= 1.7138 \times 15\angle 0° \text{ kV} = 25.71\angle 0° \text{ kV}$$
$$V_{bc} = V_b - V_c = 0 \text{ p.u.} = 0 \times V_b \text{ kV} = 0 \text{ kV} \quad \textbf{Ans.}$$
and $$V_{ca} = V_c - V_a = -0.5712 - 1.1426$$
$$= 1.7138\angle 180° \text{ p.u.} = 1.7138\angle 180° \times V_b \text{ kV}$$
$$= 1.7138\angle 180° \times 15 \text{ kV} = 25.71\angle 180° \text{ kV} \quad \textbf{Ans.}$$

Problem 13.18. A salient pole generator without dampers is rated 25MVA, 15 kV and has a direct axis subtransient reactance 0.30 per unit. The negative and zero-sequence reactances are 0.40 and 0.10 p.u. respectively. The neutral of the generator is solidly grounded. Find the subtransient currents and the line-to-line voltages at the fault under subtransient conditions when double line-to-ground fault occurs at the terminals of the generator. Assume the generator is unloaded and operating at rated voltage when the fault occurs. Neglect resistance.

Solution: Let us choose 25 MVA, 75 kV as base quantities

The circuit diagram of the given problem is as follows.

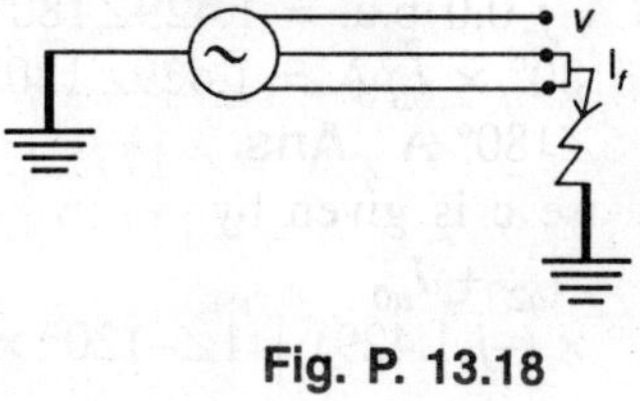

Fig. P. 13.18

Given,

$$Z_1 = 0.30 \text{ p.u.}, Z_2 = 0.40 \text{ p.u.}, Z_0 = 0.10 \text{ pu.}$$

$$V = 15 \text{ kV}, V_b = 15 \text{ kV}, MVA_b = 25 \text{ MVA}$$

At no-load, terminal voltage is equal to the internal voltage of the generator. Hence

$$E_a = V = 15 \text{ kV}$$

$$\therefore \; E_{a \text{ p.u.}} = \frac{E_a}{V_b} = \frac{15}{15} = 1\angle 0° \text{ p.u.}$$

$$\text{Base current } (I_B) = \frac{MVA_b}{\sqrt{3} \times V_b} = \frac{25 \times 10^6}{\sqrt{3} \times 15 \times 10^3} = 962.25 \text{ A}$$

For double line to ground fault,

$$I_a = 0, I_b = -I_c, V_b = V_c = 0 \;\; V_{a1} = V_{a2} = V_{a0}$$

The positive sequence current is given by

$$I_{a1} = \frac{E_{a\text{ p.u}}}{Z_1 + \dfrac{Z_2 Z_0}{Z_2 + Z_0}} = \frac{E_{a\text{ p.u}} \times (Z_2 + Z_0)}{Z_1(Z_2 + Z_0) + Z_2 Z_0} \text{ p.u.}$$

$$= \frac{(1 + j\,0.0) \times (j\,0.40 + 0.10)}{j\,0.30(j\,0.40 + 0.10) + j\,0.40 \times j\,0.10}$$

$$= -j\,2.632 \text{ p.u.}$$

For double line to ground fault,

$$V_{a1} = V_{a2} = V_{a0} = E_a - I_{a1} Z_1$$

$$= 1.0 - (-j\,2.632) \times j\,0.30 = 0.2104 \text{ p.u.}$$

The negative sequence current (I_{a2}) is given by

$$I_{a2} = -\frac{V_{a2}}{Z_2} = -\frac{0.2104}{j0.40} = j\,0.526 \text{ p.u.}$$

$$\text{and} \quad I_{a0} = -\frac{V_{a0}}{Z_0} = -\frac{0.2104}{j0.10} = j\,2.104 \text{ p.u.}$$

The current in the line *a* is given by

$$I_a = I_{a1} + I_{a2} + I_{a0} = -j\,2.632 + j\,0.526 + j\,2.104$$

$$\cong 0.0 \text{ p.u.} = 0.0 \times I_B \text{ A} = 0.0 \text{ A}$$ **Ans.**

The current in phase *b* is given by

$$I_b = \lambda^2 I_{a1} + \lambda I_{a2} + I_{a0}$$

$$= 1\angle{-120} \times -j\,2.638 + 1\angle 120° \times j\,0.526 + j\,2.104$$

$$= -2.735 + j\,3.157 \text{ p.u.} = 4.177\angle 130.9° \text{ p.u.}$$

$$= 4.177\angle 130.9° \times I_B \text{ A} = 4.177\angle 130.9° \times 962.25 \text{ A}$$

$$= 4019.32\angle 130.9° \text{ A}$$ **Ans.**

and $\quad I_c = \lambda I_{a1} + \lambda^2 I_{a2} + I_{a0}$

$= 1\angle 120° \times -j\,2.632 + 1\angle -120° \times (j\,0.526) + j\,2.104$

$= 4.177\angle 49.1°$ p.u. $= 4.177\angle 49.1 \times I_B$ A

$= 4.177\angle 49.1° \times 962.25$ A

$= 4019.32\angle 49.1°$ A **Ans.**

The fault current is given by $I_f = 3\,I_{a0} = I_b + I_c$

$= 3 \times j\,2.104 = j\,6.312$ p.u.

The line-to-ground voltages are given by

$V_a = V_{a1} + V_{a2} + V_{a0} = 3\,V_{a1} = 3 \times 0.2104$

$= 0.6312$ p.u.

and $\quad V_b = V_c = 0$ (From fault condition)

Now, the line-to-line voltages are given by

$V_{ab} = V_a - V_b = 0.6312 - 0 = 0.6312$ p.u.

$= 0.6312\angle 0° \times V_b$

Problem 13.19. One conductor of a three-phase line is open. The current flowing to the Δ-connected load through line a is 20A. With the current in line a as reference and assuming that line *c* is open. Find the symmetrical components of the line currents

Solution: The diagram of the given problem is as follows

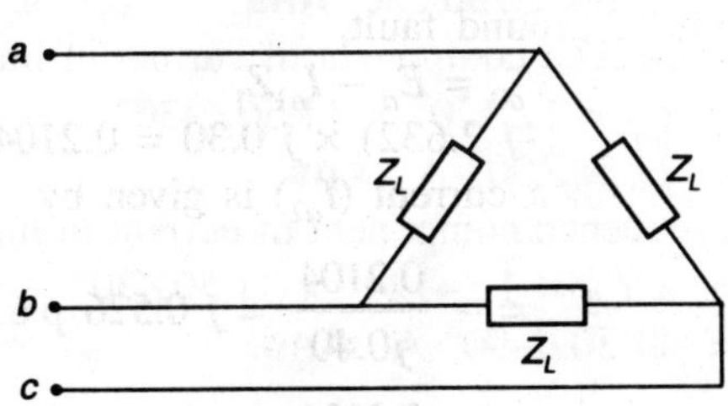

Fig. P. 13.19

Since the line c is open

$\therefore \quad I_c = 0$ A

Let us take the current is the line a as reference $I_a = 20\angle 0°$ A

Hence, $I_b = 20\angle 180°$ A

The symmetrical components of current in the line a are given by

$$I_{a0} = \frac{1}{3}(I_a + I_b + I_c)$$

$$= \frac{1}{3}(20\angle 0° + 20\angle 180° + 0) = 0 \text{ A}$$

The positive sequence component of current in line a in given by

$$I_{a1} = \frac{1}{3}[I_a + \lambda I_b + \lambda^2 I_c]$$

$$= \frac{1}{3}[20\angle 0° + 1\angle 120° \times 20\angle 180° + 1\angle -120° \times 0]$$

$$= \frac{1}{3}[20\angle 0° + 20\angle 300°] = 10 - j\,5.78$$

$$= 11.56\angle -30° \text{ A} \quad \textbf{Ans.}$$

The negative sequence component of the current in line a is given by

$$I_{a2} = \frac{1}{3}[I_a + \lambda^2 I_b + \lambda I_c]$$

$$= \frac{1}{3}[20\angle 0° + 1\angle -120° \times 20\angle 180° + 0]$$

$$= \frac{1}{3}[20\angle 0° + 20\angle 60°] = 10 + j\,5.78$$

$$= 11.56\angle 30° \text{ A} \quad \textbf{Ans.}$$

Since $I_{b0} = I_{c0} = I_{a0} = 0$ A **Ans.**

Now, The positive sequence component of current in line b is given by

$$I_{b1} = \lambda I_{a1} = 1\angle -120° \times 11.56\angle -30°$$

$$= 11.56\angle 150° \text{ A} \quad \textbf{Ans.}$$

The positive sequence component of current in line c is given by

$$I_{c1} = I_{a1} = 1\angle 120° \times 11.56\angle -30°$$

$$= 11.56\angle 90° \text{ A} \quad \textbf{Ans.}$$

The negative sequence component of current in line c is given by

$$I_{c2} = \lambda^2 I_{a2} = 1\angle -120° \times 11.50\angle 30°$$

$$= 11.50\angle -90° \text{ A} \quad \textbf{Ans.}$$

Problem 13.20. A 50 MVA, 20 kV, 3-phase alternator has a sub-transient reactance of 20% and negative and zero sequence of 20% and 10% respectively. The alternator supplies two motors over a transmission line having transformers at both ends as shown below on the one-line diagram.

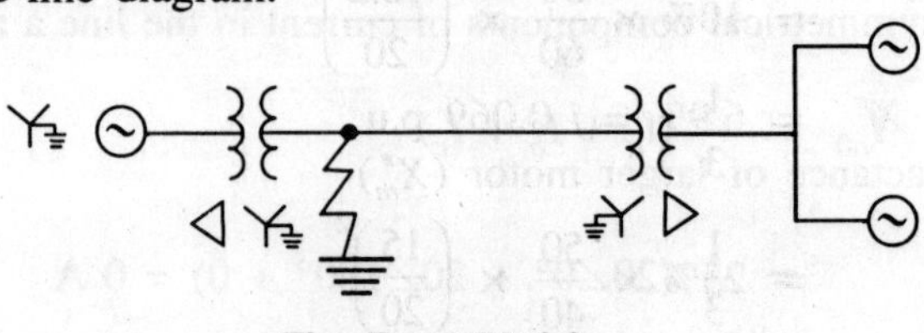

Fig. P. 13.20 (a)

The motors have rated input of 40 MVA and 30 MVA both 15 kV with 25% subtransient reactant and negative and zero sequence reactance 25% and 10% respectively. Current limiting reactors of 2.0 each are in the neutral of the alternator and larger motor. The three phase transformers are both rated 60MVA 18.2 Δ/150 Y kV with leakage reactance 10%. Series reactance of the line is 100 ohms. The zero sequence reactance of the line is 200 ohms. Determine the fault current where (i) L-G, (ii) L-L, and (iii) L-L-G fault takes place at point *p*. Assume V_f = 164.8V.

Solution: Let us choose 50 MVA and 20 kV as base quantities

Base MVA = 50 MVA

Base voltage (V_b) = 20 kV

Given,

Subtransient reactance of the alternator (X_{1g}) = j 0.20 p.u.

Negative sequence reactance (X_{2g}) = j 0.20 p.u.

Zero sequence reactance (X_{0g}) = j 0.10 pu.

The base voltage on the line side of the transformer (V_{bl})

$$= V_b \times \frac{V_H}{V_L} = 20 \times \frac{150}{18.2}$$

$$= 164.8 \text{ kV}$$

The base voltage on the motor side of the transformer

$$V_{bm} = V_{bl} \times \frac{V_L}{V_H} = 164.8 \times \frac{18.2}{150}$$

$$= 20 \text{ kV}$$

The reactances can be charged on the basis of base quantities by using the formula

$$Z_{\text{p.u. new}} = Z_{\text{p.u. old}} \times \frac{MVA_{\text{new}}}{MVA_{\text{old}}} \times \left(\frac{V_{\text{old}}}{V_{\text{new}}}\right)^2$$

Here, New quantities are base quantities. For positive sequence network

% Reactance of the transformer

$$= 10\% \times \frac{50}{60} \times \left(\frac{18.2}{20}\right)^2$$

$$X'' = 6.9\% = j\,0.069 \text{ p.u.}$$

% Reactance of larger motor (X''_m)

$$= 25\% \times \frac{50}{40} \times \left(\frac{15}{20}\right)^2$$

$$= 17.59\% = j\,0.1759 \text{ p.u.}$$

% Reactance of smaller motor (X''_m)

$$= 25\% \times \frac{50}{30} \times \left(\frac{15}{20}\right)^2$$

$$= 23.44\% = j\ 0.2344 \text{ p.u.}$$

% Reactance of the line (X''_e)

$$= \frac{X_L \times MVA_b}{(V_b)^2} = \frac{100 \times 50}{(20)^2}$$

$$= 12.5\% = j\ 0.125 \text{ p.u.}$$

Now, the positive sequence network is as follows

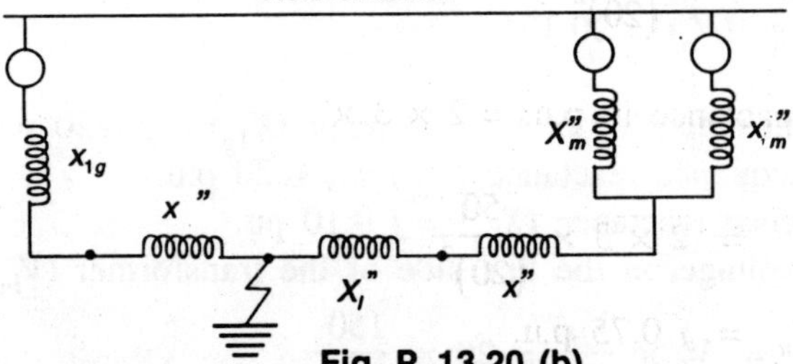

Fig. P. 13.20 (b)

From Fig. P. 13.20 (b),

Positive sequence impedance (Z_1)

$$= [X_{1g} + X''] \parallel [X''_l + X'' + (X''_m \parallel X''_m)]$$

$$= [j\ 0.20 + j\ 0.069] \parallel [j\ 0.125 + j\ 0.069 + \frac{j0.1759 \times j0.2344}{j0.1759 + j0.2344}]$$

$$= j\ 0.269 \parallel [j\ 0.194 + j\ 0.1005]$$

$$= j\ 0.269 \parallel j\ 0.2945$$

$$= \frac{j0.269 \times j0.2945}{j0.269 + j0.2945} = j\ 0.1406 \text{ p.u.}$$

The negative sequence network can be drawn as in Fig. P. 13.20 (c).

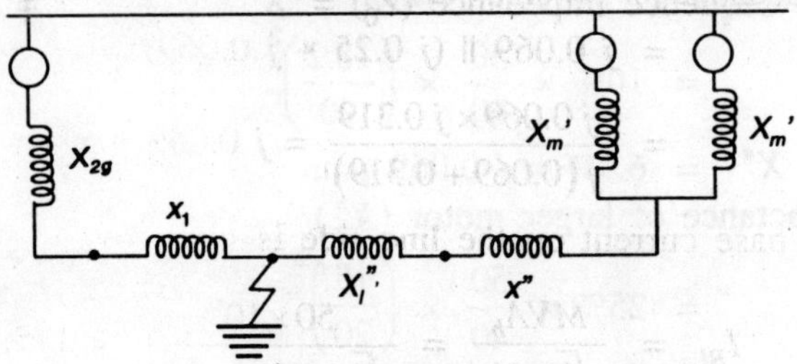

Fig. P. 13.20 (c)

From Fig. P. 13.20 (b) and Fig. P. 13.20 (c), the negative sequence network is same as positive sequence network. Hence

Negative sequence impedance (Z_2) = j 0.1406 p.u.

Calculation for zero sequence network

Given,

Zero sequence reactance of the line = 200 Ω

In p.u. zero sequence reactance of the line

$$= \frac{200 \times MVA_b}{(V_b)^2} \times 100\%$$

$$X_{l0} = \frac{200 \times 50}{(20)^2} = j\ 0.25 \text{ p.u.}$$

$$\text{Neutral reactance in p.u.} = 2 \times 3 \times \frac{MVA_b}{V_b^2}$$

$$= 2 \times 3 \times \frac{50}{(20)^2}$$

$$X_n = j\ 0.75 \text{ p.u.}$$

Now, the zero sequence reactance can be drawn as follows:

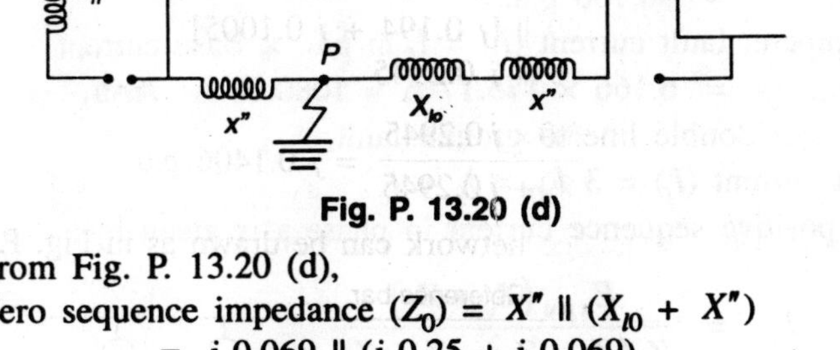

Fig. P. 13.20 (d)

From Fig. P. 13.20 (d),

Zero sequence impedance (Z_0) = $X'' \parallel (X_{l0} + X'')$

$= j\ 0.069 \parallel (j\ 0.25 + j\ 0.069)$

$$= \frac{j\,0.069 \times j\,0.319}{j(0.069 + 0.319)} = j\ 0.057 \text{ p.u.}$$

The base current on the line side is given by

$$I_{Bl} = \frac{MVA_b}{\sqrt{3} \times V_{bl}} = \frac{50 \times 10^6}{\sqrt{3} \times 164.8 \times 10^3} = 175.17 \text{ A}$$

Let us take E_a as reference phasor. $E_a = 1\angle 0°$ p.u.

(i) For line to ground fault $I_{a1} = I_{a2} = I_{a0} = \frac{I_a}{3}$

The positive sequence current in phase a is given by

$$I_{a1} = \frac{E_{a\,p.u}}{Z_1 + Z_2 + Z_3}$$

$$= \frac{1.0}{j(0.1406 + 0.1406 + 0.057)} = -j\,2.957 \text{ p.u.}$$

The fault current at point P is given by

$$I_f = I_a = I_{a1} + I_{a2} + I_{a0} = 3\,I_{a1}$$
$$= 3 \times (-j\,2.957) = -j\,8.87 \text{ p.u.}$$

In ampere, Fault current (I_f) = I_f in p.u. × I_B

= 8.87 × 175.17 A = 1553.84 A **Ans.**

(ii) For L-L fault, $I_{a1} = -I_{a2}$ and $I_{a0} = 0$

The positive sequence current in phase a is given by

$$I_{a1} = \frac{E_{a\,p.u}}{Z_1 + Z_2} = \frac{1.0}{j(2 \times 0.1406)}$$
$$= j\,3.56 \text{ p.u.}$$

Negative sequence current (I_{a2}) = $-I_{a1} = j\,3.56$ p.u.

The fault current at point p is given by

$$I_f = I_b = \lambda^2 I_{a1} + \lambda I_{a2} + I_{a0} = \lambda^2 I_{a1} - \lambda I_{a1}$$
$$= I_{a1}\,(\lambda^2 - \lambda)$$
$$= -j\,3.56\,(0.5 - j\,0.866 + 0.5 - j\,0.866)$$
$$= -6.166 \text{ p.u.}$$

In ampere, fault current (I_f) = I_f in p.u. × Base current

= 6.166 × 175.17 A = 1080.08 A **Ans.**

(iii) For double line to ground fault,

Fault current (I_f) = 3 $I_{a0} = I_b + I_c$

The positive sequence current in phase a is given by

$$I_{a1} = \frac{E_{a\,p.u}(Z_0 + Z_2)}{Z_1(Z_0 + Z_2) + Z_0 \times Z_2}$$

$$= \frac{1.0 \times (j\,0.057 + j\,0.1406)}{j\,0.1406(j\,0.057 + j\,0.1406) + j\,0.1406 \times j\,0.057}$$

$$= -j\,5.52 \text{ p.u.}$$

The zero sequence current is given by

$$I_{a0} = -\frac{E_{a\,p.u} \, I_{a1} Z_1}{Z_0} = -\frac{1.0-(-j5.52 \times j0.1406)}{j0.057}$$

$$= j\ 3.928 \text{ p.u.}$$

Fault current at point in ampere = 3 × $|I_{a0}|$ × Base current

= 3 × 3.928 × 175.17 A = 2064.13 A **Ans.**

Problem 13.21. A small generating station has a bus bar divided into three sections. Each section is connected to a tie bar with reactors each rated at 10 MVA. 0.2 p.u. reactance. A generator of 20 MVA and 0.3 p.u. reactance is connected to each section of the bus-bar. Determine the short circuit capacity of the breaker if a three-phase fault takes place on the section of the bus bar.

Solution: This power system is as follows

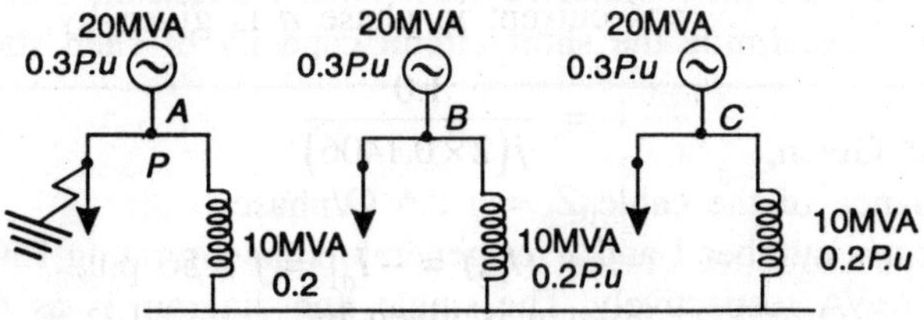

Fig. P. 13.21 (a)

Let us choose 20 MVA as base quantity

∴ Base MVA = 20 MVA

Impedance of each generator (X_g) = j 0.3 p.u.

The impedance of each reactor on the base of

$$20 \text{ MVA} = j\ 0.2 \times \frac{\text{Base MVA}}{\text{Reactor MVA}}$$

$$= j\ 0.2 \times \frac{20}{10} = j\ 0.4 \text{ p.u.}$$

Let us assume a fault occurs on the point p in Fig. P. 13.21 (a).

The circuit can be reduced Fig. P. 13.21 (b).

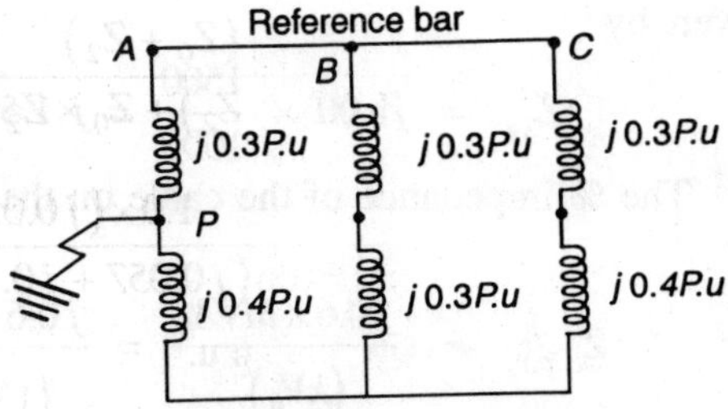

Fig. P. 13.21 (b)

The equivalent impedance between point *A* and *P* is given by

$$Z_{p.u.} = (j\ 0.3) \parallel (j\ 0.4 + \frac{j0.7}{2})$$

$$= j\ 0.3 \parallel (j\ 0.4 + j\ 0.35)$$

$$= \frac{j0.3 \times j0.75}{j(0.3+0.75)} = j\ 0.214 \text{ p.u.}$$

We know that, short circuit capacity

$$= \frac{\text{Base MVA}}{Z_{p.u}} = \frac{20 \times 10^6}{0.214} = 93.46 \text{ MVA}$$

Thus, short, circuit capacity of the breaker = 93.46 MVA **Ans.**

Problem 13.22. Two generating stations having short circuit capacities 1500 MVA and 1000 MVA respectively operating at 13.2 kV are linked by interconnected cable having a reactance of 0.6 Ω per phase. Determine the short circuit capacity of each station.

Solution: Given,

Impedance of the cable $Z_b = j\ 0.6$ Ω/phase

Let us use number 1 and 2 for generators having rating 1500 MVA and 1000 MVA respectively. The single line diagram is as follows:

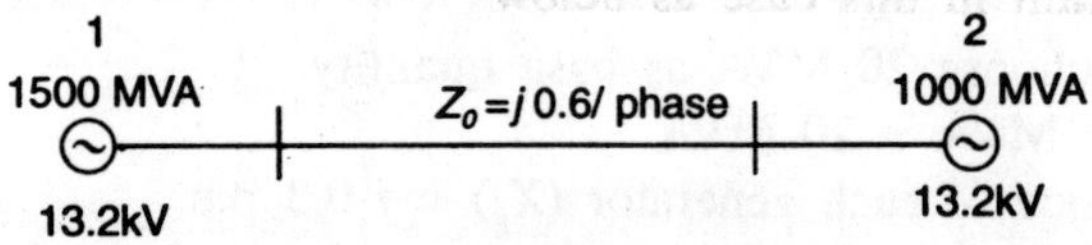

Fig. P. 13.22 (a)

Let us choose 1500 MVA as base rating and let reactance of this generator be 100%

$$\therefore \quad Z_{1g} = j\ 1.00 \text{ p.u.}$$

The impedance of the generator 2 on the base of 1500 MVA is given by

$$Z_{2g} = j1.00 \times \frac{1500}{1000} = j\ 1.5 \text{ p.u.}$$

The % impedance of the cable on the base of 1500 MVA is given by

$$Z_{c\ p.u.} = \frac{j0.6 \times MVA_b}{(kV_b)^2} = \frac{j0.6 \times 1500 \times 10^6}{(13.2 \times 10^3)^2}$$

$$= j\ 5.165 \text{ p.u.}$$

Let us assume a fault takes place at 1500 MVA capacity point. The circuit can be drawn as shown in Fig. 13.22 (b).

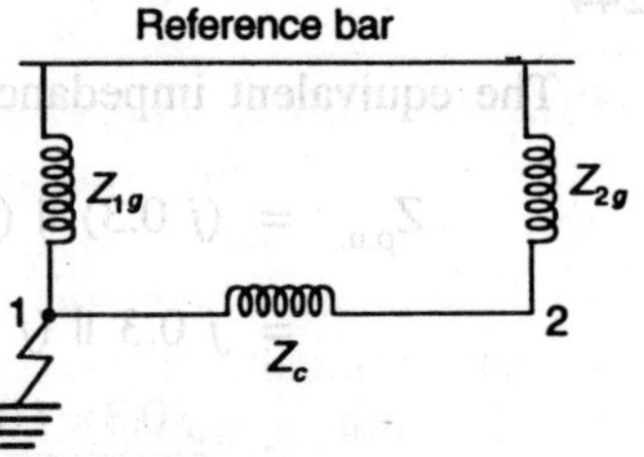

Fig. P. 13.22 (b)

The equivalent impedance between the reference bus and the node 1 is given by

$$Z_{1p.u.} = Z_{1g} \parallel (Z_c + Z_{2g})$$

$$= \frac{Z_{1g}(Zc + Z_{2g})}{Z_{1g} + Z_{2g} + Zc} = \frac{j1.0 \times (j5.165 + j1.5)}{j(1.5 + 1.0 + 5.165)}$$

$$= j\ 0.8695 \text{ p.u.}$$

The short circuit capacity of this station

$$= \frac{MVA_b}{Z_{1\text{ p.u}}} = \frac{1500 \times 10^6}{0.8695}$$

$$= 1725.13 \text{ MVA} \quad \textbf{Ans.}$$

Again, let us assume the fault takes place at 1000 MVA capacity plant. The circuit diagram in this case as below

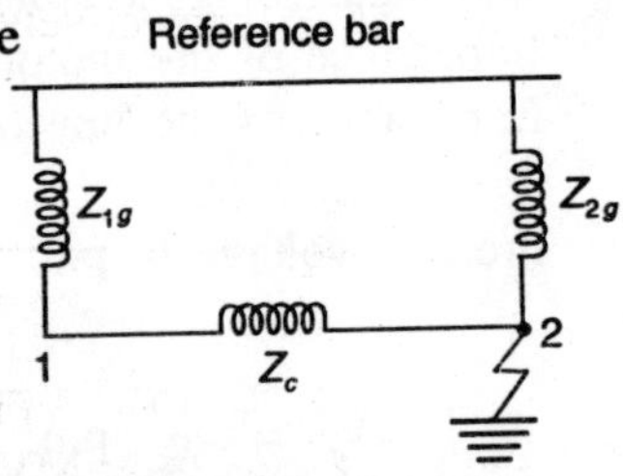

Fig. P. 13.22 (c)

The equivalent impedance between the reference bus and the node 2 is given by.

$$Z_{2\text{ p.u.}} = Z_{2g} \parallel (Z_c + Z_{1g})$$

$$= \frac{Z_{2g}(Z_c + Z_{1g})}{Z_{1g} + Z_c + Z_{2g}} = \frac{j1.5(j5.165 + j1.0)}{j(1.0 + 1.5 + 5.165)}$$

$$= j\ 1.206 \text{ p.u.}$$

The short circuit capacity of this plant

$$= \frac{MVA_b}{Z_{2\text{ p.u}}} = \frac{1500}{1.206} = 1243.78 \text{ MVA} \quad \textbf{Ans.}$$

Problem 13.23. An alternator and a synchronous motor each rated for 80 KVA, having subtransient reactance of 25% are connected through a transmission link of reactance 10% on the base of machine rating. The motor acts as a load of 50 MW at 0.8 p.f leading and the terminal voltage 15 kV when a three phase fault takes place at the motor terminals. Determine the subtransient current in the alternator, motor and the fault.

Solution: Let us choose 80 kVA and 20 kV as base quantities

Base MVA = 80 kVA

Base voltage (V_t) = 20 kV

The single-line diagram of this power system is as follows

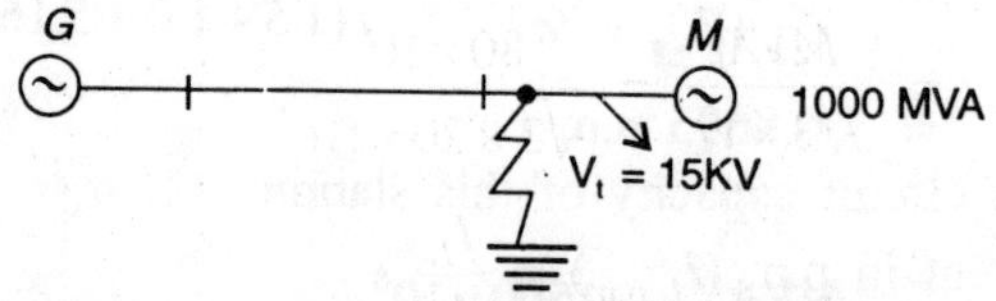

Fig. P. 13.23 (a)

Given, load power (P_r) = 50 MW at 0.8 leading p.f.

Motor terminal voltage when fault takes place (V_t) = 15 kV

Impedance of generator (Z_g) = j 0.25 p.u

Impedance of the motor (Z_m) = 10.25 p.u.

Impedance of the line (Z_l) = j 0.10 p.u.

$$\text{Prefault voltage in p.u.} = \frac{V_t}{V_b}$$

$$V_f = \frac{15}{20} \text{ p.u.} = 0.75 \text{ p.u.}$$

Let us take this voltage an reference phasor. The reactance diagram under fault condition can be drawn a shown in Fig. P. 13.23 (b).

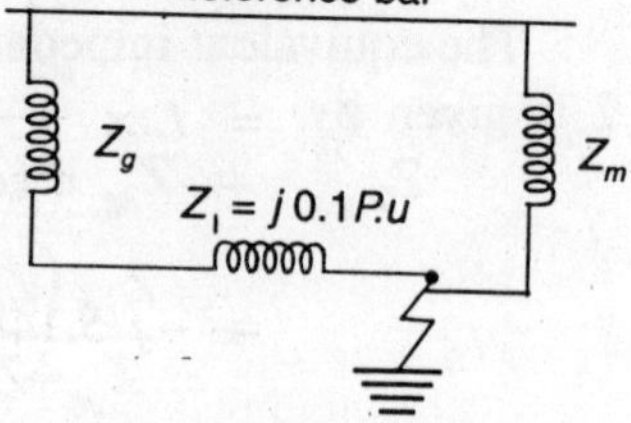

Fig. P. 13.23 (b)

The fault impedance is given by

$$Z_f = (Z_g + Z_l) \parallel Z_m = \frac{(Z_g + Z_l) Z_m}{Z_g + Z_m + Z_l}$$

$$= \frac{(j\,0.25 \times j\,0.10) \times j\,0.25}{j(0.25 + 0.25 \times 0.10)} = j\ 0.1458 \text{ p.u.}$$

$$\therefore \text{ Fault current } (I_f) = \frac{V_f}{Z_f} = \frac{0.75}{j\,0.1458} = -j\ 5.144 \text{ p.u.}$$

$$\text{Now, Load current } (I_L) = \frac{\text{Load power}}{\sqrt{3} \times V_l \times p.f}$$

$$= \frac{50 \times 10^6}{\sqrt{3} \times 15 \times 10^3 \times 0.8} = 2405.63 \text{ A}$$

In phasor form, $I_L = 2405.63\ \angle 36.8°$ A

Base current (I_b)

$$= \frac{MVA_b}{\sqrt{3} \times V_b} = \frac{80 \times 10^6}{\sqrt{3} \times 20 \times 10^3} = 2309.4 \text{ A}$$

$$\text{Load current in p.u. } (I_{L\text{ p.u.}}) = \frac{I_L}{I_b}$$

$$= \frac{2405.63\angle 36.8°}{2309.4} = 1.0416\ \angle 33.8° \text{ p.u.}$$

The p.u. fault current supplied the motor is given by

$$I_{fm} = I_f \times \frac{Z_g + Z_l}{Z_g + Z_l + Z_m}$$

$$= -j\ 5.144 \times \frac{j\,0.25 + j\,0.1}{j(0.25 + 0.25 + 0.1)}$$

$$= -j\ 3.001 \text{ p.u.}$$

The fault current supplied by generator

$$I_{fg} = I_f \times \frac{Z_m}{Z_g + Z_m + Z_l}$$

$$= -j\ 5.144 \times \frac{j\,0.25}{j(0.25 + 0.25 + 0.10)}$$

$$= -j\ 2.143 \text{ p.u.}$$

Net current supplied by the generator

$$= I_{fg} + I_{1\text{ p.u.}} = -j\ 2.143 + 1.0416\ \angle 36.8°$$

$$= 0.8340 - j\ 1.519 \text{ p.u.} = 1.733\ \angle -61.23° \text{ p.u.}$$

In ampere, net fault current supplied by the generator

= Fault current in p.u. × Base current

= 1.733 × 2309.4 A = 4002.19 A **Ans.**

The net fault current supplied by the motor is given by

$$I_{fm} = I_{fm} - I_{1\,p.u.} = -j\,3.001 - 1.0416\,\angle 36.8°$$
$$= -0.8340 - j\,3.625 \text{ p.u.}$$
$$= 3.72\,\angle -130° \text{ p.u.}$$

In ampere, net fault supplied from motor

= I_{fm} in p.u. × Base current

= 3.72 × 2309.4 A = 8590.97 A **Ans.**

In ampere, fault current = I_f × base current

= 5.144 × 2309.4 A = 11879.55 A **Ans.**

Problem 13.24. A station operating at 36 kV is divided into sections *A* and *B*. section A consists of three generators 20 MVA each having a reactance of 20% and section *B* is fed from the grid through 90 MVA transformer of 10% reactance. The circuit breakers have each a rupturing capacity of 900 MVA. Determine the reactance of a reactor to prevent the breakers being overloaded if a symmetrical short circuit occurs on the outgoing feeder connected to *A*.

Solution: The given power system is as follows

Let X_m be the reactance of the reactor for base 90 MVA

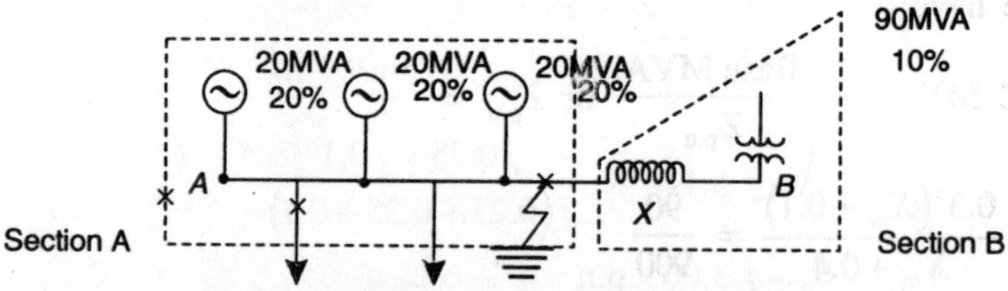

Fig. P. 13.24 (a)

Given, SC MVA = 90 MVA

Let us choose 90 MVA as base MVA.

$$\text{p.u. impedance of each generator } (Z_g) = j\,0.20 \times \frac{\text{Base MVA}}{\text{Rated MVA}}$$

$$= j\,0.20 \times \frac{90}{20} = j\,0.9 \text{ p.u.}$$

$$\text{p.u. impedance of transformer } (Z_t) = j\,0.10 \times \frac{90}{90} = j\,0.10 \text{ p.u.}$$

When the fault occurs at point on outgoing feeder, the reactance diagram can be drawn as in Fig. P. 13.24 (b).

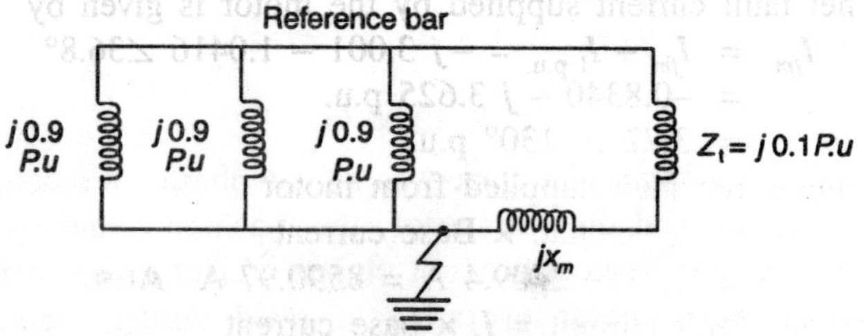

Fig. P. 13.24 (b)

The equivalent impedance between fault point and the reference bus is given by

$$Z_{p.u.} = (j\,0.9 \parallel j\,0.9 \parallel j\,0.9) \parallel (j\,X_m + Z_t)$$

$$= \frac{j\,0.9}{3} \parallel j\,(X_m + 0.10) = j\,0.3 \parallel j\,(X_m + 0.10)$$

$$= \frac{j\,0.3 \times j\left(X_m + 0.10\right)}{j\left(0.3 + X_m + 0.10\right)}$$

$$= j\,\frac{\left(0.3X_m + 0.03\right)}{X_m + 0.40}$$

We have,

$$\text{SC MVA} = \frac{\text{Base MVA}}{Z_{p.u}} \text{ or } Z_{p.u.} = \frac{\text{Base MVA}}{\text{SC MVA}}$$

$$\text{or } \frac{0.3\left(X_m + 0.1\right)}{X_m + 0.4} = \frac{90}{900}$$

$$\text{or } 3\,X_m + 0.3 = X_m + 0.4 \text{ or } 2\,X_m = 0.1$$

$$\therefore \quad X_m = \frac{0.1}{2} = 0.05 \text{ p.u.}$$

In ohms, reactance of the reactor

$$= \frac{X_m \times (V_b)^2}{MVA_b} = \frac{0.05 \times \left(36 \times 10^3\right)^2}{90 \times 10^6}$$

$= 0.72\ \Omega$ **Ans.**

Problem 13.25. A double line to ground fault occurs on phase *b* and phase *c* at point p in circuit given below.

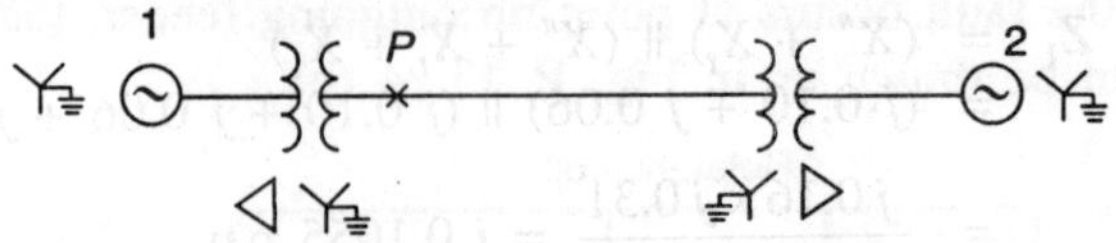

Fig. P. 13.25 (a)

Determine the subtransient current in all the phases of machine 1, the fault current and the voltage of the machine 1 and voltages at the fault point. Neglect prefault current. Assume that machine 2 is synchronous motor operating at rated voltage. Both the machines are rated at 1.5 MVA. 660 V with reactances of $X'' = X_2 = 10\%$ and $X_0 = 5\%$. Each three phase transformer is rated at 1.5 MVA 660 V Δ/ 4400 VY star with leakage reactance of 6%. The reactances of the transmission line are $X_1 = X_2 = 15\%$ and $X_0 = 50\%$ on a base of 1.5 MVA, 4400 volts.

Solution: Let us assume 1.5 MVA and 660 V as base quantities.

$MVA_b = 1.5$ MVA, $V_b = 660$ V

Base voltage referred to line side $(V_{bl}) = V_b \times \dfrac{V_H}{V_L}$

$$= 660 \times \frac{4400}{660} = 4400 \text{ V}$$

Base current on line side (I_{bl})

$$= \frac{MVA_b}{\sqrt{3} \times V_{bL}} = \frac{1.5 \times 10^6}{\sqrt{3} \times 4400} = 196.82 \text{ A} \quad \textbf{Ans.}$$

Since the transformation ratio is 660/4400 volts and the transformer is rated at 1.5 MVA, there is no change in any reactances.

The positive sequence network can be drawn shown in Fig. P. 13.25 (c).

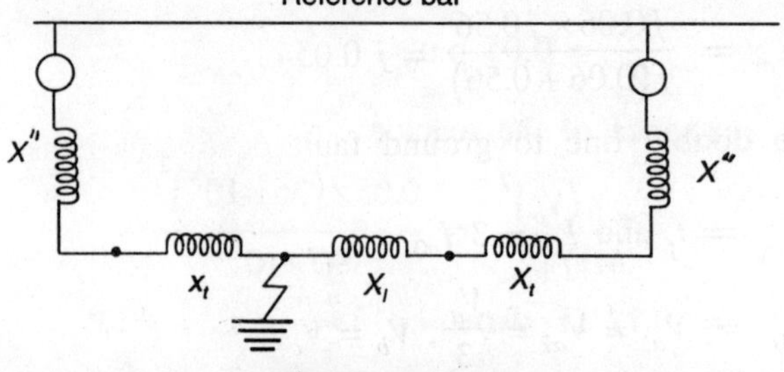

Fig. P. 13.25 (c)

The positive sequence impedance between fault-point and reference bus is given by

$$Z_1 = (X'' + X_t) \parallel (X'' + X_t + X_1)$$
$$= (j\,0.10 + j\,0.06) \parallel (j\,0.10 + j\,0.06 + j\,0.15)$$
$$= \frac{j0.16 \times j0.31}{j(0.16 + 0.31)} = j\,0.1055 \text{ p.u.}$$

The negative sequence network for fault point p is as below

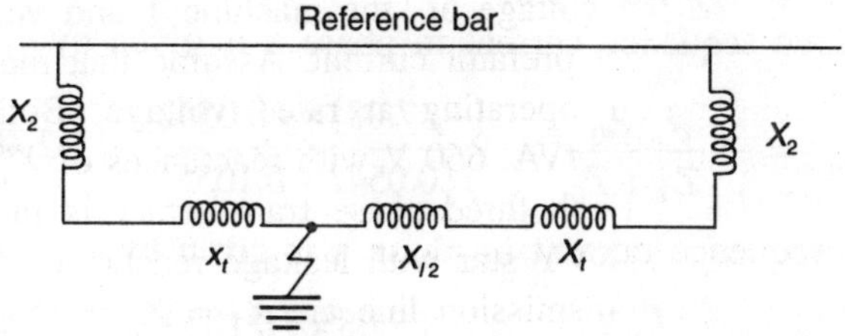

Fig. P. 13.25 (c)

The negative sequence impedance is given by

$$Z_2 = (X_2 + X_t) \parallel (X_2 + X_t + X_{l2})$$
$$= (j\,0.10 + j\,0.06) \parallel (j\,0.10 + j\,0.6 + j\,0.15)$$
$$= j\,0.1055 \text{ p.u.}$$

Now, the negative sequence impedance is as follows

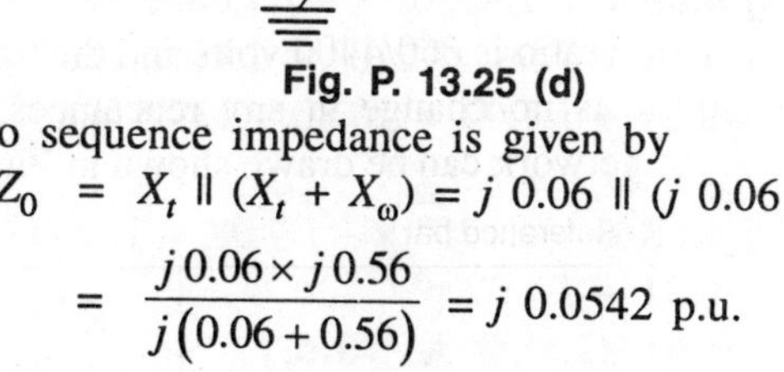

Fig. P. 13.25 (d)

The zero sequence impedance is given by

$$Z_0 = X_t \parallel (X_t + X_{\omega}) = j\,0.06 \parallel (j\,0.06 + j\,0.50)$$
$$= \frac{j0.06 \times j0.56}{j(0.06 + 0.56)} = j\,0.0542 \text{ p.u.}$$

Since the double line to ground fault occurs in phase b and phase c,

$$\therefore\ I_b + I_c = I_f \text{ and } I_f = 3\,I_{a0}$$

$$V_{a0} = V_{a1} = V_{a2} = \frac{V_a}{3},\ V_b = V_c = 0$$

Let us take E_a as reference phasor. $E_a = 1\ \angle 0°$ p.u.

The positive sequence current in phase a is given by

$$I_{a1} = \frac{E_{a\,p.u}(Z_2+Z_0)}{Z_1(Z_2+Z_0)+Z_0Z_2}$$

$$= \frac{1.0\,(j0.1055+j0.0542)}{j0.1055(j0.1055+j0.0542)+j0.0542\times j0.1055}$$

$$= -j\,7.0\,8 \text{ p.u.}$$

The negative sequence current in phase a is given by

$$I_{a2} = -\frac{I_{a1}Z_0}{Z_2+Z_0} = \frac{(-j7.08)\times j0.0542}{j(0.0542+0.1055)} = j\,2.40 \text{ p.u.}$$

The zero sequence current in phase a is given by

$$I_{a0} = -\frac{I_{a1}Z_2}{Z_1+Z_2} = \frac{-j7.08\times j0.1033}{j(0.0522+0.1055)} = j\,4.68 \text{ p.u.}$$

Now, positive sequence voltage (V_{a1})

$$= E_a - I_{a1}\cdot Z_1 = 1.0 + j\,7.08 \times j\,0.1055$$

$$= 0.2532 \text{ p.u.}$$

Negative sequence voltage (V_{a2})

$$= -I_{a2}\,Z_2 = -(j\,2.40 \times j\,0.1055) = 0.2532 \text{ p.u.}$$

Similarly, $V_{a0} = 0.2532$ p.u.

Now, fault current (I_f) $= I_{a2} = I_{a1} + I_{a2} + I_{a0}$

$$= -j\,7.08 + j\,2.40 + j\,4.68 = 0.0 \text{ p.u.}$$

$$= 0.0 \times \text{base current} = 0.0 \text{ A} \quad \textbf{Ans.}$$

Now, $I_b = I_{a0} + \lambda^2 I_{a1} + \lambda I_{a2}$

$$= j\,4.68 + 1\angle{-120^\circ} \times -j\,7.08 + 1\angle 120^\circ \times j\,2.40$$

$$= j\,4.68 + 7.08\angle{-210^\circ} + 2.40\angle 210^\circ$$

$$= -8.21 + j\,7.02 = 10.80\angle 139.5 \text{ p.u.} \quad \textbf{Ans.}$$

$$I_c = I_{a0} + \lambda I_{a1} + \lambda^2 I_{a2}$$

$$= j\,4.68 + 1\angle 120^\circ \times -j\,7.08 + 1\angle{-120} \times j\,2.40$$

$$= -8.21 - j\,7.02$$

$$= 10.80\angle 220.5^\circ \text{ A} \quad \textbf{Ans.}$$

The positive sequence current supplied by the generator is given by

$$I_{A1} = I_{a1} \times \frac{X'' + X_t + X_l}{X'' + X_t + X_1 + X_t + X''}$$

$$= -j\,7.08 \times \frac{j(0.10+0.06+015)}{j(0.10+0.06+015+0.06+0.10)}$$

$$= j\,4.67 \text{ p.u.}$$

Similarly,

$$I_{A2} = I_{a2} \times \frac{X'' + X_t + X_l}{X'' + X_t + X_1 + X_t + X''}$$

$$= j\,2.40 \times \frac{j(0.10 + 0.06 + 015)}{j(0.10 + 0.06 + 0.15 + 0.06 + 0.10)}$$

$$= j\,1.58 \text{ p.u.}$$

Now, Current supplied by generator $(I_A) = (I_{A1} + I_{A2})$

$$= 4.67 + 1.58 = 6.25 \text{ p.u.}$$

Now, $I_B = \lambda^2 I_{A1} + \lambda I_{A2} + I_{A0}$

$$= 1\angle{-120°} \times 4.67 + 1\angle 120° \times 1.58 + 0$$

$$= -3.125 - j\,2.68 = 4.117 \angle 139.4 \text{ p.u.}$$ **Ans.**

$$I_C = \lambda I_{A1} + \lambda^2 I_{A2} + I_{A0} = 4.117 \angle 220.6 \text{ p.u.}$$ **Ans.**

Voltages at fault point P

$$V_a = V_{a1} + V_{a2} + V_{a0} = 3 \times V_{a1} = 3 \times 0.2532$$

$$= 0.7596 \text{ p.u.}$$ **Ans.**

and $V_b = V_c = 0$

Now, Line-to-line voltage are given by

$$V_{ab} = V_a - V_b = 0.7596 - 0 = 0.7596 \text{ p.u.}$$

$$V_{bc} = V_b - V_c = 0 - 0 = 0 \text{ p.u.}$$

and $V_{ca} = V_c - V_a = 0 - 0.7596 = -0.7596$ p.u. **Ans.**

Chapter 14

Power System Stability

Problem 14.1. The generalized circuit constants of a nominal circuit representing a 3-phase transmission line are

$A = D = 0.980\angle 0.3°$, $B = 82.5\angle 76.0°\ \Omega$, $C = 0.0005\angle 90°\ ℧$

Determine the steady state stability limit of the line if the sending end and receiving end voltages are held at 220 kV.

Solution: Given,

$V_r = V_s = 220$ kV, $A = D = 0.980\angle 0.30°$

This yeilds, $|A| = 0.980$ and $\alpha = 0.3°$

Also given, $B = 82.5°\angle 76°\ \Omega$ and $C = 0.0005\angle 90°\ ℧$

This yeilds, $|B| = 82.5\ \Omega$ and $\beta = 76.0°$

Now, the steady state suitability limit of the line is given by

$$P_{max} = \frac{V_s V_r}{|B|} - \frac{|A| V_r^2}{|B|} \cos(\beta - \alpha)$$

$$= \frac{220 \times 220 \times 10^6}{82.5} - \frac{0.980 \times \left(220 \times 10^3\right)^2}{82.5}$$

$$\cos(76.0° - 0.3°)$$

$$= 444.66 \text{ MW}$$

Thus, steady-state stability limit of the line = 444.66 MW **Ans.**

Problem 14.2. Two power stations A and B are located close together. Station A has four identical generator sets each rated 100 MVA and each having an inertia constant of 9 MJ/MVA whereas station B has three sets each rated 200 MVA, 4 MJ/MVA. Calculate the inertia constant of the equivalent machines of both the stations on 150 MVA base.

Solution: Given,

Base MVA (G_b) = 150 MVA

For power station *A*,

Rating of each generator (G_A) = 100 MVA

Inertia constant of each generator (H_A) = 9 MJ/MVA

No. of generators (n) = 4

∴ Total energy generated at station A

$$E_A = n \times H_A \times G_A = 4 \times 9 \times 100 = 3600 \text{ MJ}$$

For power station *B*

Rating of each generator (G_B) = 200 MVA

Inertia constant for each generator (H_B) = 4 MJ/MVA

No. of generators (n) = 3

∴ Total energy generated at station *B*

$$E_B = n \times H_B \times G_B = 3 \times 200 \times 4 = 2400 \text{ MJ}$$

Total energy generated at both stations (E)

$$= E_A + E_B = 3600 + 2400 = 6000 \text{ MJ}$$

∴ Inertia constant of a single equivalent machine (H)

$$= \frac{\text{Total energy generated}}{\text{Base MVA}} = \frac{E}{G_b} = \frac{6000 \text{ MJ}}{150 \text{ MVA}}$$

$$= 40 \text{ MJ/MVA}$$

∴ Equivalent inertia constant of single machine

= 40 MJ/MVA **Ans.**

Problem 14.3. A 4 pole, 50 Hz, 22 kV turbo attenuator has a rating of 100 MVA, p.f. 0.8 lag. The moment of inertia of the rotor is 8000 kg-m^2. Determine M and H.

Solution: Given,

No. of poles (P) = 4

Frequency (f) = 50 Hz

Rating of the machine (G) = 100 MVA

Moment of inertia (I) = 8000 kgm^2

$$\text{Synchronous speed } (W_s) = \frac{4\pi f}{P} \text{ rad/sec.}$$

$$= \frac{4\pi \times 50}{4} \text{ rad/sec.} = 50\pi \text{ rad/sec.}$$

∴ Inertia constant (H)

$$= \frac{\text{K.E energy stored in rotor at synchromous speed}}{\text{Rating of the machine}}$$

$$= \frac{\frac{1}{2}I\omega^2}{G} = \frac{\frac{1}{2}\times 8\times 10^3\times(50\pi)^2}{100}$$

$= 0.98696$ MJ/MVA

To calculate M, apply the relation

$$M = \frac{GH}{180f} \text{ MJ-sec/electrical degree}$$

$$= \frac{100\times 0.98696}{180\times 50} \text{ MJ–sec/electrical degree}$$

$= 0.01097$ MJ-sec/electrical degree

Thus, $M = 0.01097$ MJ-sec/electrical degree

$H = 0.98696$ MJ/MVA **Ans.**

Problem 14.4. A turboalternator with 4-pole, 50 Hz, 80 MW, p.f. 0.8 lag and moment of inertia 40,000 kgm^2 is interconnected via a short transmission line to another alternator with 2-pole, 50 Hz, 100 MW, p.f. 0.8 lag and moment of inertia, 100,000 kgm^2. Determine the inertia constant of the single equivalent machine on a base of 100 MVA.

Solution: Given,

Base MVA (G_b) = 100 MVA

Let us use 1 and 2 for machines given for machine 1,

No. of poles (P_1) = 4

Frequency (f_1) = 50 Hz

$$\text{Rating of this machine } (G_1) = \frac{\text{Active power}}{\text{p.f.}}$$

$$= \frac{80}{0.8} \text{ MVA} = 100 \text{ MVA}$$

Moment of inertia of this machine (I_1) = 40×10^3 kgm^2

$$\text{Synchronous speed } (W_s) = \frac{4\pi f_1}{P_1} \text{ rad/sec.}$$

$$= \frac{4\pi\times 50}{4} \text{ rad/sec.} = 50\,\pi \text{ rad/sec.}$$

∴ K.E. stored at synchronous speed in this machine (E_1)

$$= \frac{1}{2}\, I\omega^2 = \frac{1}{2} \times 40 \times 10^3 \times (50\pi)^2 = 493.48 \text{ MJ}$$

For machine 2,
No. of poles $(P_2) = 2$
Frequency $(f_2) = 50$ Hz
Moment of inertia $(I_2) = 10 \times 10^3$ kg-m^2

$$\therefore \text{ Synchronous speed } (W_s) = \frac{4\pi \times f_2}{P_2} \text{ rad/sec.}$$

$$= \frac{4\pi \times 50}{2} \text{ rad/sec.} = 100\ \pi \text{ rad/sec.}$$

$\therefore$ K.E stored in the rotor at synchronous speed (E_2)

$$= \frac{1}{2} I_2 \omega_s^2 = \frac{1}{2} \times 10 \times 10^3 (100\ \pi)^2$$
$$= 493.48 \text{ MJ}$$

$\therefore$ Total energy stored in the rotors of both machines

$$(E) = E_1 + E_2 = (493.48 + 493.48) \text{ MJ} = 986.96 \text{ MJ}$$

At base 100 MVA, Inertia constant (H)

$$= \frac{\text{Total K.E stored}}{\text{Base MVA}} = \frac{E}{G_b} = \frac{986.96}{100} \text{ MJ/MVA}$$

$$= 9.87 \text{ MJ/MVA} \quad \textbf{Ans.}$$

Problem 14.5. Station *A* transmits 50 MW of power to station *B* through a tie line. The maximum steady state capacity of the line is 100 MW. Determine the allowable sudden load that can be switched on without loss of stability.

Solution: Given,

$P_{max} = 100$ MW

Initial operating power

$(P_1) = 50$ MW

The power-angle equation under this case is given by

$$P_e = P_{max} \times \sin \delta$$

or $\quad P_e = 100 \sin \delta \quad ...(1)$

The power-angle curve is as given in Fig. P. 14.5

Let, δ_1 be the load-angle at point *a*

For operating point *a*,

$P_e = 50$ MW

or $100 \sin \delta_1 = 50$

$\therefore \sin \delta_1 = 0.5$

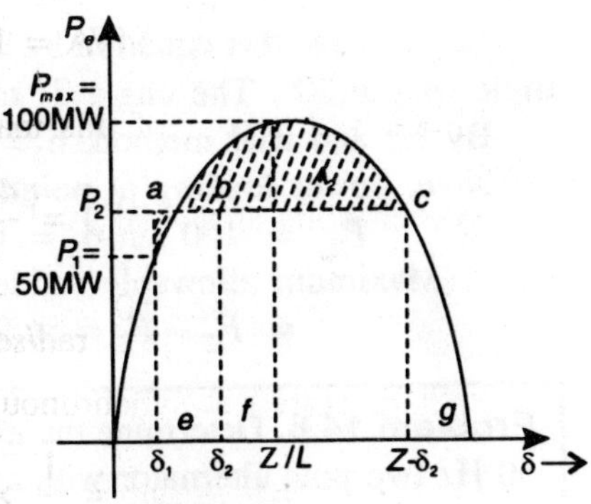

Fig. P. 14.5

This yeilds $\delta_1 = 30°$
$= 0.5236$ radian.

When additional load is switched on, the rotor accelerates due to accelerating power $(P_2 - P_1)$.

Let δ_2 be the load angle at point *b*.

At point *b*,

$\therefore \quad P_e = P_2$ or $100 \sin \delta_2 = P_2$ ....(ii)

For system to be remained instability.

Area A_1 = Area A_2

$$\text{or} \quad \int_{\delta_1}^{\delta_2} (P_2 - P_1 \sin \delta)\, d\delta = \int_{\delta_2}^{\pi-\delta_2} (100 \sin \delta - P_2)\, d\delta$$

$$\text{or} \quad \int_{\delta_1}^{\delta_2} (P_2 - 100 \sin \delta)\, d\delta = -\int_{\delta_2}^{\pi-\delta_2} (P_2 - 100 \sin \delta)\, d\delta$$

Using integration property, this yields

$$\int_{\delta_1}^{\pi-\delta_2} (P_2 - 100 \sin \delta)\, d\delta = 0$$

$$\text{or} \quad \left[(P_2\delta + 100 \cos \delta)\right]_{\delta_1}^{\pi-\delta_2} = 0$$

or $P_2 (\pi - \delta_2 - \delta_1) + 100 [\cos (\pi - \delta_2) - \cos \delta_1] = 0$

since $P_2 = 100 \sin \delta_2$

$$\text{or } 100 \sin \delta_2 \left(\pi - \frac{\pi}{6} - \delta_2\right) + 100 [-\cos \delta_2 - \cos 30°] = 0$$

$$\text{or} \quad 100 \sin \delta_2 \left(\frac{5\pi}{6} - \delta 2\right) = 100 [\cos \delta_2 + 0.866]$$

$$\text{or} \quad \sin \delta_2 \left(\frac{5\pi}{6} - \delta_2\right) = \cos \delta_2 + 0.866$$

Let us solve this equation by hit and trial method since initial load angle (δ_1) is 30°. The angle δ_2 must be greater than δ_1.

By hit and trial method, $\delta_2 = 60.4° = 1.054$ radian

Now, power loading at point *b* is given by

$$P_2 = 100 \sin \delta_2 = 100 \sin 60.4° = 86.95 \text{ MW}$$

$\therefore$ Maximum allowable sudden loading on the machine

$= P_2 - P_1 = 86.95 - 50 = 36.95$ MW **Ans.**

Problem 14.6. Determine the kinetic energy stored by a 50 MVA, 50 Hz two pole alternator with an inertia constant (H) of 5 kW sec per kVA. If the machine is running steadily at synchronous speed with a shaft input (minus rotational losses) of 65000 HP when the electrical power developed suddenly changes from its normal value to a value of 40 MW, determine the acceleration or deceleration of the rotor. If the acceleration computed for the generator is constant for a period of 10 cycles, determine the change in torque angle in that period and the rpm at the end of the 10 cycles.

Solution: Given,

Rated MVA (G) = 50 MVA

No. of poles (P) = 2

Operating frequency (f) = 50 Hz

Inertial constant (H) = 5 kW-sec/MVA

Mechanical power input (P_i) = 6500 HP

$= 65{,}000 \times 0.735$ kW = 47,775 kW

Electrical power output (P_e) = 40 MW = 40,000 kW.

Now, Synchromous speed of the roter (W_s)

$$= \frac{4\pi f}{P} \text{ rad/sec.} = \frac{4\pi \times 50}{2}$$

$$= 100\,\pi \text{ rad/sec} = 314 \text{ rad/sec}$$

The kinetic energy stored in the rotor at synchronous speed is given by

K.E. stored = $H \times$ rating of the machine

$= H \times G = 5 \times 50 = 250$ MJ **Ans.**

We know that, accelerating power $(P_a) = P_i - P_e$

$= 47{,}775 - 40{,}000 = 7{,}775$ MW

$$\text{Here, } M = \frac{G_H}{\pi f} \text{ MJ-sec/rad} = \frac{50 \times 5}{\pi \times 50} \text{ MJ-sec/rad}$$

$$= 1.592 \text{ MJ-sec/rad}$$

To calculate acceleration of the rotor, apply the swing equation

$$M\frac{d^2\delta}{dt^2} = P_a$$

$$\therefore \text{ Acceleration } \left(\frac{d^2\delta}{dt^2}\right) = \frac{P_a}{M}$$

$$= \frac{7.775}{1.592} = 4.884 \text{ rad/sec}^2$$

$$= 279.82 \text{ electrical dgree/sec}^2$$ **Ans.**

Now, we have, $\dfrac{d^2\delta}{dt^2} = 4.884$

Multiplying both sides by $2\dfrac{d\delta}{dt}$, we get

$$2\frac{d\delta}{dt} \cdot \frac{d^2\delta}{dt^2} = 9.768\ \frac{d\delta}{dt}$$

Integrating both sides with respect to t,

$$\left(\frac{d\delta}{dt}\right)^2 = 9.768\ \delta + A$$

At $t = 0$, $\delta = 0$, $\therefore A = 0$

$$\therefore \left(\frac{d\delta}{dt}\right)^2 = 9.768\ \delta$$

This yields $\dfrac{d\delta}{dt} = \sqrt{9.768} \times \sqrt{\delta}$

$$\text{or} \quad \frac{d\delta}{dt} = 3.125\sqrt{\delta} \qquad \text{...........(1)}$$

Again $\dfrac{d\delta}{dt} = 3.125\sqrt{\delta}$ or $\dfrac{d\delta}{\sqrt{\delta}} = 3.125\ t$

Integrating both sides, $2 \times \sqrt{\delta} = 3.125\ t$

Squaring both sides, we get

$$\delta = 2.441\ t^2 \qquad \text{................(2)}$$

Again, Given,

No. of cycle (n) = 10, f = 50 Hz

$\therefore$ Time taken to complete 10 cycles $t = \dfrac{10}{50} = 0.2$ sec

During 10 sec,

Change in rotor angle ($\Delta\delta$) = $2.414 \times (0.2)^2$

= 0.0977 rad = 5.595 degree **Ans.**

To calculate rpm at the end of 10 cycles,

From Eq. (1) $\dfrac{d\delta}{dt} = 3.125\ \sqrt{\delta}$

$= 3.125\ \sqrt{0.0977}$

$= 0.9768$ rad/sec $= \dfrac{0.9768 \times 60}{2\pi}$ rpm = 9.33 rpm

Revolution per minute at the end of 10 cycles

$$= \frac{120f}{P} + 9.33 = \frac{120 \times 50}{2} + 9.33$$

$$= 3009.33 \text{ rpm} \quad \textbf{Ans.}$$

Problem 14.7. Given a generator and a system with reactance of X_a = 1.5 p.u. and X_e = 0.2 p.u. both on 100 MVA base. Assume a generator terminal voltage of 0.96 p.u. Determine the centre and radius for the pull out curve and also the minimum permissible output vars when the output power is zero.

Solution: Given,

X_d = 1.5 p.u., X_e = 0.2 p.u.

Generator terminal voltage (V) = 0.96 p.u.

Base MVA = 100 MVA

The coordinates of the centre is given by

$$= (P_c, \theta_c) = \left(0, \frac{V^2}{2}\left[\frac{1}{X_e} - \frac{1}{X_d}\right]\right)$$

$$= \left(0, \frac{(0.96)^2}{2}, \left[\frac{1}{0.2} - \frac{1}{1.5}\right]\right) = (0,\ 1.997) \quad \textbf{Ans.}$$

Now, Radius of circle (r) = $\frac{V^2}{2}\left[\frac{1}{X_d} + \frac{1}{X_e}\right]$

$$= \frac{(0.96)^2}{2}\left[\frac{1}{1.5} + \frac{1}{0.2}\right]$$

$$= 2.61 \text{ p.u.} \quad \textbf{Ans.}$$

Problem 14.8. A generator operating at 50 Hz delivers 1 p.u. power to an infinite bus when a fault occurs which reduces the maximum power transferable to 0.4 p.u. where as the maximum power transferable before the fault was 1.75 p.u. and is 1.25 p.u. after the fault is cleared. Determine the critical clearing angle. If the inertia constant (H) of the generator is 4 p.u., determine the critical clearing time of the breakers.

Solution: The power angle curve for the given situations is as given in Fig. P. 14.8.

For point a an curve A,

$$P_2 = P_{max} \sin \delta_0 \text{ .....(i)}$$

or $\quad 1 = 1.75 \sin \delta_0$

or $\quad \text{Sin } \delta_0 = \dfrac{1}{1.75}$

This yeilds

$$\delta_0 = \text{Sin}^{-1}\left(\frac{1}{1.75}\right)$$

$$= 0.6082 \text{ radian}$$

$$= 34.85°$$

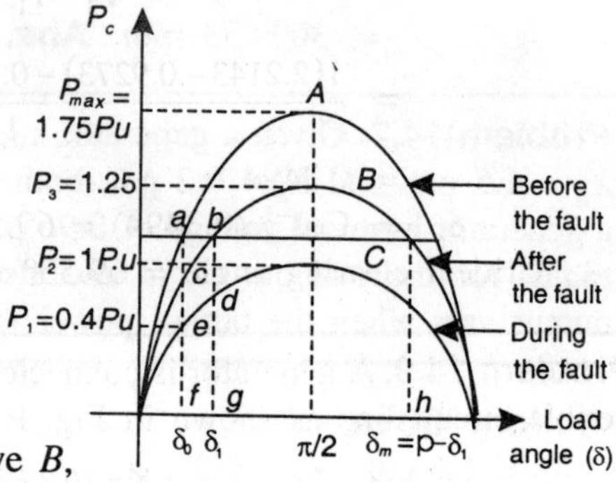

Fig. P. 14.8

Again, At point *b* on curve *B*,

$$P_2 = P_3 \sin \delta_1$$

or $P_{max} \sin \delta_0 = P_3 \sin \delta_1$

or $1.75 \times \dfrac{1}{1.75} = 1.25 \sin \delta_1$

or $\quad \sin \delta_1 = \dfrac{1}{1.25}$

This yeilds $\delta_1 = \sin^{-1}\left(\dfrac{1}{1.25}\right) = 53.10° = 0.9273$

Now, $\sin = \pi - \delta_1 = 3.1416 - 0.9273 = 2.2143 \text{ rad} = 126.87°$

To calculate minimum permissible output vars, apply the relation

$$P^2 + \left\{Q - \frac{V^2}{2}\frac{(X_d - X_e)}{X_d \times Xe}\right\}^2 = \left\{\frac{V^2}{2}\frac{(X_d + X_e)}{X_d \times X_e}\right\}^2$$

It is given, $\quad P = 0$

$$\left\{Q - \frac{V^2}{2}\frac{(X_d - X_e)}{X_d \times Xe}\right\}^2 = \left\{\frac{V^2}{2}\frac{(X_d + X_e)}{X_d \times X_e}\right\}^2$$

$(\theta - 1.997)^2 = (2.61)^2$

This yields

$\theta - 1.997 = \pm 2.61$ or $\theta = \pm 2.61 + 1.997$

$= 4.607$ or -0.613

For minimum vars, Neglecting the vars

$\therefore$ Minimum permissible vars = −0.613 p.u. **Ans.**

The critical clearing angle can be calculated by using the relation

$$\text{Cos } \delta_c = \frac{P_5(\delta_m - \delta_0) - P_1 \cos \delta_0 + P_3 \cos \delta_m}{P_3 - P_1}$$

$$= \frac{1(2.2143 - 0.9273) - 0.4 \cos 53.13° + 1.25 \cos 126.87°}{1.25 - 0.4}$$

$$= 0.3494$$

$\therefore \quad \delta_c = \text{Cos}^{-1}(0.3494) = 69.55°$

$\therefore$ Critical clearing angle = 69.55° **Ans.**

Problem 14.9. A generator is connected to an infinite bus through double circuit line as shown in Fig. P. 14.9 (a).

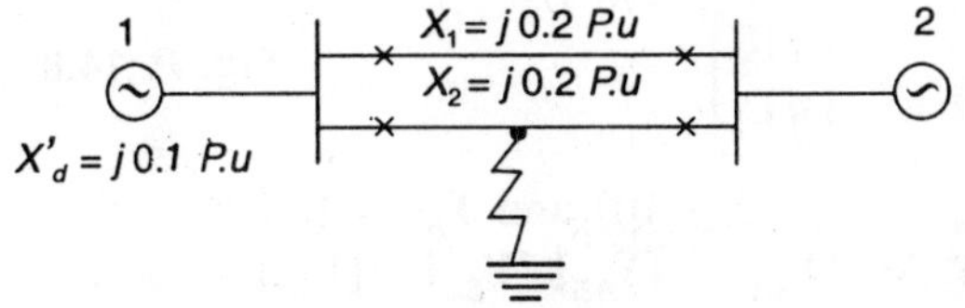

Fig. P. 14.9 (a)

Complete the prefault, faulted and post-fault reduced Y matrices. Suppose a three-phase fault takes place at the mid point of one feeder and the breaker operates after certain time.

Solution: Given, reactance of the generator (X'_d) = 0.1 p.u.

$\therefore$ Admittance of the generator $= \dfrac{1}{0.1} = 10$ ℧

Reactance of the each feeder ($X_1 = X_2$) = 0.2 p.u.

$\therefore$ Admittance of each feeder $= \dfrac{1}{0.2} = 5$ ℧

(i) Prefault condition:

The equivalent circuit before fault is as shown in Fig. P. 14.9 (b).

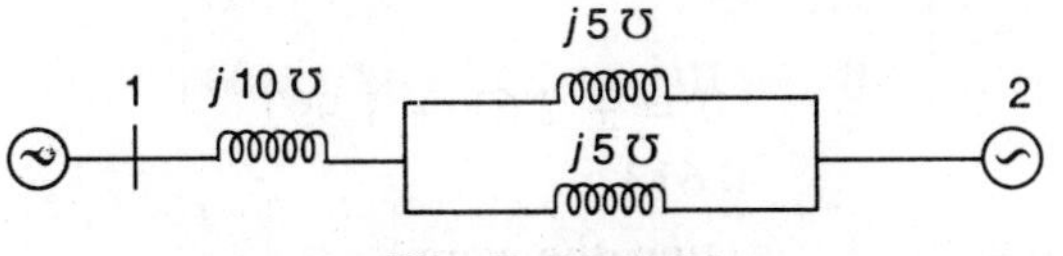

Fig. P. 14.9 (b)

The full Y– matrix is given by

$$= \left[\begin{array}{cc|c} -j10 & 0 & j10 \\ 0 & -j10 & j10 \\ \hline j10 & j10 & -j20 \end{array}\right]$$

Comparing this with $\begin{bmatrix} Y_{AA} & Y_{AB} \\ Y_{BA} & Y_{BB} \end{bmatrix}$

This yields

$$Y_{AA} = \begin{bmatrix} -j10 & 0 \\ 0 & -j10 \end{bmatrix}$$

$$Y_{AB} = \begin{bmatrix} j10 \\ j10 \end{bmatrix}$$

$$Y_{BA} = [j\ 10, j\ 10] \text{ and } Y_{BB} = [j\ 20]$$

Now, $Y = [Y_{AA}] - [Y_{AB}]\,[Y_{BB}]^{-1}\,[Y_{BA}]$

$$= \begin{bmatrix} -j10 & 0 \\ 0 & -j10 \end{bmatrix} - \begin{bmatrix} j10 \\ j10 \end{bmatrix} [-j\ 20]^{-1}\ [j\ 10, j\ 10]$$

$$= \begin{bmatrix} -j10 & 0 \\ 0 & -j10 \end{bmatrix} - \begin{bmatrix} j10 \\ j10 \end{bmatrix} \left[\frac{1}{j20}\right] [j\ 10, j\ 10]$$

$$= \begin{bmatrix} -j10 & 0 \\ 0 & -j10 \end{bmatrix} - \begin{bmatrix} (+j10)\times(+j10) & (j10)\times(j10) \\ (+j10)\ \ (+j10) & (j10)\times(j10) \end{bmatrix}$$

$$\times \left[\frac{1}{j20}\right]$$

$$= \begin{bmatrix} -j10 & 0 \\ 0 & -j10 \end{bmatrix} - \begin{bmatrix} -100 & -100 \\ -100 & -100 \end{bmatrix} \times \frac{j}{20}$$

$$= \begin{bmatrix} -j10 & 0 \\ 0 & -j10 \end{bmatrix} + \frac{1}{20} \begin{bmatrix} j100 & j100 \\ j100 & j100 \end{bmatrix}$$

$$= \begin{bmatrix} -j10 & 0 \\ 0 & -j10 \end{bmatrix} + \begin{bmatrix} j5 & j5 \\ j5 & j5 \end{bmatrix} = \begin{bmatrix} -j5 & j5 \\ j5 & -j5 \end{bmatrix} \quad \textbf{Ans.}$$

(ii) When the fault takes place, the equivalent circuit is as follow

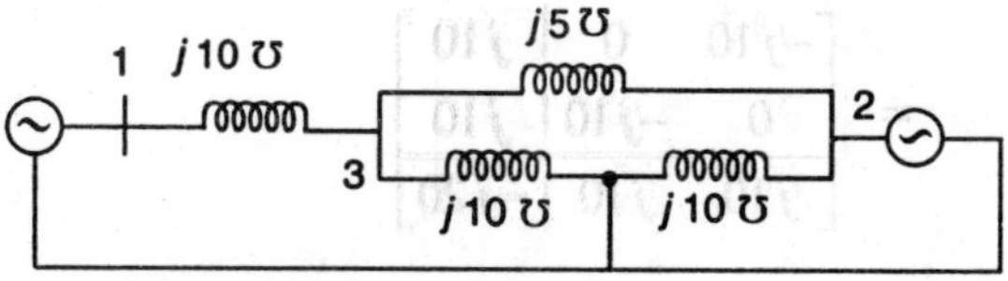

Fig. P. 14.9 (c)

The full Y–matrix is given by = $\begin{bmatrix} -j10 & 0 & j10 \\ 0 & -j15 & j5 \\ j10 & j5 & -j25 \end{bmatrix}$

Companying this with = $\begin{bmatrix} Y_{AA} & Y_{AB} \\ Y_{BA} & Y_{BB} \end{bmatrix}$

This yields

$$[Y_{AA}] = \begin{bmatrix} -j10 & 0 \\ 0 & -j15 \end{bmatrix}, [Y_{AB}] = \begin{bmatrix} j10 \\ j5 \end{bmatrix}$$

$$[Y_{BA}] = [j\ 10, j\ 5], [Y_{BB}] = [-j\ 25]$$

Now, Reduced Y matrix is given by

$$[Y] = [Y_{AA}] - [Y_{AB}]\ [Y_{BA}] = [Y_{BB}]^{-1}$$

$$= \begin{bmatrix} -j10 & 0 \\ 0 & -j15 \end{bmatrix} - \begin{bmatrix} j10 \\ j5 \end{bmatrix} [j\ 10, j\ 5]\ [-j\ 25]^{-1}$$

$$= \begin{bmatrix} -j10 & 0 \\ 0 & -j15 \end{bmatrix} - \begin{bmatrix} -100 & -50 \\ -50 & -25 \end{bmatrix} \begin{bmatrix} \frac{j}{25} \end{bmatrix}$$

$$= \begin{bmatrix} -j10 & 0 \\ 0 & -j15 \end{bmatrix} + \frac{1}{25} \begin{bmatrix} j100 & j50 \\ j50 & j25 \end{bmatrix}$$

$$= \begin{bmatrix} -j10 & 0 \\ 0 & -j15 \end{bmatrix} + \begin{bmatrix} j4 & j2 \\ j2 & j1 \end{bmatrix} = \begin{bmatrix} -6j & j2 \\ j2 & -j14 \end{bmatrix}$$ **Ans.**

(iii) For post fault

The equivalent circuit under this condition is given by

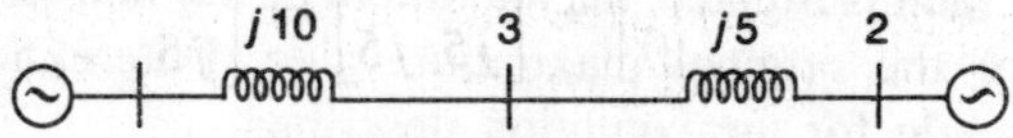

Fig. P. 14.9 (d)

The full matrix for this circuit is given by

$$\begin{bmatrix} -j10 & 0 & j10 \\ 0 & -j5 & j5 \\ j10 & j5 & -j15 \end{bmatrix}$$

Companying this with the matrix $\begin{bmatrix} Y_{AA} & Y_{AB} \\ Y_{BA} & Y_{BB} \end{bmatrix}$

This yields

$$[Y_{AA}] = \begin{bmatrix} -j10 & 0 \\ 0 & -j5 \end{bmatrix}, \ [Y_{AB}] = \begin{bmatrix} j10 \\ j5 \end{bmatrix}$$

$$[Y_{BA}] = [j10, j5], \ Y_{BB} = [-j\ 15]$$

The reduced Y matrix is given by

$$[Y] = [Y_{AA}] - [Y_{AB}]\ [Y_{BA}]\ [Y_{BB}]^{-1}$$

$$= \begin{bmatrix} -j10 & 0 \\ 0 & -j5 \end{bmatrix} - \begin{bmatrix} j10 \\ j5 \end{bmatrix} [j\ 10, j\ 5]\ [-j\ 15]^{-1}$$

$$= \begin{bmatrix} -j10 & 0 \\ 0 & -j5 \end{bmatrix} - \begin{bmatrix} j10 \times j10 & j10 \times j5 \\ j5 \times j10 & j5 \times j5 \end{bmatrix} \begin{bmatrix} \frac{j}{15} \end{bmatrix}$$

$$= \begin{bmatrix} -j10 & 0 \\ 0 & -j5 \end{bmatrix} - \begin{bmatrix} -100 & -50 \\ -50 & -25 \end{bmatrix} \times \frac{j}{15}$$

$$= \begin{bmatrix} -j10 & 0 \\ 0 & -j5 \end{bmatrix} + \frac{1}{15} \begin{bmatrix} j100 & j50 \\ j50 & j25 \end{bmatrix}$$

$$= \begin{bmatrix} -j10 & 0 \\ 0 & -j5 \end{bmatrix} + \begin{bmatrix} -j6.67 & j3.33 \\ j3.33 & -j1.67 \end{bmatrix}$$

$$= \begin{bmatrix} -j3.33 & j3.33 \\ j3.33 & -j3.33 \end{bmatrix}$$ **Ans.**

Problem 14.10. A 50 Hz generator is delivering 50% of the power that it is capable of delivering through a transmission line to an infinite bus. A fault occurs that increases the reactance between the generator and the infinite bus to 400% of the value before the fault. When the fault is isolated, the maximum power that can be delivered is 80% of the original maximum value. Determine the critical clearing angle for the condition described.

Solution: The power – angle relation is

$$P_e = \frac{V_1V_2}{\pi} \sin \delta \qquad \text{............(1)}$$

Let P_m be the maximum power that can be transferred from generator to bus.

$$\therefore \quad P_e = P_m \sin \delta_0 \qquad \text{............ (2)}$$

Let δ_0 be the load angle when the power transferred is 50% of the maximum power, ie. P_s = 50% of P_m

$\therefore$ 50% $\times P_m = P_m \sin \delta_0$

or $0.5\ P_m = P_m \sin \delta_0$

This yields

$$\delta_0 = 30° = 0.5236 \text{ rad}$$

During the fault, $P_e = \frac{P_m}{400} \sin \delta = 0.25\ P_m \sin \delta$

This relation yields

$$\gamma_1 = 0.25$$

Let δ_m be the load angle when power transferred is 80% of the maximum power

$$\therefore \quad P_e = 0.80\ P_m \sin \delta_m$$

This yields $\gamma_2 = 0.80$

Also, We have, At point b

$$0.5\ P_m = 0.80\ P_m \sin \delta_1$$

or $\sin \delta_1 = \frac{5}{8}$

This yields

$$\delta_1 = \sin^{-1}\left(\frac{5}{8}\right)$$

$$= 0.6751 \text{ rad} = 38.68°$$

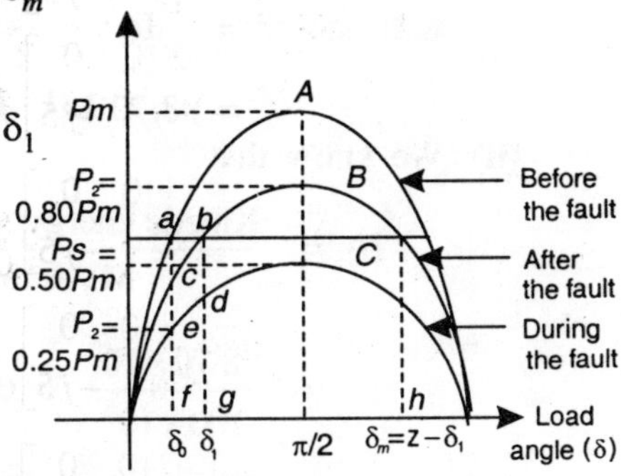

Fig. P. 14.10

Here, $\delta_m = 180° - 38.68°$

$$= 141.32 = 21.465 \text{ radian}$$

Now, the critical clearing angle δ_C is given by

$$\delta_C = \cos^{-1} \frac{(P_s / P_m)(\delta_m - \delta_0) + \gamma_2 \cos \delta_m - \gamma_1 \cos \delta_0}{r_2 - r_1}$$

$$= \cos^{-1} \left[\frac{(0.5P_m / 0.8P_m)(2.465 - 0.5236) + 0.80 \cos 141.32° - 0.25 \cos 30°}{0.80 - 0.25} \right]$$

$$= \cos^{-1} (0.684) = 46.89°$$

$\therefore$ Critical clearing angle (δ_c) = 46.89 **Ans.**

Problem 14.11. A 4 pole 50 Hz 100 MVA turbo generator has a moment of inertia of 8×10^3 kg-m^2. calculate
(ii) The kinetic energy is MJ at rated speed
(iii) The inertia constants M and H
(iv) and the inertia constant on 50 MVA Base.

Solution: Given,

MVA rating = 100 MVA
Frequency (f) = 50 Hz
No of poles (p) = 4
Moment of inertia (I) = 8×10^3 kg-m^2

$$\text{Speed of the rotor } (\omega) = \frac{4\pi f}{p} \text{ rad/sec}$$

$$= \frac{4 \times \pi \times 50}{4} \text{ rad/sec} = 50\,\pi \text{ rad/sec}$$

$$= 157.08 \text{ rad/sec}$$

(i) To calculate kinetic energy at rated speed, apply the relation:

$$\text{K.E. stored} = \frac{1}{2} I\omega^2 = \frac{1}{2} \times 8 \times 10^3 \times (157.08)^2$$

$$= 98.70 \text{ MJ} \quad \textbf{Ans.}$$

(ii) We know that

$$H = \frac{\text{Kinetic energy stored in MJ}}{\text{MVA rating}}$$

$$= \frac{98.70 \times 10^6}{100 \times 10^6} = 0.987 \text{ MJ/MVA} \quad \textbf{Ans.}$$

$$M = \frac{\text{Kinetic energy stored in MJ}}{180 \times f}$$

$$= \frac{98.70}{180 \times 50} = 0.0109 \text{ MJS/electrical degree} \quad \textbf{Ans.}$$

(iii) The inertia constant on the base 50 MVA is given by

$$\text{p.u. inertia constant} = \frac{M}{\text{Base MVA}}$$

$$= \frac{0.0109}{50} \text{ p.u.}$$

$$= 2.193 \times 10^{-4} \text{ p.u.} \quad \textbf{Ans.}$$

Problem 14.12. A loss-free generator supplies 60 MW to an infinite bus, the steady state limit of the system being 120 MW. Determine whether the generator will remain in synchronism if the prime mover input is abruptly increased by 40 MW.

Solution: The power–angle curve of this problem can be drawn as in Fig. P. 14.12.

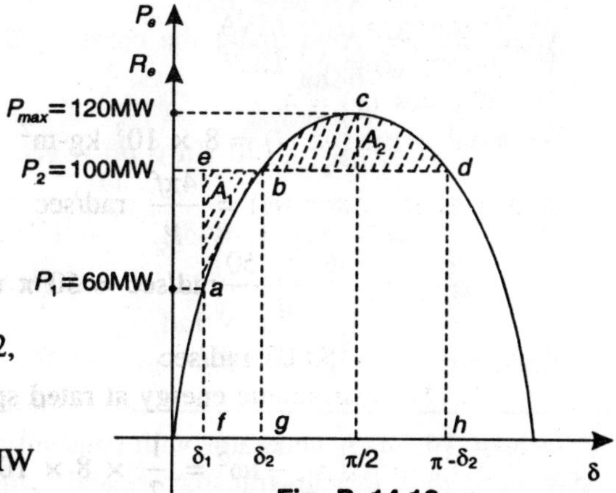

Fig. P. 14.12

From Fig. P. 14.12,

P_1 = 60 MW

P_2 = 100 MW

and P_{max} = 120 MW

The equation of the power–angle curve is given by

$P_e = P_{max} \sin \delta$, $P_e = 120 \sin \delta$ .... (1)

From curve, the initial operating point is a. Let δ_1 be the initial load angle.

At point *a*,

$P_1 = 120 \sin \delta$ or $60 = 120 \sin \delta$

This yields

$$\delta_1 = \sin^{-1}\left(\frac{1}{2}\right),\ \delta_1 = 30° \text{ or } \delta_1 = 0.523 \text{ rad}$$

Let δ_2 be the load only for operating point b,

Hence, $P_2 = 120 \sin \delta_2$ or $100 = 120 \sin \delta_2$

This yields

$$\delta_2 = \sin^{-1}\left(\frac{100}{120}\right) = 56.4° = 0.9851 \text{ rad}$$

For the generator to be in synchronism, area A_1 should be less than or equal to area A_2.

Now, Let us calculate areas A_1 and A_2.

Area A_1 = Area of reactangle befg – Area under the curve from *a* to *b*

$$= eg \times ef - \int_{\delta_1}^{\delta_2} P_e d\delta$$

$$= 100 \times (\delta_2 - \delta_1) - \int_{\delta_1}^{\delta_2} 120 \sin \delta \, d\delta$$

$$= 100 \times (\delta_2 - \delta_1) + 120 \left[\cos \delta\right]_{\delta_1}^{\delta_2}$$

$$= 100 \times (\delta_2 - \delta_1) + 120 [\cos \delta_2 - \cos \delta_1]$$

$$= 100 \times (0.9851 - 0.523) + 120 [\cos 56.4° - \cos 30°]$$

$$= 8.7 \text{ MW rad}$$

Area A_2 = Area under the curve b to d – Area of rectangle bdhg

$$= \int_{\delta_2}^{180-\delta_2} 120 \sin \delta \, d\delta - gH \times bg$$

$$= -120 \left[\cos \delta\right]_{56.4°}^{123.6°} - 100 (\pi - \delta_2 - \delta_2)$$

$$= -120 [\cos 123.6° - \cos 56.4] - 100 [\pi - 2 \times 0.9851]$$

$$= 15.83 \text{ MW rad.}$$

Since area A_2 is greater than area A_1, the system is stable **Ans.**

Problem 14.13. A generator with constant excitation supplies 50 MW through a step-up transformer and a high voltage line to an infinite bus bar. If the steady-state stability of the system is 100 MW. Estimate the maximum permissible sudden increase of generator, if the stability is to be maintained. The resistance of the generator, transformer and the line may be neglected.

Solution: Given,

P_{max} = 100 MW, P_1 = 50 MW

The power–angle curve of this problem is as shown in Fig. P. 14.13.

Let δ_1 be the initial load angle

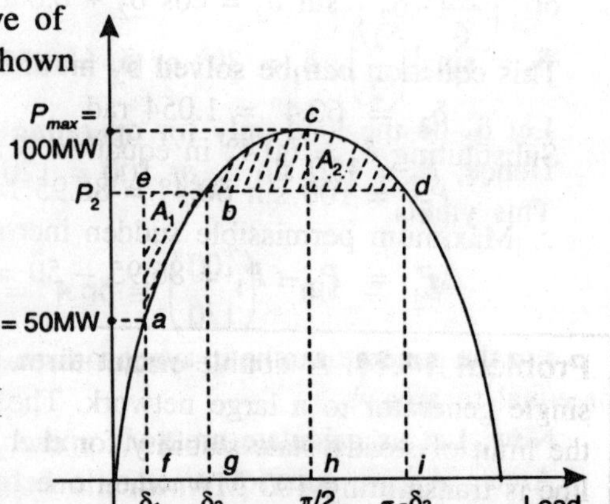

Fig. P. 14.13

At parts a,

$P_1 = P_{max} \times \sin \delta_1$

or $50 = 100 \sin \delta_1$

This yields

$\delta_1 = 30° = 0.523$ rad

The equation of the power-angle curve is given by

$$P_e = 100 \sin \delta \quad \text{.................(1)}$$

Let P_2 be the maximum load on the generator due to sudden increase of generator output and δ_2 be the corresponding load angle.

$$P_2 = 100 \sin \delta_2 \quad \text{..............} (2)$$

For stability to be maintained,

Area A_1 = Area A_2

or $$\int_{\delta_1}^{\delta_2} (P_2 - 100 \sin \delta)\, d\delta = \int_{\delta_2}^{\pi-\delta_2} (100 \sin \delta - P_2)\, d\delta$$

or $$\int_{\delta_1}^{\delta_2} (P_2 - 100 \sin \delta)\, d\delta = -\int_{\delta_2}^{\pi-\delta_2} (P_2 - 100 \sin \delta)\, d\delta$$

Using integration property,

$$\int_{\delta_1}^{\pi-\delta_2} (P_2 - 100 \sin \delta)\, d\delta = 0$$

$$[P_2\, \delta_2 + 100 \cos \delta]_{\delta_1}^{\pi-\delta_2} = 0$$

$$P_2 (\pi - \delta_2 - \delta_1] + 100 [\cos (\pi - \delta_2) - \cos \delta] = 0$$

$$P_2 (\pi - \delta_2 - \delta_1] + 100 [-\cos \delta_2 - \cos \delta_1] = 0$$

or $$[P_2\, \delta + 100 \cos \delta]_{\delta_2}^{\pi-\delta_2} = 0$$

From equation (2)

or $$100 \sin \delta_2 \left(\pi - \delta_2 - \frac{\pi}{6}\right) - 100 (\cos \delta_2 + \cos \delta_1) = 0$$

or $$\sin \delta_2 \left(\frac{5\pi}{6} - \delta_2\right) - \cos \delta_2 - 100 \cos \delta_1 = 0$$

or $$\left(\frac{5\pi}{6} - \delta_2\right) \sin \delta_2 = \cos \delta_2 + 1 \times 0.866$$

or $$\left(\frac{5\pi}{6} - \delta_2\right) \sin \delta_2 = \cos \delta_2 + 0.866$$

This equation can be solved by hit and trial method. This yields

$$\delta_2 = 60.4° = 1.054 \text{ rad.}$$

Substituting $\delta_2 = 60.4°$ in equation (2), this yields

$$P_2 = 100 \sin 60.4° = 86.95 \text{ MW}$$

$\therefore$ Maximum permissible sudden increase of load

$$\Delta P = P_2 - P_1 = 86.95 - 50 = 36.95 \text{ MW}$$ **Ans.**

Problem 14.14. A double-circuit three phase feeder connects a single generator to a large network. The power corresponding to the limit of steady state stability for each circuit is 150 MW. The line is transmitting 100 MW when one of the circuits is switched out suddenly. Determine with reference to appropriate diagram whether the generator is likely to vermin in synchronism.

Solution: Given

Steady-state stability limit for each circuit (P_{2max}) = 150 MW

$\therefore$ Steady-state stability limit for both circuit

$$P_{max} = 2 \times 150 = 300 \text{ MW}$$

When one of the circuit is switched out, the power—angle curve is given by

$$P_{e2} = P_{2max} \sin \delta, P_{e2} = 150 \sin \delta \qquad \text{.............(1)}$$

When both circuits are being operated.

$$P_{e1} = 300 \sin \delta \qquad \text{................ (2)}$$

The power—angle curve for P_e both the conditions can be drawn as in Fig. 14.14.

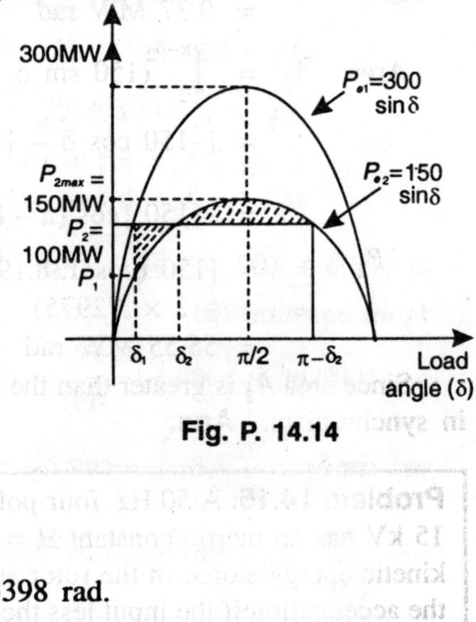

Fig. P. 14.14

Let δ_1 be the initial load angle when both the circuits are being operated

$$\therefore \quad P_2 = 300 \sin \delta_1$$

$$\text{or} \quad 100 = 300 \sin \delta_1$$

This yields

$$\delta_1 = \text{Sin}^{-1}\left(\frac{1}{3}\right)$$

$$= 19.47° = 0.3398 \text{ rad.}$$

From curve,

$$P_2 = P_{2max} \sin \delta_2$$

$$\text{or} \quad 100 = 150 \sin \delta_2$$

This yields

$$\delta_2 = \sin^{-1}\left(\frac{10}{15}\right)$$

$$= 0.7297 \text{ rad} = 41.81°$$

$$\text{and} \quad \pi - \delta_2 = 180° - 41.81° = 138.19°$$

$$= 3.14 - 0.7297 = 2.403 \text{ rad}$$

Again, from curve 2, P_1 can be calculated by using the relation

$$P_{e2} = P_{max} \sin \delta \text{ or } P_1 = 150 \sin \delta_1$$

$$= 150 \sin 19.47° = 50 \text{ MW}$$

For the system to remain in synchronism, Area $A_1 \leq$ Area A_2

Now,

$$\text{Area } A_1 = \int_{\delta_1}^{\delta_2} (P_2 - P_{2max} \sin \delta)\, d\delta$$

$$= \int_{\delta_1}^{\delta_2} (100 - 150 \sin \delta)\, d\delta$$

$$= [100\, \delta + 150 \cos \delta]_{\delta_1}^{\delta_2}$$

$$= 100\, [\delta_2 - \delta_1] + 150\, [\cos \delta_2 - \cos \delta_1]$$

$$= 100\, (0.7297 - 0.3398) + 150\, [\cos 41.81^\circ - \cos 19.47^\circ]$$

$$= 9.37 \text{ MW rad}$$

$$\text{Area } A_2 = \int_{\delta_2}^{\pi-\delta_2} (150 \sin \delta - 100)\, d\delta$$

$$= [-150 \cos \delta - 100\, \delta]_{\delta_2}^{\pi-\delta_2}$$

$$= -[150\, \{\cos (\pi - \delta_2) - \cos \delta_2\}] + 100\, (\pi - \delta_2 - \delta_2)$$

$$= [150\, (\cos 138.19^\circ - \cos 41.81^\circ) + 100\, (3.14 - 2 \times .72975)$$

$$= 55.55 \text{ MW rad}$$

Since area A_2 is greater than the area A_1, the generator will remain in synchronism. **Ans.**

Problem 14.15. A 50 Hz four-pole turbogenerator rated 25 MVA, 15 kV has an inertia constant H = 10 kW-sec/kVA. Determine the kinetic energy stored in the rotor at synchronous speed. Determine the acceleration if the input less the rotational loss is 30000 HP and the electric power developed in 20000 kW. If the acceleration computed for the generator is constant for a period of 16 cycles, determine the change in torque angle in that period and the rpm at the end of 16 cycles. Assume that the generator is synchronized with a large system and has no accelerating torque before the 16 cycle period begins.

Solution: Given,

Rated MVA (G) = 25 MVA

No. of poles of the generator (P) = 4

Operating frequency (f) = 52 Hz, H = 10 kW-Sec/MVA

Mechanical power input (P_i) = 30000 Hp

= 30,000 × 0.735 kW = 22050 kW

Electrical power output (P_e) = 20000 kW

Speed of the rotor = $\frac{4\pi f}{p}$ rad/sec

$$= \frac{4 \times \pi \times 50}{4} \text{ rad/sec} = 157.08 \text{ rad/sec}$$

Now, kinetic energy stored in the rotor

= H × rated MVA of the generator

= 10 × 25 = 250 megajoules **Ans.**

We know that accelerating power is given by

$$P_a = P_i - P_e = 22050 - 20000 = 2050 \text{ kW} = 2.050 \text{ MW}$$

Here, $M = \frac{GH}{\pi f} = \frac{25 \times 10}{\pi \times 50}$ megajoules-sec/radian

= 1.592 megajoules-sec/radian

Now, to calculate acceleration, apply the swing equation.

$$M\frac{d^2\delta}{dt^2} = p_a$$

or $$\frac{d^2\delta}{dt^2} = \frac{Pa}{M}$$

or $$\frac{d^2\delta}{dt^2} = \frac{2.050}{1.592} = 1.288 \text{ rad/sec}^2$$ **Ans.**

Multiplying both sides by $2\frac{d\delta}{dt}$, we get

$$2\frac{d\delta}{dt} \times \frac{d^2\delta}{dt^2} = 2.575\frac{d\delta}{dt}$$

Integrating both sides with respect to t, we get

$$\int 2\frac{d\delta}{dt} \cdot \frac{d^2\delta}{dt^2} \cdot dt = \int 2.575\frac{d\delta}{dt} . dt$$

This yields

$$\left(\frac{d\delta}{dt}\right)^2 = 2.575\,\delta + A \qquad \text{.......... (1)}$$

From boundary condition, at $t = 0$, $\frac{d\delta}{dt} = 0$

From equation (1) A = 0

Hence,

$$\frac{d\delta}{dt} = \sqrt{2.575\delta} \text{ or } \frac{d\delta}{\sqrt{\delta}} = \sqrt{2.575}\, dt$$

Integrating both sides, we get $2 \times \sqrt{\delta} = 1.605\, t$

Squaring both sides

$$\delta = \frac{(1.605)^2}{4} t^2 \text{ or } \delta = 0.644\, t^2$$

Given, no of cycle (n) = 16, f = 50 Hz

∴ Time taken to complete 16 cycles

$$= \frac{n}{f} = \frac{16}{50} = 0.32 \text{ sec}$$

During this period,

Change in load angle = $0.644 \times t^2$

$$\Delta\delta = 0.644 \times (0.32)^2 = 0.066 \text{ radian}$$

$$= 3.782 \text{ electrical degrees} \quad \textbf{Ans.}$$

To calculate rpm at the end of 16 cycle,

Now, we have

$$\frac{d\delta}{dt} = 1.605\sqrt{\Delta\delta} = 1.605\sqrt{0.066}$$

$$= 0.412 \text{ rad/sec}$$

$$= \frac{0.412}{2\pi} \text{ rev/sec} = \frac{0.412}{2\pi} \times 60 \text{ rpm}$$

$$= 3.93 \text{ rpm}$$

∴ Revolution per minute at the end of 16 cycles

$$= \frac{120f}{p} + 3.93$$

$$= \frac{120 \times 50}{4} + 3.93 = 1503.93 \text{ rpm} \quad \textbf{Ans.}$$

Problem 14.16. A 50 Hz synchronous generator is connected to an infinite bus through a line. The p.u. reactance of generator and the line are j 0.4 p.u. and j 0.2 respectively

The no-load voltage of the generator is 1.2 p.u. and that of infinite bus is 1.0 p.u. The inertia constant of the generator is 4 MW-sec/MVA. Determine the frequency of natural oscillations if the generator is loaded to (i) 50% and (ii) 80% of its maximum power transfer capacity and small perturbation in power is given.

Solution: The single-line diagram of the given problem is as given in Fig. P. 14.16.

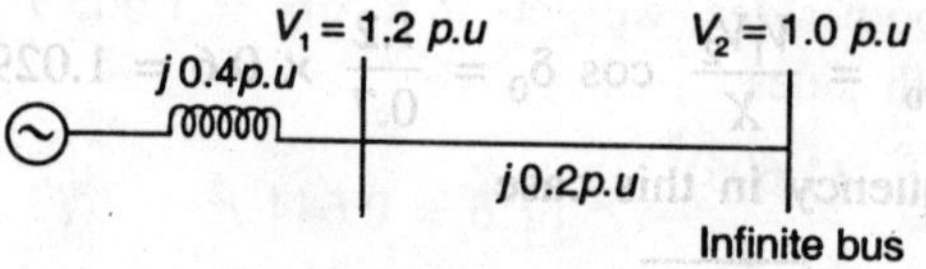

Fig. P. 14.16

Given, H = 4 MW-sec/MVA

Reactance of the generator $(X_g) = j\ 0.4$ p.u.

Reactance of the line $(X_l) = j\ 0.3$ p.u.

Total reactance between generator and infinite bus

$$X = X_g + X_l = j\ 0.4 + j\ 0.3 = j\ 0.7$$

Since the system is operating initially under steady state condition, a small perturbation in power will make the rotor oscillate. The natural frequency of oscillation is given by

$$f_n = \left\{\left(\frac{\partial pe}{\partial \delta}\right)_{\delta_0} \Big/ M\right\}\frac{1}{2}$$

(i) For 50% loading, sin δ_0 = 0.50 yields δ_0 = 30°

Now, the power—angle equation is given by

$$P_e = \frac{V_1 V_2}{X} \sin \delta$$

Differentiating it partially with respects to δ,

$$\frac{\partial p_e}{r\delta} = \frac{V_1 V_2}{X} \cos \delta_0$$

At δ_0 = 30°

$$\left(\frac{\partial p_e}{r\delta}\right) \delta_0 = \frac{1.2 \times 1.0}{0.7} \cos 30° = 1.485$$

$$\text{Here, } M = \frac{GH}{\pi f} = \frac{4 \times 1}{50\pi} = \frac{4}{50\pi}$$

$$= 0.0255 \text{ MJ-sec/rad}$$

$$\text{Natural frequency } (f_n) = \sqrt{\frac{\left.\frac{\partial p_e}{r\delta}\right|_{\delta_0}}{M}}$$

$$= \sqrt{\frac{1.485}{0.0255}} = 7.63 \text{ rad/sec.}$$

$$= \frac{7.63}{2\pi} \text{ rev/sec.} = 1.214 \text{ Hz}$$ **Ans.**

(ii) For 80% loading, $\sin \delta_0 = 0.8$ yields $\cos \delta_0 = 0.6$

Here, $\left.\dfrac{\partial p_e}{r\delta}\right|_{\delta_0} = \dfrac{V_1 V_2}{X} \cos \delta_0 = \dfrac{1.2}{0.7} \times 0.6 = 1.029$

Natural frequency in this case

$$(f_n) = \sqrt{\frac{\left.\dfrac{\partial p_e}{r\delta}\right|_{\delta_0}}{M}} = \sqrt{\frac{1.029}{0.0255}}$$

$$= 6.35 \text{ rad /sec}$$

$$= \frac{6.35}{2\pi} \text{ rev/sec.} = 1.01 \text{ Hz}$$ **Ans.**

Chapter 15

Economic Operation of Power System and Load Characteristics

Problem 15.1. Calculate the power loss in the system given below by the loss coefficients and by $I_1^2 R$ for $I_1 = 1.5 \angle 0°$ p.u. and $I_2 = 1.2 \angle 0°$ Assume $V_3 = 1.0 \angle 0°$ p.u.

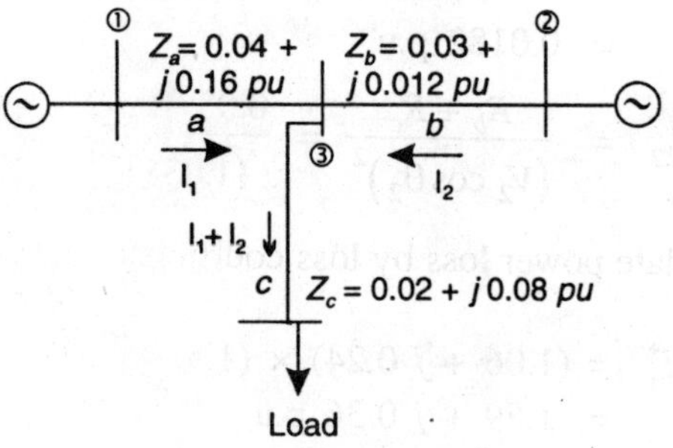

Fig. 15.1

Solution: Given

$$I_1 = 1.5\angle 0° = 1.5 + j\ 0.0 \text{ p.u.}$$
$$I_2 = 1.2\angle 0° = 1.2 + j\ 0.0 \text{ p.u.}$$
$$V_3 = 1.0\angle 0° = 1.0 + j\ 0.0 \text{ p.u.}$$
$$Z_a = 0.04 + j\ 0.16 \text{ p.u.} = R_a + j\ X_a \text{ p.u.}$$
$$Z_b = 0.03 + j\ 0.12 \text{ p.u.} = R_b + j\ X_b \text{ p.u.}$$

and $Z_c = 0.02 + j\ 0.08 \text{ p.u.} = R_c + j\ X_c$ p.u.

Now, Let us calculate voltage at buses 1 and 2.

The voltage at bus 1 is given by

$$V_1 = V_3 + I_1 Z_a$$
$$= 1.0 + j\ 0.0 + (1.5 + j\ 0.0)\ (0.04 + j\ 0.16)$$
$$= 1.06 + j\ 0.24 \text{ p.u.}$$

and
$$V_2 = V_3 + I_2 Z_b$$
$$= 1.0 + j\ 0.0 + (1.2 + j\ 0.0)\ (0.03 + j\ 0.12)$$
$$= 1.036 + j\ 0.144 \text{ p.u.}$$

Since all the currents have phase angles zero, the power factor at each source node is the cosine of the angle of the voltage at the node, and the voltage magnitude times the power factor equals the real part of the complex expression for the voltages V_1 and V_2.

Therefore from the expression of V_1

$V_1 \cos \theta_1 = 1.06$ p.u.

And from the expression of V_2

$V_2 \cos \theta_2 = 1.036$

Now, the loss coefficients are given by

$$B_{11} = \frac{R_a + R_c}{(V_1 \cos \theta_1)^2} = \frac{0.04 + 0.02}{(1.06)^2}$$

$$= 0.0534 \text{ p.u.}$$

$$B_{12} = \frac{R_c}{(V_1 \cos \theta_1) \times V_2 \cos \theta_2} = \frac{0.02}{1.06 \times 1.036}$$

$$= 0.0182 \text{ p.u.}$$

and $$B_{22} = \frac{R_b + R_c}{(V_2 \cos \theta_2)^2} = \frac{0.03 + 0.02}{(1.036)^2} = 0.0466 \text{ p.u.}$$

To calculate power loss by loss coefficient method, let us calculate P_1 and P_2

$$V_1 I_x^* = (1.06 + j\,0.24) \times (1.5 - j\,0.0)$$
$$= 1.59 + j\,0.36 \text{ p.u.}$$

This yields

$$P_1 = 1.59 \text{ p.u.}$$

and $$V_2 I_2^* = (1.036 + j\,0.144 \times (1.2 - j\,0.0)$$
$$= 1.243 + j\,0.173 \text{ p.u.}$$

This yields

$$P_2 = 1.243 \text{ p.u.}$$

The power loss by loss coefficient method is given by

$$P_L = P_1^2 B_{11} + 2 P_1 P_2 B_{12} + P_2^2 B_{22}$$
$$= (1.59)^2 \times 0.0534 + 2 \times 1.59 \times 1.243 \times 0.0182 + (1.243)^2 \times 0.0466$$
$$= 0.02789 \text{ p.u.}$$ **Ans.**

Again, the power loss by $I^2 R$ method is given by

$$P_L = 3\,[I_1^2 (R_a + R_c) + 2 I_1 I_2 R_c + I_2^2 (R_b + R_c)]$$
$$= 3\,[(1.5)^2 (0.04 + 0.02) + 2 \times 1.5 \times 1.2 \times 0.02 + (1.2)^2 (0.03 + 0.02]$$
$$= 0.279 \text{ p.u.}$$

Problem 15.2. On a system consisting of two generating plants, the incremental costs in dollars per megawatt hour with P_1 and P_2 in megawatts are

$$\frac{dF_1}{dP_1} = 0.008\ P_1 + 8.0, \quad \frac{dF_2}{dP_2} = 0.012\ P_2 + 9.0$$

The system is operating on economic dispatch with $P_1 = P_2 = 500$ MW and $\frac{\partial P_1}{\partial P_2} = 0.2$. Find the penalty factor of plant 1.

Solution: Let L_1 and L_2 be the penalty factor for plants 1 and 2 respectively

Given,

$$P_1 = P_2 = 500 \text{ MW}$$

$$\frac{\partial P_L}{\partial P_2} = 0.2 = \frac{dF_1}{dP_1} = 0.008\ P_1 + 8.0$$

$$\frac{dF_2}{dP_2} = 0.012\ P_2 + 9.0$$

The penalty factor for plant 2 is given by

$$L_2 = \frac{1}{1-\frac{\partial P_L}{\partial P_2}} = \frac{1}{1-0.2} = \frac{1}{0.8} = 1.25$$

For economic dispatch

$$L_1 \frac{dF_1}{dP_1} = L_2 \frac{dF_2}{dP_2}$$

or $L_1\ (0.008\ P_1 + 8.0) = 1.25\ (0.012\ P_2 + 9.0)$

Substituting $P_1 = P_2 = 500$ MW in the above equation, this yeilds

$$L_1 = \frac{1.25\,(0.012 \times 500 + 9)}{0.008 \times 500 + 8} = 1.563$$

Penalty factor for plant 1 = 1.563 **Ans.**

Problem 15.3. A power system has only two generating plants, and power is being dispatched economically with P_1 = 140 MW and P_2 = 250 MW. The loss coefficients are:

$$B_{11} = 0.10 \times 10^{-2}\ \text{MW}^{-1}$$

$$B_{12} = -0.01 \times 10^{-2}\ \text{MW}^{-1}$$

$$B_{22} = 0.13 \times 10^{-2}\ \text{MW}^{-1}$$

To raise the total load on the system by 1 MW will cost an additional \$12 per hour. Find (a) the penalty factor or plant 1, and (b) the additional cost per hour to increase the output of this plant by 1 MW.

Solution: Given,

$$P_1 = 140 \text{ MW},$$
$$P_2 = 250 \text{ MW}, B_{11} = 0.10 \times 10^{-2} \text{ MW}^{-1}$$
$$B_{12} = -0.01 \times 10^{-2} \text{ MW}^{-1}$$
$$B_{22} = 0.13 \times 10^{-2} \text{ MW}^{-1}$$
and $\lambda = \$ 12$ per MWh

The transmission loss of the two plants power system in given by

$$P_L = P_1^2 B_{11} + 2 P_1P_2B_{12} + P_2^2B_{22} \qquad \text{........ (1)}$$

Differentiating equation (1) partially wrt P_1

$$\frac{\partial P_L}{\partial P_1} = 2P_1B_{11} + 2P_2B_{12}$$
$$= 2 \times 140 \times 0.10 \times 10^{-2} + 2 \times 250 \times (-0.01 \times 10^{-2})$$
$$= 0.23$$

(a) We have $\frac{\partial P_L}{\partial P_1} = 0.23$

The penalty factor for plant 1 is given by

$$L_1 = \frac{1}{1-\frac{\partial P_L}{\partial P_1}} = \frac{1}{1-0.23} = \frac{1}{0.77} = 1.299$$

$\therefore$ Penalty factor for plant 1 = 1.299 **Ans.**

(b) Given = \$ 12 /MWh

Let $\frac{dc_1}{dp_1}$ be additional cost per hour.

To calculate $\frac{dc_1}{dp_1}$, apply the relation $L \frac{dc_1}{dp_1} = \lambda$

or $1.299 \times \frac{dc_1}{dp_1} = 12$

$$\therefore \quad \frac{dc_1}{dp_1} = \frac{12}{0.299} = 9.238$$

Incremental cost per hour to increase the power output by 1 MW

= \$ 9.238 /MWh **Ans.**

Problem 15.4. Incremental fuel costs in Rs. per megawatt hour for two units in a plant are given by

$$\frac{dF_1}{dP_1} = 0.1\ P_1 + 20$$

$$\frac{dF_2}{dP_2} = 0.12\ P_2 + 16$$

The minimum and maximum loads on each unit are to be 20 MW and 125 MW respectively. Determine the incremental fuel cost and the allocation of load between units for the minimum cost when loads are (i) 100 MW, (ii) 150 MW. assume both the units are operating.

Solution: Given,

$$\frac{dF_1}{dP_1} = 0.1\ P_1 + 20$$

and $$\frac{dF_2}{dP_2} = 0.12\ P_2 + 16$$

For minimum cost $\frac{dF_1}{dP_1} = \frac{dF_2}{dP_2}$

or $0.1\ P_1 + 20 = 0.12\ P_2 + 16$

Multiplying both sides by 10, $P_1 + 200 = 1.2\ P_2 + 160$

or $P_1 = 1.2\ P_2 + 160 - 200$

or $P_1 = 1.2\ P_2 - 40$ .... (1)

(i) When load is 100 MW,

$P_1 + P_2 = 100$ .... (2)

Substituting $P_1 = 1.2\ P_2 - 40$ in (2)

$1.2\ P_2 + P_2 = 100 + 40$ .... (3)

Again, from equation (3)

$P_1 = 150 - P_2 = 150 - 86.36 = 63.64$ WM

Thus, the power delivered by plant 1 is 63.64 MW

Now, the incremental fuel cost is given by

$$\frac{dF_1}{dP_1} = \frac{dF_2}{dP_2} = 0.1\ \mathrm{P}_1 + 20 = 0.12\ \mathrm{P}_2 + 10$$

$$= 0.1 \times 63.64 + 20 = \text{Rs } 26.364/\text{MWh}$$

Thus, Incremental fuel cost = Rs 26.364 MWh **Ans.**

Problem 15.5. A system consists of two plants connected by a tie line and a load is located at plant 2. When 100 MW are transmitted from plant 1, a loss of 10 MW takes place on the tie-line. Determine the generation schedule at both the plants and the power received by the load when λ for the system is Rs. 25 per megawatt hour and the incremental fuel costs are given by the equation.

$$\frac{dF_1}{dP_1} = 0.03\ P_1 + 17 \text{ Rs./MWhr}$$

$$\frac{dF_2}{dP_2} = 0.06\ P_2 + 19 \text{ Rs./MWhr}$$

Solution: Let P_1 an P_2 be the power generated by plants 1 and 2 respectively

Given

$$P_L = 10 \text{ MW}, \quad \lambda = \text{Rs } 25\text{/MWh}$$

$$\frac{dF_1}{dP_1} = 0.03\ P_1 + 17 \text{ Rs./MWh}$$

$$\frac{dF_2}{dP_2} = 0.06\ P_1 + 19 \text{ Rs./MWh}$$

The transmission loss for two plant system is given by

$$P_L = B_{11}\ P_1^2 + 2\ B_{12}\ P_1\ P_2 + B_{22}\ P_2 \quad \text{....... (1)}$$

or $2.2\ P_2 = 140 \therefore P_2 = 63.64$ MW **Ans.**

From equation (2)

$$P_1 = 100 - 63.64 = 36.36 \text{ MW} \quad \textbf{Ans.}$$

Now, the incremental fuel cost is given by

$$\frac{dF_1}{dP_1} = \frac{dF_2}{dP_2} = 0.1\ P_1 + 20$$

$$= 0.12\ P_2 + 16 = 0.1 \times 36.36 + 20$$

$$= 0.12 \times 63.64 + 16 = 23.636$$

$$= \text{Rs } 3.636\text{/MWhr} \quad \textbf{Ans.}$$

(ii) When load is 150 MW

$$P_1 + P_2 = \text{`}150 \quad \text{......... (2)}$$

Substituting $P_1 = 1.2\ P_2 - 40$ in equations (2)

$1.2\ P_2 - 40 + P_2 = 150$ or $2.2\ P_2 = 190$

$$\therefore \quad P_2 = \frac{190}{2.2} = 86.36 \text{ MW} \quad \textbf{Ans.}$$

Since the plant 2 is not supplying power. Therefore

$$B_{22} = 0 \text{ and } B_{12} = 0$$

From equation 1, When $P_1 = 100$ MW,

$$B_{11} = \frac{P_L}{P_1^2} = \frac{10}{(100)^2} = 0.001 \text{ MW}^{-1}$$

and

$$\frac{\delta P_L}{\delta P_1} = 2P_1 B_{11} + 2B_{12}P_2$$

$$= 2P_1 \times 0.001 + 2 \times 0 \times P_2 = 0.002\ P_1$$

Now, the penalty factor for plant 1 is given by

$$L_1 = \frac{1}{1 - \dfrac{\partial P_L}{\partial P_1}} = \frac{1}{1 - 0.002 P_1}$$

Again, from equation (1)

$$\frac{\delta P_L}{\delta P_2} = 2P_2\ B_{22} + 2\ P_1 B_{12}$$

$$= 2\ P_2 \times 0 + 2\ P_1 \times 0 = 0$$

Penalty factor for plant 2 (L_2) $= \dfrac{1}{1 - \dfrac{\partial P_L}{\partial P_2}} = \dfrac{1}{1-0} = 1.0$

As we know that

Penalty factor × Incremental fuel cost = λ

For plant 1

$$\frac{1}{1 - 0.002\, P_1} \times (0.03\ P_1 + 17) = 25$$

or $0.03\ P_1 + 17 = 25 \times (1 - 0.002\ P_1)$

or $0.03\ P_1 + 17 = 25 - 0.05\ P_1$

or $0.08\ P_1 = 25 - 17$

$$\therefore \quad P_1 = \frac{8}{0.08} = 100 \text{ MW}$$

Power generated by plant 1 (P_1) = 100 MW

Again for plant 2,

$$L_2 \times \frac{dF_2}{dP_2} = \lambda$$

$1 \times (0.006\ P_2 + 19) = 25$

or $0.06\ P_2 = 6$

$$\therefore \quad P_2 = \frac{6}{0.06} = 100 \text{ MW}$$

$\therefore$ Power generated by plant 2 = 100 MW **Ans.**

The power received at the load is given by

$$P_R = P_1 + P_2 - P_L = 100 + 100 - 10 = 190 \text{ MW}$$

Power received by load = 190 MW **Ans.**

Problem 15.6. For the system whose one line diagram is shown is Fig. P. 15.6, assume $I_1 = 2.0 \angle 0°$ per unit and $I_2 = 0.9 \angle 0°$ per unit. If the voltage at the bus 3 is $V_3 = 1.0 \angle 0°$ per unit, find the loss coefficient line impedances are $0.02 + j\, 0.14$ per unit, $0.04 + j\, 0.16$ per unit, and $0.01 + j\, 0.06$ per unit for sections a,b and c respectively

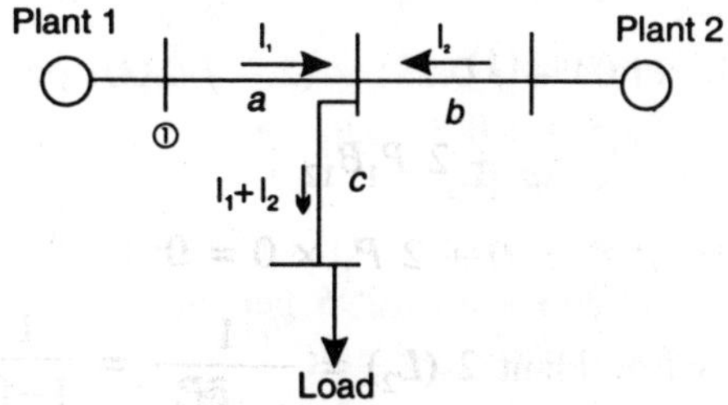

Fig. P. 15.6

Also calculate the transmission loss using loss coefficient formula.

Solution: Given,

$I_1 = 2\angle 0° = 2 + j\, 0.0$ per unit

$I_2 = 0.9\angle 0° = 0.9 + j\, 0.0$ per unit

$Z_a = 0.02 + j\, 0.14$ per unit $= R_a + j\, X_a$

$Z_b = 0.04 + j\, 0.16$ per unit $= R_b + j\, X_b$

$Z_c = 0.01 + j\, 0.06$ per unit $= R_c + j\, X_c$

and voltage at bus 3 $(V_3) = 1\angle 0°$ per unit

Let us calculate voltages at bus 1 and 2

The voltage at bus 1 is given by

$$V_1 = V_3 + I_1 Z_a = 1\angle 0° + (2 + j\, 0.0)\,(0.02 + j\, 0.14)$$
$$= 1.04 + j\, 0.28 \text{ per unit.}$$
$$V_2 = V_3 + I_2 Z_b = 1\angle 0° + (0.9 + j\, 0.0)\,(0.04 + j\, 0.16)$$
$$= 1.036 + j\, 0.144 \text{ per unit}$$

Since all the currents have phase angles zero, the power factor at each source node is the cosine of the angle of the voltage at the node, and the voltage magnitude times power factor equals the real part of the complex expression for the voltage V_1 and V_2

From expression of V_1, $V_1 \cos\theta_1 = 1.04$ p.u. and

From expression of V_2, $V_2 \cos\theta_2 = 1.036$ p.u.

Now, the loss coefficients are given by

$$B_{11} = \frac{R_a + R_c}{(V_1 \cos\theta_1)^2} = \frac{0.02 + 0.01}{(1.04)^2} = 0.0277 \text{ p.u.} \quad \textbf{Ans.}$$

$$B_{22} = \frac{R_b + R_c}{(V_2 \cos\theta_2)^2} = \frac{0.04 + 0.01}{(1.036)^2} = 0.04658 \text{ p.u.} \quad \textbf{Ans.}$$

and $$B_{12} = \frac{Rc}{V_1 \cos\theta_1 \times V_2 \cos\theta_2} = \frac{0.01}{1.04 \times 1.036}$$

$$= 0.0093 \text{ p.u.} \quad \textbf{Ans.}$$

To calculate transmission loss, let us calculate P_1 and P_2

Now,

$$V_1 I_1^* = (1.04 + j\,0.28) \times (2 - j\,0.00) \text{ pu}$$

$$= 2.08 + j\,0.56 \text{ p.u.}$$

This yields, $P_1 = 2.08$ P.u.

$$V_2 I_2^* = (1.036 + j\,0.144) \times (0.9 - j\,0.0)$$

$$= 0.9296 + j\,0.1296 \text{ per unit}$$

This yields, $P_2 = 0.9296$ per unit

The transmission loss is given by

$$P_L = B_{11}P_1^{\,2} + 2\,P_1\,P_2\,B_{12} + P_2^{\,2}\,B_{22}^{\,2}$$

$$= 0.0277 \times (2.08)^2 + 2 \times 2.08 \times 0.9296 \times 0.0093 + (0.9296)^2 \times 0.04658$$

$$= 0.1961 \text{ p.u.}$$

Problem 15.7. A system consists to of two plants connected by a transmission line. The only load is located at plant 2. When 250 MW is transmitted from plant 1 to plant 2, the power loss in the line is 20 MW. Find the required generator for each plant and the power received by the load when λ for the system is Rs 50 per megawatt hour. Assume that the incremental fuel costs can be approximated by the following equation

$$\frac{dF_1}{dP_1} = 0.020\,P_1 + 16 \text{ Rs/MWh}$$

and $$\frac{dF_2}{dP_2} = 0.030\,P_2 + 20 \text{ Rs/MWh}$$

Solution: Let P_1 and P_2 be the generated power for plants 1 and 2 respectively.

Given, $P_1 = 250$ MW, $P_L = 20$ MW, λ = Rs 50 /MWh

The transmission loss for a two-plant system is

$$P_L = B_{11} P_1^2 + B_{22} P_2^2 + 2 P_1 P_2 B_{12} \quad \text{.......... (1)}$$

Since plant 2 is not supplying power through transmission line, Therefore

$$B_{22} = 0 \text{ and } B_{12} = 0$$

From equation (1)

$$B_{11} = \frac{P_L}{P_1^2} = \frac{20}{(250)^2} = 0.0032 \text{ MW}^{-1}$$

and

$$\frac{\partial P_L}{\partial P_1} = 2 P_1 B_{11} + 2 P_2 B_{12}$$

$$= 2 P_1 \times 0.00032 + 2 P_2 \times 0 = 0.00064 P_1$$

The penalty factor for plant 1 is given by

$$L_1 = \frac{1}{1 - \frac{\partial P_L}{\partial P_1}} = \frac{1}{1 - 0.00064 P_1}$$

Again, From equation (1)

$$\frac{\partial P_L}{\partial P_2} = 2 P_2 B_2^2 + 2 P_1 B_{12} = 2 P_2 \times 0 + 2 P_1 \times 0 = 0$$

$\therefore$ Penalty factor for plant 2 $(L_2) = \dfrac{1}{1 - \dfrac{\partial P_L}{\partial P_2}} = \dfrac{1}{1-0} = 1.0$

As we know that

Penalty factor × Incremental fuel cost = λ

For plant 1,

$$L_1 \times \frac{\partial F_1}{\partial P_1} = 50$$

or

$$\frac{0.020 P_1 + 16}{1 - 0.00064 P_1} = 50$$

or $0.020 P_1 + 16 = 50 - 0.032 P_1$

or $0.020 P_1 + 0.032 P_1 = 50 - 16$ or $0.052 P_1 = 34$

$$\therefore \quad P_1 = \frac{34}{0.052} = 653.85 \text{ MW } \textbf{Ans.}$$

For plant 2,

$$L_2 \times \frac{dF_2}{dP_2} = 50$$

or $1 \times 0.030\ P_2 + 20 = 50$ or $0.030\ P_2 = 30$

$$P_2 = 1000 \text{ MW} \quad \textbf{Ans.}$$

Now, the power received by the load is given by

$$P_R = P_1 + P_2 - P_L = P_1 + P_2 - B_{11}\ P_1^2$$
$$= 653.85 + 1000 - 0.00032 \times (653.85)^2$$
$$= 1517.01 \text{ MW} \quad \textbf{Ans.}$$

$\therefore$ Power received by load = 1517.04 MW **Ans.**

Problem 15.8. The fuel costs of two-unit plant are given by

$$C_1 = 160 + 4\ P_1 + 0.008\ P_1^2$$
$$C_2 = 250 + 4\ P_2 + 0.02\ P_2^2$$

Where P_1 and P_2 are in MW, the plant supplies a load of 600 MW. Find the economic load scheduling of the two units and the incremental fuel cost, neglect losses.

Solution: Given,

$$P_1 + P_2 = 600 \qquad \text{.................} (1)$$
$$C_1 = 160 + 4\ P + 0.008\ P_1^2$$

and $$C_2 = 250 + 4\ P_2 + 0.02\ P_2^2$$

Let P_1 and P_2 be the power generated by plants, and 2 respectively.
The incremental fuel cost of the plant 1 is given by

$$I_{C1} = \frac{dc_1}{dp_1} = 4 + 2 \times 0.008\ P_1$$
$$= 4 + 0.016\ P_1$$

The incremental fuel cost of the plant 2 is given by

$$I_{C2} = \frac{dc_2}{dp_2} = 4 + 0.02 \times 2\ P_2 = 4 + 0.04\ P_2$$

For economic load scheduling

$$I_{C1} = I_{C2}$$

or $4 + 0.016\ P_1 = 4 + 0.04\ P_2$

or $16\ P_1 = 40\ P_2$

$$\therefore \quad P_2 = 0.4\ P_1 \qquad \text{.................} (2)$$

Substituting $P_2 = 0.4\ P_1$ in equation (1)

$P_1 + 0.4\ P_1 = 600$ or $1.4\ P_1 = 600$

$$\therefore \quad P_1 = 428.57 \text{ MW} \quad \textbf{Ans.}$$

From equation (2)

$$P_2 = 0.4 \times 428.57 = 171.43 \text{ MW} \quad \textbf{Ans.}$$

The incremental fuel cost for plant 1 is given by

$$\frac{dc_1}{dP_1} = 4 + 0.016\ P_1 = 4 + 0.016 \times 428.57$$
$$= 10.86 \text{ Rs/MWh}$$

Again, the incremental fuel cost for plant 2 is given by

$$\frac{dc_2}{dP_2} = 4 + 0.04\, P_2 = 4 + 0.04 \times 171.43$$

$$= 10.86 \text{ Rs/MWh}$$

Hence the incremental fuel cost of the plant for most economic operation is Rs 10.86 per MWh **Ans.**

> **Problem 15.9.** The fuel inputs per hour of plants 1 and 2 are given as
>
> $$F_1 = 0.4\, P_1^2 + 80\, P_1 + 160 \text{ Rs/hr}$$
> $$F_2 = 0.5\, P_2^2 + 60\, P_2 + 200 \text{ Rs/hr}$$
>
> Determine the economic operating scheduled and the corresponding cost of generation of the maximum and minimum loading 150 MW and 30 WM. The demand is 200 MW and transmission losses are neglected

Solution: Given

$$F_1 = 0.4\, P_1^2 + 80\, P_1 + 160 \text{ Rs./hr} \quad \text{......... (1)}$$

and $$F_2 = 0.5\, P_2^2 + 60\, P_2 + 200 \text{ Rs./hr} \quad \text{......... (2)}$$

The incremental production costs of plant 1 is given by

$$\frac{dF_1}{dP_1} = 0.8\, P_1 + 80$$

The incremental production costs of plant 2 is given by

$$\frac{dF_2}{dP_2} = 1.0\, P_2 + 60$$

For economic operation $\dfrac{dF_1}{dP_1} = \dfrac{dF_2}{dP_2}$

or $0.8\, P_1 + 80 = 1.0\, P_2 + 60$

or $P_2 = 0.8\, P_1 + 20$ .... (3)

When load is 200 MW

$P_1 + P_2 = 200$

From equation (3)

$P_1 + 0.8\, P_1 + 20 = 200$ or $1.8\, P_1 = 180$

$$\therefore \quad P_1 = \frac{180}{1.8} = 100 \text{ MW}$$

Substituting $P_1 = 100$ MW in equation (3)

$P_2 = 0.8 \times 100 + 20 = 100$ MW **Ans.**

Now, total cost of generation = $F_1 + F_2$

$= 0.4\ P_1^2 + 80\ P_1 + 160 + 0.5\ P_2^2 + 60\ P_2 + 200$

$= 0.4 \times 100^2 + 80 \times 100 + 160 + 0.5 \times 100^2 + 60 \times 100 + 20$

= Rs 23.180 /hr

Problem 15.10. Assume that the fuel input in btu per hour for unit 1 and 2 are given by

$$F_1 = (10\ P_1 + 0.025\ P_1^2 + 90)\ 10^6$$
$$F_2 = (8\ P_2 + 0.05\ P_2^2 + 125)\ 10^6$$

The maximum and minimum loads on the units are 150 MW and 20 MW respectively. Determine the minimum cost of generation when the following load is supplied as shown in Fig. P. 15.10.

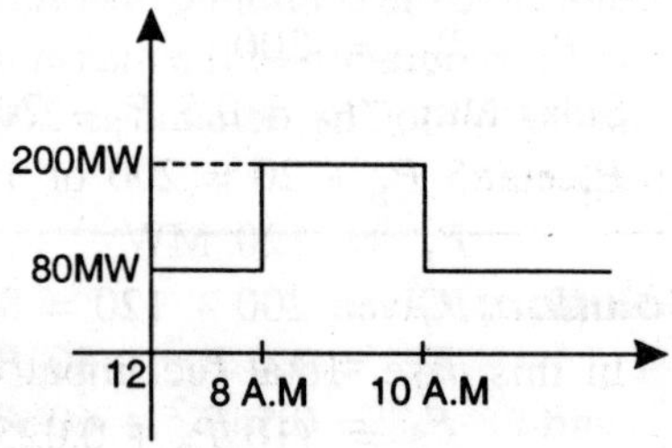

Fig. P. 15.10

The cost of fuel is Rs. 4 per million btu.

Solution: Given,

$$F_1 = (10\ P_1 + 0.025\ P_1^2 + 90) \times 10^6$$
$$F_2 = (8\ P_2 + 0.05\ P_2^2 + 125) \times 10^6$$

The incremental production cost of plant 1 is given by

$$\frac{dF_1}{dP_1} = (10 + 0.5\ P_1) \times 10^6$$

The incremental production cost of plant 2 is given by

$$\frac{dF_2}{dP_2} = (8 + 0.1\ P_2) \times 10^6$$

For economic operation.

$$\frac{dF_1}{dP_1} = \frac{dF_2}{dP_2}$$

or $10 + 0.05\ P_1 = 8 + 0.1\ P_2$

Multiplying both sides by 10, we get

$100 + 0.5\ P_1 = 80 + P_2$ or $P_2 = 0.5\ P_1 + 20$ ....(1)

(i) When load is 80 MW, we have

$1 + P_2 = 80$ ....(2)

Putting $P_2 = 0.5\ P_1 + 20$ in equation (2)

$P_1 + 0.5\ P_1 + 20 = 80$

or $1.5\ P_1 = 60,$

$\therefore\ P_1 = 40$ MW **Ans.**

Again from equation (2), $P_2 = 80 - P_1 = 80 - 40 = 40$ MW

Total fuel input $= F_1 + F_2$

$= (10\ P_1 + 0.025\ P_1^2 + 90) \times 10^6 + (8\ P_2 + 0.05\ P^2_2 + 125) \times 10^6$

$= (10 \times 40 + 0.025\ (40)^2 + 90) \times 10^6 + (8 \times 40 + 0.05 \times (40)^2 + 125) \times 10^6$

$= 1055 \times 10^6$ btu/hr

(ii) When the load is 200 MW

$P_1 + P_2 = 200$ .... (3)

Substituting $P_2 = 0.5\ P_1 + 20$ in equation (3)

$P_1 + 0.5\ P_1 + 20 = 200$ or $1.5\ P_1 = 180$

$P_1 = 120$ MW

and $P_2 = 200 - 120 = 80$ MW

In this case, Total fuel input $= F_1 + F_2$

$= (10\ P_2 + 0.025\ P_1^2 + 90) \times 10^6 + (8\ P_2 + 0.05\ P_2^2 + 125) \times 10^6$

$= (10 \times 120 + 0.025 \times (120)^2 + 90) \times 10^6$ $(8 \times 80 + 0.05 \times (80)^2 + 125) \times 10^6$

$= 2735 \times 10^6$ btu per/hr

Total fuel input in both cases

$F = (2735 + \text{loss})$ million btu per hr

$= 3790$ million btu per hr

Operating duration = 10 hr

$\therefore$ Cost of 1 million Btu = Rs 4 for 1 hr

$\therefore$ Total cost $= F \times t \times$ cost of 1 million btu

$= 10 \times 4 \times 3790 =$ Rs 1,51,600 **Ans.**

Problem 15.11. The load—duration curve for a plant is shown below. Determine the load factor from load factor

from load—duration curve

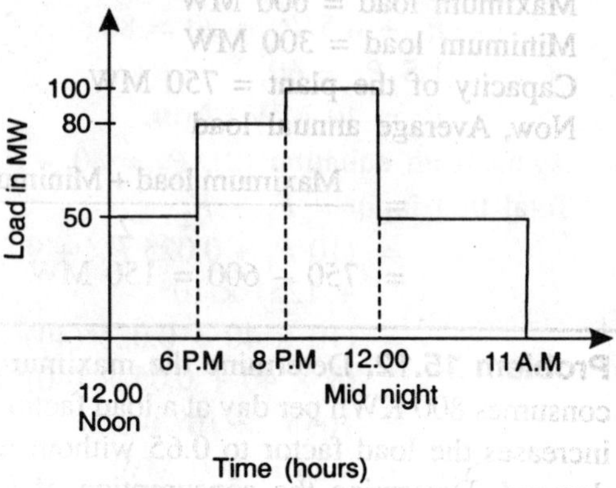

Fig. P. 15.11

Solution: For $P_1 = 50$ MW, $t_1 = 6$ h

For $P_2 = 80$ MW, $t_1 = 2$ h

For $P_3 = 100$ MW, $t_3 = 4$ h

For $P_4 = 50$ MW, $t_4 = 11$ h

Total time duration $= t_1 + t_2 + t_3 + t_4$

$= 6 + 2 + 4 + 11 = 23$ h

Actual energy consumed

$= P_1 \times t_1 + P_2 \times t_2 + P_3 \times t_3 + P_4 \times t_4$

$= 50 \times 6 + 80 \times 2 + 100 \times 4 + 50 \times 11$

$= 30 + 160 + 400 + 550 = 1140$ MWh

$\therefore$ Average load

$$= \frac{\text{Actual energy consumed}}{\text{Time duration}} = \frac{1140}{23} = 49.57 \text{ MW}$$

Again from load—duration curve

$\therefore$ Maximum demand = 100 MW

$$\text{Load Factor} = \frac{\text{Average load}}{\text{Maximum demond}} = \frac{49.57}{100} = 0.4957 \quad \textbf{Ans.}$$

Problem 15.11. The yearly load duration curve of a power plant is a straight line, The maximum load is 600 MW and the minimum load is 300 MW. The capacity of the plant is 75 MW.

Find (i) Plant capacity factor

(ii) Load factor

(iii) Utilization factor

(iv) Reserve capacity

Solution: Given,

Maximum load = 600 MW

Minimum load = 300 MW

Capacity of the plant = 750 MW

Now, Average annual load

$$= \frac{\text{Maximum load} + \text{Minimum load}}{2}$$

$= 750 - 600 = 150$ MW **Ans.**

Problem 15.12. Determine the maximum value of a load which consumes 800 KWh per day at a load factor of 0.45. If the consumer increases the load factor to 0.65 without increasing the maximum demand. Determine the consumption of energy in kWh.

Solution: Given,

Initial load factor = 0.45

Final load factor = 0.65

Initial energy consumption = 800 kW/h per day

Let X be the maximum demand. For both initial and final cases

For initial operation,

$$\text{Initial load factor} = \frac{\text{Initial energy consumption in 24 hr}}{(\text{Maximum demand in kW}) \times 24}$$

$$\text{or Maximum demand } (X) = \frac{\text{Initial energy consumption}}{\text{Initial load factor in 24 hr} \times 24}$$

$$= \frac{800}{0.45 \times 24} \text{ MW} = 74.07 \text{ MW}$$

$$= \frac{600 + 300}{2} = \frac{900}{2} = 450 \text{ MW}$$

(a) Plant capacity factor

$$= \frac{\text{Average annual load}}{\text{Capacity of the plant}} = \frac{450}{750} = 0.6 \quad \textbf{Ans.}$$

(b) Load factor

$$= \frac{\text{Average load}}{\text{Maximum demand}} = \frac{450}{600} = 0.75 \quad \textbf{Ans.}$$

(c) Utilization factor

$$= \frac{\text{Maximum demand}}{\text{Plant capacity}} = \frac{600}{750} = 0.8 \quad \textbf{Ans.}$$

(d) We know that, reserve capacity

= plant capacity – maximum demand

For final operation, energy consumed in kWh

= (load factor) × (maximum demand) × 24

= 0.65 × 74.07 × 24 = 1155.56 kWh **Ans.**

Problem 15.13. A power station supplied the peak loads of 40 MW, 25 MW and 30 MW to three localities the annual load factor is 0.80 and the diversity of the load at the station is 1.80. Calculate

(i) The maximum demand on the station

(ii) Installed capacity

(iii) Energy supplied in a year

Solution: Given,

load factor = 0.80

Diversity factor = 1.80

The peak loads are

$P_1 = 40$ MW, $P_2 = 25$ MW, $P_3 = 30$ MW

(a) We know that diversity factor

$$= \frac{\text{Sum of individual maximum demands}}{\text{Maximum demand on the station}}$$

Maximum demand on the station

$$= \frac{P_1 + P_2 + P_3}{\text{Diversity factor}} = \frac{40 + 25 + 30}{1.80}$$

= 52.78 MW

(b) Installed capacity

= Sum of individual maximum demands

= $P_1 + P_2 + P_3$ = 40 + 25 + 30 = 95 MW **Ans.**

(c) We have,

$$\text{Load factor} = \frac{\text{Average load}}{\text{Maximum demand}}$$

$$= \frac{\dfrac{\text{Total energy produced in a year}}{\text{Time duration}}}{\text{Maximum demond}}$$

Total energy produced in a year

= Time duration × maximum demand × load factor

= 87.60 × 52.78 × 0.80

= 389866.67 MWh **Ans.**

Problem 15.14. A hydroelectric station has to operate with a mean head of 40 m and is supplied from a reservoir lake which drains catchments area of 300 square km over which the average rain fall is 120 cm per annum. If the 70% of the rain is utilized and the expected load factor for the station is 85%, calculate the power in KW far which the station should be designed . Assume that no - head is lost in pipes, penstocks, etc. The mechanical efficiency of turbine is 90% and efficiency of generator is 94%.

Solution: Given

Water head (h) = 40 m

Catchment area (A) = 300 Sq.km = 300 × 10^6 Sq.m

Rain fall 120 cm per annum = 1.20 m

Run off = 70% = 0.7

Load factor = 0.85, η_t = 0.90 and η_g = 0.94

Now, water stored in the reservoir

= Rainfall × catchments area × run off

= 300 × 10^6 × 1.20 × 0.70 = 252 × 10^6 cubic meter

Discharge of the water from the reservoirs is given by

$$Q = \frac{\text{Total water stored in the reservoir}}{365 \times 24 \times 60 \times 60} \text{ m}^3\text{/sec}$$

$$= \frac{252 \times 10^6}{365 \times 24 \times 60 \times 60} \text{ m}^3\text{/sec} = 7.991 \text{ m}^3\text{/sec.}$$

Density of water (δ) = 100 kg/m^3

Overall efficiency (η) = $\eta_t \times \eta_g \times \eta_a$

= 0.90 × 0.94 × 1 = 0.846

The power generated in kW can be given by

$$P = \frac{0.736}{75} Q\rho h\eta \text{ kW}$$

$$= \frac{0.736}{75} \times 7.991 \times 1000 \times 40 \times 0.846 \text{ kW}$$

= 2,653.67 kW **Ans.**

Chapter 16

Load Frequency Control and Load Power Flow

Problem 16.1. Two generating stations *A* and *B* have full load capacities of 500 MW and 210 MW respectively. The interconnector connecting the two stations has a motor-generator set (Plant-*C*) near station *A* of full load capacity 50 MW. Percentage changes of speeds of *A*, *B*, and *C* are 5, 4 and 2.5 respectively. The loads on bus bars *A* and *B* are 250 MW and 100 MW respectively. Determine the load taken by set *C* and indicate the direction in which the energy is flowing.

Solution: The circuit diagram of the given problem is as follows

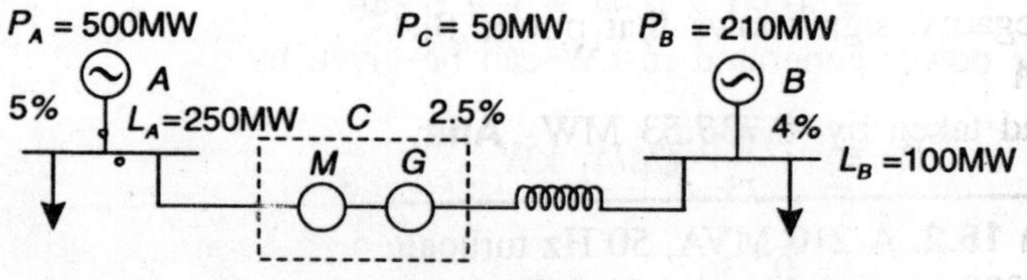

Fig. P. 16.1

Let α be the power in MW flows from generator *A* to generator *B*.

Load on station *A*

$$= \text{Load on bus } A + \text{power flow from } A \text{ to } B$$

$$= 250 + \alpha$$

and Load on station *B*

$$= \text{Load on bus } B - \alpha = 100 - \alpha$$

Given, % change in speed of generator $A = 5$

∴ % Drop in speed of generator A

$$= \frac{5}{P_A} \times (250 + \alpha) = \frac{5}{500} (250 + \alpha)$$

$$= \frac{250+\alpha}{100}$$

Similarly, % Drop in speed of generator

$$B = \frac{4}{P_B} \times (100 - \alpha)$$

$$= \frac{4}{210} (100 - \alpha) = \frac{2(100-\alpha)}{105}$$

Since the output of A corresponding to the power flow through the interconnector is the input to C, therefore, there is further reduction in frequency which is given by $\frac{2.5d}{50}$, i.e. $\frac{d}{20}$. The total reduction in frequency from B when referred to bus B, should be same.

$$\therefore \frac{250+\alpha}{100} + \frac{\alpha}{20} = \frac{2(100-\alpha)}{105}$$

or $\frac{250+\alpha+5\alpha}{100} = \frac{2(100-\alpha)}{105}$

or $(250 + 6\alpha) \times 105 = 200\ (100 - \alpha)$

or $250 \times 105 + 6 \times 105 \times \alpha = 200 \times 100 - 200\ \alpha$

or $830\ \alpha = -6250$

$\therefore\ \alpha = -7.53$ MW

The negative sign shows that power flows from generator B to generator A

∴ Load taken by C = 7.53 MW **Ans.**

Problem 16.2. A 210 MVA, 50 Hz turboalternator operates at no load at 3000 rpm. A load of 75 MW is suddenly applied to the machine and the steam valves to the turbine commence to open after 1 sec due to the time lag in the governor system. Assuming inertia constant H of 5 kW–sec per KVA of generator capacity, calculate the frequency to which the generated voltage drops before the steam flow commences to increase to meet the new load.

Solution: Given,

Generator rating (G) = 210 MVA

Inertia constant (H) = 5 kW-Sec/kVA

Time (t) = 1 sec
Load power (P_L) = 75 MW
We have,

$$H = \frac{\text{Kinetic energy stored in the rotor at synchnous speed}}{G}$$

$\therefore$ K.E. stored in the rotor (k.E_S)
$= H \times G = 5 \times 210 = 1050$ MJ

Before the steam valve open,
Energy lost by the rotor (k.E_L)
$= P_L \times t = 75 \times 1 = 75$ MJ

Due to energy lost, there is reduction in speed of the rotor and therefore, reduction in frequency

The new frequency is given by

$$f_n = \sqrt{\frac{k.E_S - k.E_L}{k.E_S}} \times 50 \text{ Hz} = \sqrt{\frac{1050 - 75}{1050}} \times 50 \text{ Hz}$$

$= 48.18$ Hz **Ans.**

Problem 16.3. An interconnector with inductive reactance of 25 ohm and negligible resistance has two units of generations with voltage of 33 kV and 30 kV at its ends. A load of 6 MW is to be transferred from 33 kV to 30 kV side of the interconnector. Determine the power factor of the power transmitted and other necessary conditions between the two ends

Solution: Given,

Reactance of the line (X) = 25 ohm
Sending end voltage (V_S) = 33 kV
Receiving end voltage (V_r) = 30 kV
Power transferred to the load through the line (P) = 6 MW.
The circuit diagram is as follows

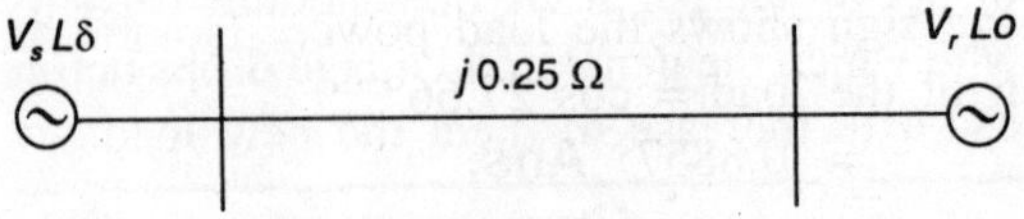

Fig. P. 16.3

The power transferred through a line is given by

$$P = P_{max} \sin \delta = \frac{V_S V_r}{X} \sin \delta$$

$\therefore$ Maximum power transferred $(P_{max}) = \frac{33 \times 30}{25} = 39.6$ MW

Let δ be the load angle when 6 MW is to be transferred.

$$\therefore \quad P = 6 \text{ or } P_{max} \sin \delta = 6$$

$$\therefore \quad \delta = \sin^{-1}\left(\frac{6}{39.6}\right) = 8.7147.$$

$\therefore$ Load angle $(\delta) = 8.7147$

Let $\cos \theta$ be the load power factor.

I be the current flowing through the line

$$\therefore \quad \bar{I} = I\angle\theta$$

Let us take receiving end voltage as reference phase

$$\bar{V}_r = 30\angle 0° \text{ kV } (L - L)$$

$$= \frac{30}{\sqrt{3}} \angle 0° \text{ kV (phase)}$$

Now, the current through the line is given by

$$I\angle\theta = \frac{V_S\angle\delta - V_r\angle 0°}{Z\angle\theta}$$

$$= \frac{33\angle 8.7147° - 30\angle 0°}{\sqrt{3} \times 25\angle 90°} \times 10^3$$

$$= \frac{1}{\sqrt{3}}[1.32\angle{-81.2853°} - 1.2\angle{-90°}] \times 10^3$$

$$= [1.32\angle{-81.2853°} + j\,1.2] \times \frac{1}{\sqrt{3}} \times 10^3$$

$$= [0.1999 - j\,0.1048]\,\frac{1}{\sqrt{3}} \times 10^3$$

$$= 130.31\ \angle{-27.66°}\text{ A}$$

The –ve sign shows the load power factor is logging.

$\therefore$ p.f. of the load $= \cos 27.66°$

$= 0.8857$ **Ans.**

Problem 16.4. Two generating stations *A* and *B* are of capacities 20 MW and 10 MW and speed regulation of 2% and 3% respectively. The two stations are connected through an interconnector and a motor-generator set. The set is connected to bus bars of A and has a capacity of 3 MW and full load slip of 4%. Determine the load on the interconnector when there is a load of 8 MW on B bus bars due to its own consumer but A has no external load.

Solution: The circuit diagram of this problem is as follows

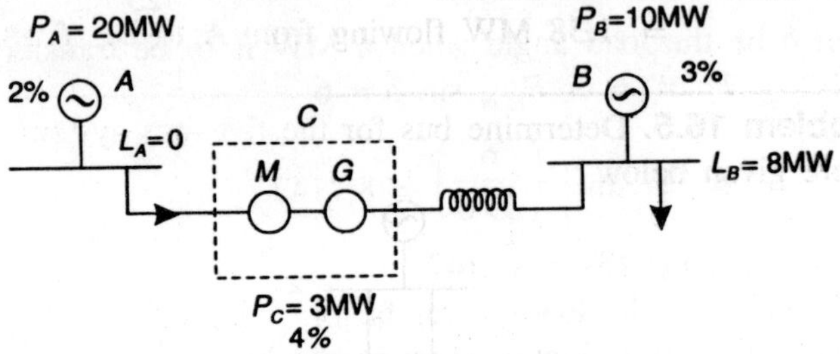

Fig. P. 16.4

Let α be the power in MW flows from generator to generator B

$\therefore$ Load on station A

$$= \text{Load on bus } A + \text{power transmitted from } A \text{ to } B$$

$$= 0 + \alpha = \alpha$$

Similarly,

Load on station B = Load on bus $B - \alpha$

$$= 8 - \alpha$$

Now,

$$\% \text{ Drop in speed of generator A} = \frac{2}{P_A} \times \text{load}$$

$$= \frac{2}{20} \times \alpha = \frac{\alpha}{10}$$

$$\% \text{ Drop in speed of generator } B = \frac{3}{P_B} \times \text{load}$$

$$= \frac{3}{10}(8 - \alpha)$$

Since the output of A corresponding to the power flow through the inter connector is the input to C, therefore, there is further reduction in frequency which is given by $\frac{4\alpha}{P_C}$, ie $\frac{4\alpha}{3}$, the total reduction in frequency both from A and C and the reduction in frequency from B, where referred to bus B should be same

$$\therefore \quad \frac{\alpha}{10} + \frac{4\alpha}{3} = \frac{3}{10}(8 - \alpha)$$

or $\dfrac{3\alpha + 40\alpha}{30} = \dfrac{3}{10}(8 - \alpha)$ or $43\alpha = 9\,(8 - \alpha)$

or $43\alpha + 9\alpha = 72$ or $52\,\alpha = 72$

$\therefore \quad \alpha = \dfrac{72}{52} = 1.38$

∴ Load on interconnector

= 1.38 MW flowing from A to B. **Ans.**

Problem 16.5. Determine bus for the five–bus system shown in figure given below

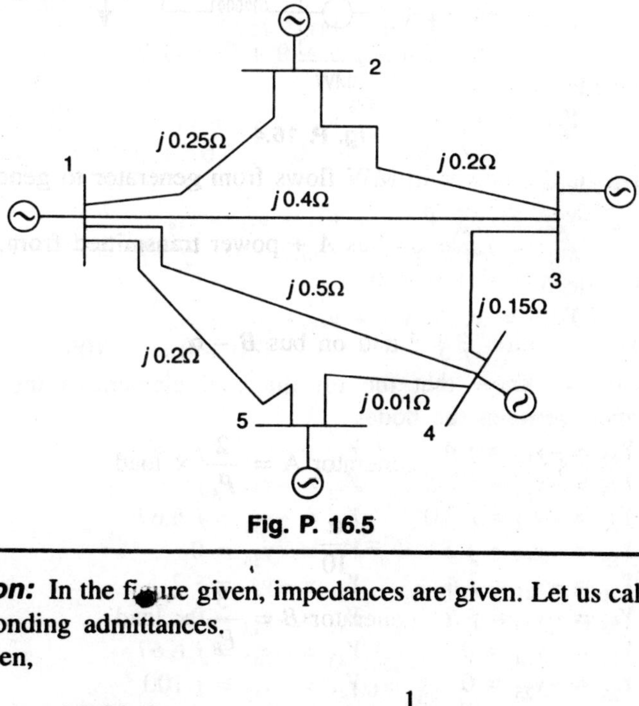

Fig. P. 16.5

Solution: In the figure given, impedances are given. Let us calculate corresponding admittances.

Given,

$Z_{12} = j\,0.25\ \Omega \quad \therefore\ y_{12} = y_{21} = \dfrac{1}{j0.25} = -\,j\,4\ ℧$

$Z_{23} = j\,0.2\ \Omega \quad \therefore\ y_{23} = y_{32} = \dfrac{1}{Z_{23}} = \dfrac{1}{j0.2} = -\,j\,5\ ℧$

$Z_{34} = j\,0.15\ \Omega \quad \therefore\ y_{34} = y_{43} = \dfrac{1}{Z_{34}} = \dfrac{1}{j0.15} = -\,j\,6.67\ ℧$

$Z_{45} = j\,0.01\ \Omega \quad \therefore\ y_{45} = y_{54} = \dfrac{1}{Z_{45}} = \dfrac{1}{j0.01} = -\,j\,100\ ℧$

$Z_{51} = j\,0.2\ \Omega \quad \therefore\ y_{51} = y_{15} = \dfrac{1}{Z_{51}} = \dfrac{1}{j0.2} = -\,j\,5\ ℧$

$$Z_{14} = j\,0.5\ \Omega \qquad \therefore\ y_{14} = y_{41} = \frac{1}{Z_{14}} = \frac{1}{j0.5} = -j\,2\ \mho$$

$$Z_{13} = j\,0.4\ \Omega \qquad \therefore\ y_{13} = y_{31} = \frac{1}{Z_{13}} = \frac{1}{j0.4} = -j\,2.5\ \mho$$

As we know that in admittance matrix, the diagonal element of each node is the sum of the admittances connected to it.

At node 1

$$\begin{aligned} Y_{11} &= y_{11} + y_{12} + y_{13} + y_{14} + y_{15} \\ &= 0 - j\,4 - j\,2.5 - j\,2 - j\,5 = -j\,13.5 \end{aligned}$$

at node 2

$$\begin{aligned} Y_{22} &= y_{22} + y_{21} + y_{23} + y_{24} + y_{25} \\ &= 0 - j\,4 - j\,5 + 0 + 0 = -j\,9 \end{aligned}$$

At node 3

$$\begin{aligned} Y_{33} &= y_{31} + y_{32} + y_{33} + y_{34} + y_{35} \\ &= -j\,2.5 - j\,5 + 0 - j\,6.67 + 0 = -j\,14.17 \end{aligned}$$

At node 4

$$\begin{aligned} Y_{44} &= y_{41} + y_{42} + y_{43} + y_{44} + y_{45} \\ &= -j\,2 + 0 - j\,6.67 + 0 - j\,100 = -j\,180.67 \end{aligned}$$

At node 5

$$\begin{aligned} Y_{55} &= y_{51} + y_{52} + y_{53} + y_{54} + y_{55} \\ &= -j\,5 + 0 + 0 - j\,100 + 0 = -j\,105 \end{aligned}$$

Also, we know that the off diagonal element is the negated admittance between the nodes

$Y_{12} = -y_{12} = j\,4$ $\qquad Y_{31} = -y_{31} = j\,2.5$

$Y_{13} = -y_{13} = j\,2.5$ $\qquad Y_{32} = -y_{32} = j\,5$

$Y_{14} = -y_{14} = j\,2.0$ $\qquad Y_{34} = -y_{34} = j\,6.67$

$Y_{15} = -y_{15} = j\,5$ $\qquad Y_{35} = -y_{35} = 0$

$Y_{21} = -y_{21} = j\,4$ $\qquad Y_{41} = -y_{41} = j\,2$

$Y_{23} = -y_{23} = j\,5$ $\qquad Y_{42} = -y_{42} = 0$

$Y_{24} = -y_{24} = 0$ $\qquad Y_{43} = -y_{43} = j\,6.67$

$Y_{25} = -y_{25} = 0$ $\qquad Y_{45} = -y_{45} = j\,100$

$Y_{52} = -y_{52} = 0$ $\qquad Y_{51} = -y_{51} = j\,5$

$Y_{53} = -y_{53} = 0$

$Y_{54} = -y_{54} = j100$

Now, the Y_{bus} matrix is given by

$$Y_{bus} = \begin{bmatrix} Y_{11} & Y_{12} & Y_{13} & Y_{14} & Y_{15} \\ Y_{21} & Y_{22} & Y_{23} & Y_{24} & Y_{25} \\ Y_{31} & Y_{32} & Y_{33} & Y_{34} & Y_{35} \\ Y_{41} & Y_{42} & Y_{43} & Y_{44} & Y_{45} \\ Y_{51} & Y_{52} & Y_{53} & Y_{54} & Y_{55} \end{bmatrix}$$

$$= \begin{bmatrix} -j\,13.5 & j\,4 & j\,2.5 & j\,2.0 & j\,5 \\ j\,4 & -j\,9 & j\,5 & 0 & 0 \\ j\,2.5 & j\,5 & -j\,14.17 & j\,6.67 & j\,0 \\ j\,2 & j\,0 & j\,6.67 & -j\,108.67 & j\,100 \\ j\,5 & j\,0 & j\,0 & j\,100 & -j\,105 \end{bmatrix}$$ **Ans.**

Problem 16.5. The line series impedances are as follows

Line (Bus to Bus)	*Impedances p.u.*
1-2	0.03 + j 0.04
1-3	0.04 + j 0.03
2-3	0.12 + j 0.05

Neglect the short capacitances of the lines.

Solution: Let us first draw equivalent circuit as shown below

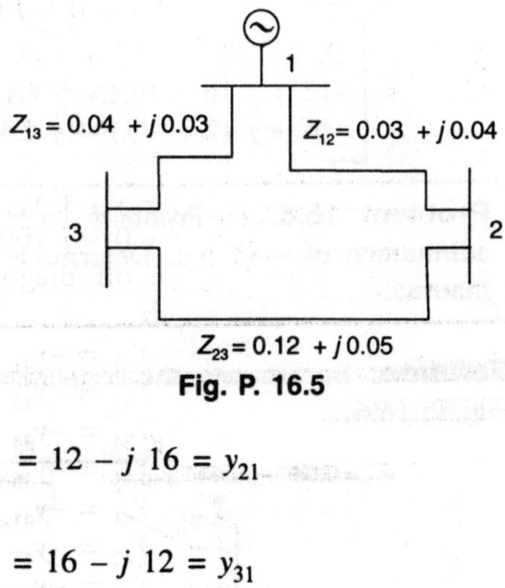

Fig. P. 16.5

Given,
$Z_{12} = 0.03 + j\,0.04$
$Z_{13} = 0.04 + j\,0.03$
$Z_{23} = 0.12 + j\,50.05$

$$y_{12} = \frac{1}{Z_{12}} = \frac{1}{0.03 + j0.04} = 12 - j\,16 = y_{21}$$

$$y_{13} = \frac{1}{Z_{13}} = \frac{1}{0.04 + j0.03} = 16 - j\,12 = y_{31}$$

$$y_{23} = y_{32} = \frac{1}{Z_{23}} = \frac{1}{0.12 + j0.05} = 7.1 - j\,2.959$$

As we know that in admittance matrix, the diagonal elements of each node is the sum of the admittances connected to it.

$$\begin{aligned} Y_{11} &= y_{11} + y_{12} + y_{13} \\ &= 0 + 12 - j\,16 + 16 - j\,12.6 \\ &= 28 - j\,28 \\ Y_{22} &= y_{21} + y_{22} + y_{23} \\ &= 12 - j\,16 + 0 + 7.1 - j\,2.959 \\ &= 19.1 - j\,18.959 \end{aligned}$$

$$Y_{33} = Y_{31} + Y_{32} + Y_{33}$$
$$= 16 - j\,12 + 7.1 - j\,2.959$$
$$= 23.1 - j\,14.959$$

Also we know that off – diagonal element is the negated admittance between the nodes.

$$\therefore\ Y_{12} = -Y_{12} = -12 + j\,16$$
$$Y_{13} = -Y_{13} = -16 + j\,12$$
$$Y_{23} = -Y_{23} = -7.1 + j\,2.959$$
$$Y_{21} = -Y_{21} = -12 + j\,16$$
$$Y_{31} = -Y_{31} = -16 + j\,12$$
$$Y_{32} = -Y_{32} = -7.1 + j\,2.959$$

The bus admittance matrix is given by

$$Y_{bus} = \begin{bmatrix} Y_{11} & Y_{12} & Y_{13} \\ Y_{21} & Y_{22} & Y_{23} \\ Y_{31} & Y_{32} & Y_{33} \end{bmatrix}$$

$$= \begin{bmatrix} 28 - j\,28 & -12 + j\,16 & -16 + j\,12 \\ -12 + j\,16 & 19.1 - j\,18.959 & -7.1 + j\,2.959 \\ -16 + j\,12 & -7.1 + j\,2.959 & 23.1 - j\,14.959 \end{bmatrix}$$

Problem 16.6. In Problem 16.5 each line has a total shunt admittance of $-j40$ p.u. Determine the modified bus admittance matrix.

Solution: In this case, the equivalent circuit diagram is as given in Fig. P. 16.6.

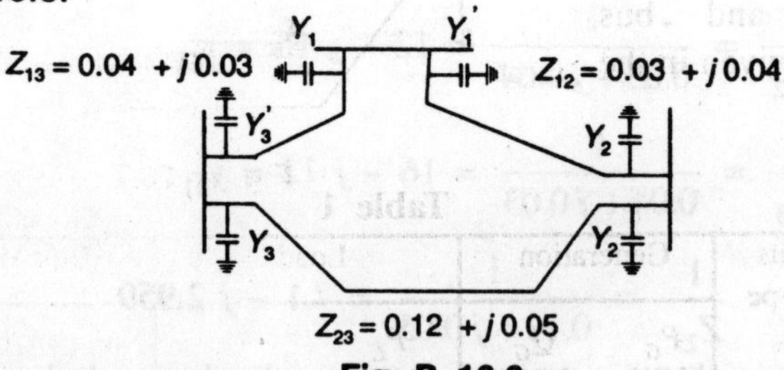

Fig. P. 16.6

For the line between bases 1 and 2

Total shunt admittance = $Y'_1 + Y'_2 = -j\,4.0$ and $Y'_1 = Y'_2$

For the line between bases 2 and 3,

Total shunt – admittance = $Y'_2 + Y'_3 = -j\,4.0$ and $Y'_2 = Y'_3$

For the line between buses 3 and 1

Total shunt admittance = $Y'_3 + Y'_1 = -j\,4.0$ and $Y'_3 = Y'_1$

$\therefore\ Y'_1 = Y'_1 = Y'_2 = Y'_2 = Y_3 Y'_3 = -j\,2.0$

The effect of shunt admittances is only for the diagonal elements in Y_{bus}.

Hence, the new diagonal admittances are

$$Y_{11n} = Y_{11}(\text{old}) + Y_1 + Y'_1 = 28 - j\,28 - j\,2.0 - j2.0$$
$$= 1.0 = 28 - j32$$
$$Y_{22n} = Y_{22}(\text{old}) + Y_2 + Y'_2 = 19.1 - j\,18.959 - j\,4.0$$
$$= 19.1 - j\,22.959$$
$$Y_{33n} = Y_{33}(\text{old}) + Y_3 + Y'_3 = 23.1 - j2.959 - j4$$
$$= 23.1 - j6.959$$

Now the modified bus admittance matrix is given by

$$= \begin{bmatrix} Y_{11n} & Y_{12} & Y_{13} \\ Y_{21} & Y_{22} & Y_{23} \\ Y_{31} & Y_{32} & Y_{33} \end{bmatrix}$$

$$\begin{bmatrix} 28 - j32 & -12 + j16 & -16 + j12 \\ -12 + j16 & 19.1 - j22.595 & -7.1 + j2.959 \\ -16 + j12 & -7.1 + j2.959 & 23.1 - j6.959 \end{bmatrix}$$

Problem 16.7. A three bus power system is shown in Fig. P. 16.7.

The relevant per unit line admittances on 200 MVA base are indicated on the diagram and bus data are given in the Table 1

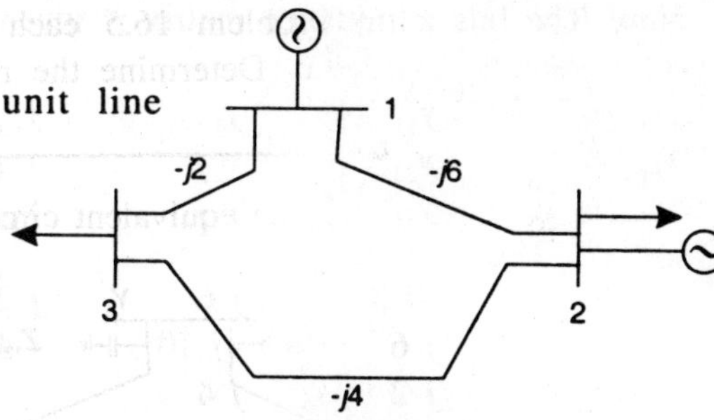

Fig. P. 16.7

Table 1

Bus number	Bus type	Generation		Load		Bus Voltage	
		P_G (MW)	Q_G (MVA_r)	P_L (MVA)	Q_L (MVA_r)	$V_{p.u}$	δ (degree)
1	Slack	?	?	0	0	1.01	0
2	PQ	30	20	60	30	?	?
3	PQ	0	0	70	35	?	?

Form Y_{bus} and determine the voltages at the bus 2 and 3 after the second iteration using Gauss-Seidel Method. Take acceleration factor = 1.6

Solution: Given

$$\gamma_{11} = 0 \quad \gamma_{12} = -j6 \quad \gamma_{13} = -j2$$
$$\gamma_{22} = 0 \quad \gamma_{21} = -j6 \quad \gamma_{23} = -j4$$
$$\gamma_{31} = -j2, \quad \gamma_{32} = -j4 \quad \gamma_{33} = 0$$

As we know that in admittance matrix, the diagonal element of each node is the sum of the admittances connected to it.

$$\begin{aligned} Y_{11} &= y_{11} + y_{12} + y_{13} \\ &= 0 - j\,6 - j\,2 = -j\,8 \\ Y_{22} &= y_{21} + y_{22} + y_{23} \\ &= -j\,6 + 0 - j\,4 \\ &= -j\,10 \\ Y_{33} &= y_{31} + y_{32} + y_{33} \\ &= -j\,2 - j\,4 + 0 = -j\,6 \end{aligned}$$

Also we know that off-diagonal element is negative admittance between the nodes.

$$\begin{aligned} Y_{12} &= -y_{12} = j\,6 \\ Y_{13} &= -y_{13} = j\,2 \\ Y_{21} &= -y_{21} = j\,6 \\ Y_{23} &= -y_{23} = j\,4 \\ Y_{31} &= -y_{31} = j\,2 \\ Y_{32} &= -y_{32} = j\,4 \end{aligned}$$

Now, The bus admittance matrix is given by

$$Y_{bus} = \begin{matrix} y_{11} & y_{12} & y_{13} \\ y_{21} & y_{22} & y_{23} \\ y_{31} & y_{32} & y_{33} \end{matrix}$$

$$= \begin{matrix} -j\,8 & j\,6 & j\,2 \\ j\,6 & -j\,10 & j\,4 \\ j\,2 & j\,4 & -j\,6 \end{matrix} \quad \textbf{Ans.}$$

Now, Let us calculate bus power.

As we know that powers for generator bases are taken as positive while that for load bus are taken as negative

At bus 2

$$P_2 = \frac{P_G - P_L}{\text{base } MVA} = \frac{30 - 60}{200} = \frac{-60}{200} = -0.3 \text{ p.u.}$$

$$Q_2 = \frac{Q_G - Q_L}{\text{base MVA}} = \frac{20 - 30}{200} = \frac{-10}{200} = -0.05 \text{ p.u.}$$

At bus B,

$$P_3 = \frac{P_G - P_L}{\text{base MVA}} = \frac{0-70}{200} = \frac{-70}{200} = -0.35 \text{ p.u.}$$

$$Q_3 = \frac{Q_G - Q_L}{\text{base MVA}} = \frac{0-35}{200} = -0.175 \text{ p.u.}$$

According to Guass-Seidel Method,

The voltage at bus K at the $(r + 1)$ th iteration is given by

$$V_{k1}^{(r+1)} = \frac{1}{Y_{kk}}\left[\frac{P_k - j\theta_k}{\left(V_k^{(r)}\right)} - Y_{k1}V_1^{(r+1)} \ldots\ldots Y_{kk-1}V_{k-1}^{(r+1)} \ldots\ldots \gamma_{kn}V_n^{(r)}\right]$$

For first iteration

Here, $V_1 = 1.01\angle 0°$, $V_K^{(0)} = 1\angle 0°$ for $K = 2, 3$

Now, the voltage at bus 2 in given by

$$\text{V}_2^{(1)} = \frac{1}{Y_{22}}\left[\frac{P_k - j\theta_2}{\left(V_k^{(0)}\right)} - Y_{21}V_1 - Y_{23}V_3^{(0)}\right]$$

$$= -\frac{1}{j10}\ \frac{-0.3 + j\,0.05}{(1\angle 0°)} - (j\ 6) \times 1.01\angle 0° - (j\ 4\ 1\angle 0°)$$

$$= -\frac{1}{j10}\ [-0.3 + j\ 0.05 - j\ 6.06 - j\ 4]$$

$$= \frac{j}{10}\ [-0.3 - j\ 10.01] = 1.001 - j\ 0.03$$

$$\Delta V_2^{(1)} = V_2^{(1)} - V_2^{(0)} = 1.001 - j\ 0.03 - (1 + j\ 5.0)$$

$$= 0.001 - j\ 0.03$$

The accelerated valve of $V_2^{(1)}$ is given by

$$V^{(1)}_{2(acc)} = V_2^{(0)} + \Delta V_2^{(1)} = 1\angle 0° + 1.6\ [0.001 - j\ 0.03]$$

$$= 1 + j\ 0.0 + 0.0016 - j\ 0.048$$

$$= 1.0016 - j\ 0.048$$

The voltage at bus 3 is given by

$$V_3^{(1)} = \frac{1}{Y_{33}}\left[\frac{P_3 - j\theta_k}{\left(V_3^{(r)}\right)} - Y_{31}V_1 - Y_{32}V_2^{(1)}\right]$$

$$= \frac{1}{-j6}\left[\frac{-0.35 + j\,0.175}{(1\angle 0°)} - (j\ 2) \times 1.01\angle 0° - (j\ 4) \times (1.0016 - 10.043)\right]$$

$$= \frac{j}{6}\ [-0.35 + j\ 0.175 - j\ 2.02 - j\ 4.0064 - 0.192]$$

$$= 1.170 - j\,0.0903$$

$$\therefore \quad \Delta V_3^{(1)} = V_3^{(1)} - V_3^{(0)}$$

$$= 1.170 - j\,0.0903 - (1 + j\,0.0)$$

$$= 0.170 - j\,0.0903$$

The accelerated value of $V_3^{(1)}$ value of $V_3^{(1)}$ is given by

$$V_{3ace}^{(1)} = V_3^{(0)} + \alpha \Delta V_3^{(1)}$$

$$= (1 + j\,0.0) + 1.6\,(0.170 - j\,0.0903)$$

$$= 1 + j\,0.0 + 0.272 - j\,0.1445$$

$$= 1.272 - j\,0.1445$$

The voltages at the end of first iteration are

$$V_1^{(1)} = 1.01\angle 0^\circ \text{ p.u.}$$

$$V_2^{(1)} = 1.0016 - j\,0.048 \text{ p.u.}$$

$$V_3^{(1)} = 1.272 - j\,0.1445 \text{ p.u.}$$

For second iteration

The voltage at bus 2 is given by

$$V_2^{(2)} = \frac{1}{y_{22}}\left[\frac{P_2 - j\theta_2}{\left(V_2^{(1)}\right)} - y_{21}V_1 - y_{23}V_3^{(1)}\right]$$

$$= \frac{1}{-j10}\left[\frac{-0.3 + j\,0.05}{(1.0016 + j\,0.048)} - (j\,6) \times (1 + 100) - (j\,4) \times (1.272 - j0.1445)\right]$$

$$= \frac{j}{10}\left[\frac{0.3041\angle +170.54^\circ}{1.0027\angle 2.74} - j\,6 \times j\,6.068 - 0.5680\right]$$

$$= \frac{j}{10}\,[0.3033\angle 16.78 - j\,6 - j\,6.088 - 0.5680]$$

$$= 1.2024 - j\,0.0864$$

$$\therefore \quad \Delta V_2^{(2)} = V_2^{(2)} - V_2^{(1)}$$

$$= 1.2024 - j\,0.0864 - (1.0016 - j\,0.048)$$

$$= 0.2008 - j\,0.0384$$

$$= 0.2008 - j\,0.0384$$

The accelerated value of V^2_2 is given by

$$V_{2acc}^{(2)} = V_2^{(1)} + \alpha \Delta V_2^{(2)}$$

$$= 1.0016 - j\,0.048 + 1.6\,(0.2008 - j\,0.0348)$$

$$= 1.3229 - j\,0.10368$$

The voltage at bus 3 is given

$$V_3^{(2)} = \frac{1}{Y_{33}}\left[\frac{P_2 - j\theta_3}{\left(V_3^{(1)}\right)} - y_{31}V_1 - y_{32}V_2^{(2)}\right]$$

$$= \frac{1}{-j6}\left(\frac{-0.35+j0.175}{1.272+j0.1445}\right)-(j\ 2) \times (1 + j\ 0) - (j\ 4)$$
$$\times (1.3229 - j\ 0.1037)]$$
$$= \frac{j}{6}\left(\frac{0.3913\angle 153.43}{1.2802\angle 6.48°}\right) - j\ 2 \times j\ 5.2916 - 0.4148]$$
$$= \frac{j}{6}[0.3057\angle 146.95 - j7.2916 - 0.4148]$$
$$= 1.1875 - j\ 0.1118$$
$$\therefore \quad \Delta V_3^{(2)} = V_3^{(2)} - V_3^{(1)}$$
$$= 1.1875 - j\ 0.1118 - (1.272 - j0.1445)$$
$$= -0.0845 - j\ 0.2563$$

The accelerated value of V_3 is given by

$$V_{3acc}^{(2)} = V_3^{(1)} + \alpha(\Delta V_3)^{(2)}$$
$$= 1.272 - j\ 0.1445 + 1.6\ (-0.0845 - j\ 0.2563)$$
$$= 1.1368 - j\ 0.5546$$

Thus, The bus voltage at the end of 2nd iteration are given by

$$V_1^{(2)} = 1.01 + j\ 0.0$$
$$V_2^{(2)} = 1.3229 - j\ 0.10368$$
$$V_3^{(2)} = 1.1368 - j\ 0.5546 \quad \textbf{Ans.}$$

Problem 16.8. Two generating stations A and B have full load capacities of 250 MW and 100 MW. The interconnector connecting the two stations has an induction motor/synchronous generator (Plant-C) of full load capacity 50 MW. The percentage change of speeds of *A*, *B* and C are 6, 5 and 4 respectively. The loads on the bus bars *A* and *B* are 100 MW and 60 MW respectively. Determine the load taken by the set C and indicate the direction in which the energy is flowing.

Solution: The circuit diagram of the given problem can be drawn as given in Fig. P. 16.8.

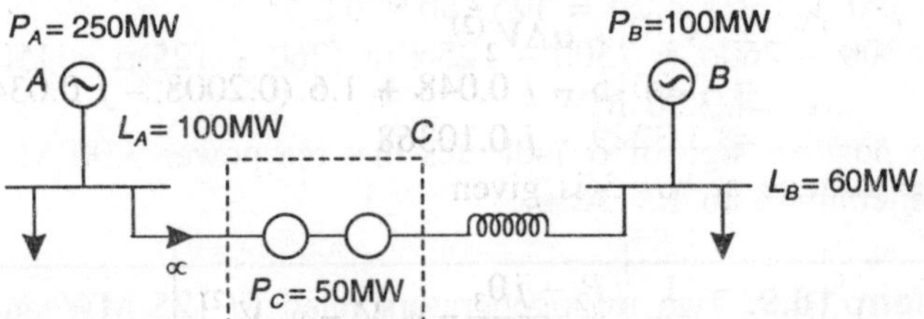

Fig. P. 16.8

Let α be the power in MW flows from generator A to generator B.

$\therefore$ Load on station A

$$= \text{Load on bus } A + \text{Power flow from } A \text{ to } B$$
$$= 100 + \alpha$$

$\therefore$ Load an station B

$$= \text{Load on bus } B - \alpha$$
$$= 60 - \alpha$$

Given, % change in speed of generator $A = 6$

$$\% \text{ Drop or in speed of generator } A = \frac{6}{P_A} \times (1000 + \alpha)$$

$$= \frac{6}{250}(100 + \alpha) = \frac{3}{125}(100 + \alpha)$$

$$\text{Similarly, \% drop in speed of generator } B = \frac{5}{P_B}(60 - \alpha)$$

$$= \frac{5}{100}(60 - \alpha) = \frac{60 - \alpha}{20}$$

Since the output of A corresponding to the power flow through the interconnector is the input to C, therefore, there is further reduction in frequency which is given by $\frac{4\alpha}{P_C}$, i.e. $\frac{4\alpha}{50}$ the total reduction in frequency both from A and C and the reduction in frequency from B when referred to bus B, should be same.

$$\therefore \quad \frac{3}{125}(100 + \alpha) + \frac{4\alpha}{50} = \frac{(60 - \alpha)}{20}$$

$$\text{or} \quad \frac{3}{125}(100 + \alpha) + \frac{2\alpha}{25} = \frac{60 - \alpha}{20}$$

$$\text{or} \quad \frac{3(100 + \alpha) + 10\alpha}{125} = \frac{60 - \alpha}{20}$$

or $(300 + 13\alpha) \times 20 = 125\,(60 - \alpha)$

or $6000 + 260\alpha = 7500 - 125\alpha$ or $(260 + 125)\alpha = 1500$

$\therefore \quad \alpha = 3.90$ MW

The positive sign of α indicates that the power 3.90 MW flows from generator A to B. **Ans.**

Problem 16.9. Two turbogenerators rated to 125 MW and 250 MW have governor drop characteristics of 10% from no-load to full load. They are connected imparalled to share a loud of 300 MW.

Determine the load shared by each machine assuming free governer action.

Solution: The circuit diagram of the problem is as below's

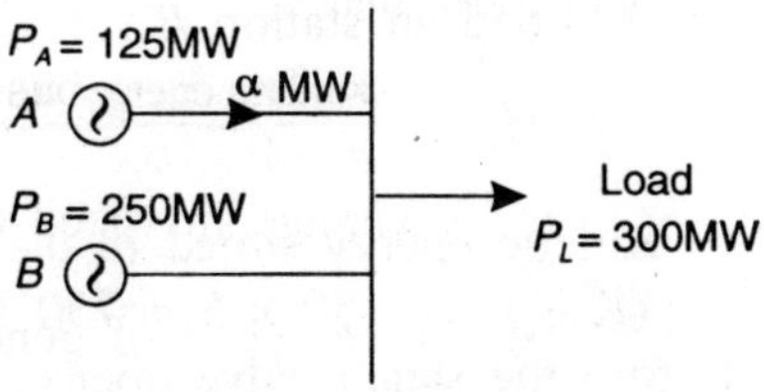

Fig. P. 16.9

Let α be the power in MW supplied to load by the generator A

∴ 300 – α be the power in MW supplied by the generator B

Given, % drop = 10,

∴ % Drop in speed of generator *A*

$$= \frac{10}{P_A} \times \alpha = \frac{10\alpha}{125}$$

∴ % Drop in speed of generator *B*

$$= \frac{10}{P_B}(300 - \alpha) = \frac{10}{250}(300 - \alpha)$$

Since the two machines are working in parallel, the percentage drop in frequency from both the machines due to different loadings must be same.

$$\therefore \quad \frac{10\alpha}{125} = \frac{10}{250}(300 - \alpha)$$

$$\text{or } 300 - \alpha = \left(\frac{250}{125}\right)\alpha \text{ or } 2\alpha + \alpha = 300$$

$$\therefore \quad \alpha = 100 \text{ MW}$$

Power shared by Generator *B* = 300 – 100 = 200 MW **Ans.**

and, Power shared by generator *A* = 200 MW **Ans.**

Problem 16.10. A 150 MVA 50 Hz turbogenerator operates at no load at 3000 rpm. A load of 40 MW is suddenly applied to the machine and the steam valves to the turbine commence to open after 0.8 see due to time – log in the governor system. Assuming inertia constant H of 5 kW – sec/MVA of the generating capacity, calculate the frequency to which the generated voltage drops before the steam flow commences to increase to meet the new load.

Solution: Given

Generator rating (G) = 150 MVA

Inertia constant (H) = 5 MW-sec/KVA

Time (t) = 0.8 sec

Load power (P_L) = 40 MW

As we know that

$$H = \frac{\text{Kinetic energy stored in the rotor at synchromous speed}}{\text{Rating}}$$

$\therefore$ Kinetic energy stored in the rotor = $G \times H$

$$(\text{K.}E_s) = 150 \times 5 = 750 \text{ MJ}$$

Before the steam valve opens. Energy lost by the rotor

$$= P_L \times t$$

$$\text{K.}E_L = 0.8 \times 40 = 32 \text{ MW}$$

Due to energy lost. There is reduction in speed of the rotor and therefore, reduction in frequency.

The new frequency is given by

$$f_n = \sqrt{\frac{k.E_S - k.E_L}{kE_S}} \times f \text{ Hz}$$

$$= \sqrt{\frac{750 - 32}{750}} \times 50 \text{ Hz}$$

$$= 48.92 \text{ Hz}$$ **Ans.**

Chapter 17

Distribution System

Problem 17.1. In a d.c. two wire system, a feeder is working on 240V supplying a constant load. If the supply voltage is increased to 440V with the same power transmitted, calculate percentage saving in conductor material.

Solution: Let P be the power transmitted in each case.

(i) When supply voltage is 240 V

Let l be the length of each conductor and j be the current density

Since power transmitted is P

$$\therefore \text{ Line current } (I_1) = \frac{P}{V} = \frac{P}{240}$$

$$\text{We know that, current density } (j) = \frac{\text{current}}{\text{Area}}$$

$$\therefore \text{ Area of the conductor } (A_1) = \frac{I_1}{j} = \frac{P}{240j}$$

∴ Volume of the conductor material required

$$V_1 = A_1 \times \text{Total length of the conductor}$$

$$= A_1 \times 2l = 2lA_1 = \frac{2lP}{240j}$$

(ii) When supply voltage is 440 V,

Power transmitted = P

$$\text{Line current } (I_1) = \frac{P}{V} = \frac{P}{440}$$

$$\therefore \text{ Area of the conductor } (A_2) = \frac{I_1}{j} = \frac{P}{440j}$$

$\therefore$ Volume of the conductor material $(V_2) = A_2 \times 2l = \dfrac{2lP}{440Vj}$

Now,

$$\frac{V_2}{V_1} = \frac{2Pl/440j}{2Pl/240j} \text{ or } \frac{V_2}{V_1} = \frac{240}{440}$$

$$\text{or } 1 - \frac{V_2}{V_1} = \frac{440-240}{440} = 0.4545$$

The % saving of conductor material is given by

$$= \left(\frac{V_2 - V_1}{V_1}\right) \times 100\% = \left(1 - \frac{V_2}{V_1}\right) \times 100\%$$

$$= 0.4545 \times 100\% = 45.45\%$$

$\therefore$ Required % saving of conductor material = 45.45% **Ans.**

Problem 17.2. A given amount of power is to be transmitted by an overhead line. Compare the diameter and weight of aluminium conductor with those of copper for the same power loss in the line. The specific data may be

Specific resistance of aluminium (δ_a) = 2.80 μΩ-cm

Specific resistance of copper (ρ_c) = 1.69 μΩ-cm

Specific gravity of aluminium (ρ_a) = 2.71

Specific gravity of copper (ρ_c) = 8.89

Solution: Let A_C be the area of copper conductor

A_a be the area of aluminium conductor and l be the length conductor. Again, let D_a and D_c be the diameter of aluminium and copper conductor respectively.

As we know that the resistance of a conductor is given by

$$R = \rho \frac{l}{A}$$

$\therefore$ Resistance of copper conductor $(R_c) = \rho_C \dfrac{l}{A_c}$

and that of aluminium conductor $(R_a) = \rho_a \dfrac{l}{A_a}$

For the same power to be transmitted,

$$R_a = R_c$$

$$\text{or } \rho_C \frac{l}{A_c} = \rho_a \frac{l}{A_a} \text{ or } \frac{A_a}{A_c} = \frac{\rho_a}{\rho_c} \text{ or } \frac{\pi D_a^2/4}{\pi D_a^2/4} = \frac{2.80}{1.69}$$

$$\therefore \quad \frac{D_a}{D_c} = \sqrt{\frac{2.80}{1.69}} = 1.287 \quad \textbf{Ans.}$$

Now, $\dfrac{\text{Weight of the aluminium}}{\text{Weight of the copper}}$

$$= \frac{\text{Specific gravityof } A_1 \times \text{volume of the aluminium conductor}}{\text{Specific gravity of cu} \times \text{volume of coppes conductor}}$$

$$= \frac{\rho a \times A_a \times 1}{\rho c \times A_c \times 1} = \frac{\rho a}{\rho c} \times \frac{\pi D_a^2/4}{\pi D_a^2/4} = \frac{\rho a}{\rho c} \times \frac{Da^2}{Dc^2}$$

$$= \frac{2.71}{8.89} \times \frac{2.80}{1.69} = 0.5069 \quad \textbf{Ans.}$$

Problem 17.3. A 100 km long transmission line supplies a load of 10 MVA at 0.8 pf. lagging at 32 kV. The efficiency of transmission line is 92%. Calculate the volume of the conductor aluminium required for the line when

(i) Single phase, 2-wire system is used.

(ii) Three phase, 3-wire system is used.

Take the resistively of aluminium (ρ_a) = 2.85 × 10^{-8} Ω-m

Solution: Given, power transmitted to load in MVA = 10 MVA

Load p.f. = 0.8 (lagging)

Efficiency of transmission line = 92%

∴ Power transmitted to the load = MVA × p.f

= 10 × 0.8 = 8 MW = 8 × 10^6 W

Length of the conductor (l) = 100 km = 100 × 10^3 m

Line loss (P_L) = (100 – 92) % of the power transmitted

= 8% of 8 × 10^6 W = 64 × 10^4 W

(i) When single-phase, 2-wire system is used.

Now,

$$\text{MVA} = V \times I \times 10^{-6} \text{ or } 10 = 32 \times 10^3 \times I \times 10^{-6}$$

$$\therefore \quad I = \frac{10 \times 10^6}{32 \times 10^3} = 312.5 \text{ A}$$

Let A_1 be the area of cross-section of each conductor, line loss,

$$(P_L) = 2I^2R = 2I^2 \times \rho_l \times \frac{l}{A_1}$$

$$\therefore \quad A_1 = \frac{2I^2\rho_l \times l}{P_L}$$

Hence, Volume of the conductor required

$$= A_1 \times 2l = \frac{4I^2\rho_l \times l^2}{P_L}$$

$$= \frac{4\times(312.5)^2\times 2.85\times 10^{-8}\times(10\times 10^3)^2}{64\times 10^4}$$

$$= 1.75 \text{ m}^3 \quad \textbf{Ans.}$$

(ii) When three phase, 3-wire system is used.

$$\text{MVA} = \sqrt{3}\times V_L \times I_L \times 10^{-6}$$

$$I_L = \frac{MVA}{\sqrt{3}\times I_L}$$

$$\therefore \quad I_L = \frac{10\times 10^6}{\sqrt{3}\times 32\times 10^3} = 180.42 \text{ A}$$

Let A_2 be the area of each phase conductor.

$$\therefore \text{ Total line loss } (P_L) = 3I_1^2R = 3I_L^2 \times \rho_l \times \frac{l}{A_2}$$

$$\therefore \quad A_2 = \frac{3I_L^2\rho_l \times l}{P_L}$$

$\therefore$ Volume of conductor material required

$$= 3l \times A_2 = 3l \times \frac{3I_L^2\rho_l \times l}{P_L}$$

$$= \frac{9I^2\rho_l \times l^2}{P_L}$$

$$= \frac{9\times(180.42)^2\times 2.85\times 10^{-8}\times(10\times 10^3)^2}{64\times 104}$$

$$= 1.305 \text{ m}^3 \quad \textbf{Ans.}$$

> **Problem 17.4.** An existing single phase line ac system comprising of two overhead conductors is to be converted into a 3-phase, 3-wire system by providing an additional similar conductor. Calculate the percentage of additional load that can be transmitted by the 3-phase system if the operating line voltage is increased to 150% and percentage line losses remain the same in both the cases.

Solution: Let V be supply voltage in the single phase system

$\therefore$ operating voltage for 3-phase system = 150% × V = 1.5 V

Again, Let I_1 be the single-phase current

I_3 be the three- phase current

R be the resistance of each conductor

Cos θ be the power factor in both the system.

(i) For single-phase system,

Power transmitted $(P_1) = VI_1 \cos\theta$ .... (1)

Line loss $(P_L) = 2\, I_1^2\, R$

$$\therefore \%\text{ line loss} = \frac{\text{Line loss}}{\text{Power transmitted}} \times 100\%$$

$$= \frac{P_L}{P_1} \times 100\%$$

(ii) For 3-phase system

power transmitted $(P_3) = \sqrt{3}\; V_L\, I_3 \cos\theta$

$$= \sqrt{3} \times 1.5\, VI_3 \cos\theta \qquad \text{.......... (2)}$$

Line loss $(P_{L3}) = 3\, I^2_3\, R$

$$\therefore \%\text{ Line loss} = \frac{P_{L3}}{P_3} \times 100\%$$

$$= \frac{3I_3^2 R}{\sqrt{3} \times 1.5V \times I_3 \cos\theta} \times 100\%$$

For the same percentage line loss

$$\frac{3I_3^2 R}{\sqrt{3} \times 1.5V \times I_3 \cos\theta} \times 100 = \frac{2I_1^2 R}{VI_1 \cos\theta} \times 100$$

$$\text{or} \quad \frac{3I_3}{\sqrt{3} \times 1.5} = 2\, I_1$$

$$\therefore I_3 = \frac{2 \times 1.5 \times \sqrt{3}}{3}\, I_1 = \sqrt{3}\, I_1$$

From equation (2)

Power transmitted

$$(P_3) = \sqrt{3} \times 1.5 \times V \times \sqrt{3}\ I_1 \cos\theta$$

$$= 4.5\ V\ I_1 \cos\theta$$

Now, % additional load

$$= \frac{P_3 - P_1}{P_1} \times 100\%$$

$$= \frac{4.5 VI_1 \cos\theta - VI_1 \cos\theta}{VI_1 \cos\theta} \times 100\%$$

$$= 3.5 \times 100\% = 350\ \%$$

Thus, 350% addition load can be transmitted by converting single phase to 3-phase with 150% of operating voltage

Problem 17.5. Estimate the weight of copper required for a three phase transmission system supplying 360 kV between the lines a load of 150 MW at a lagging power factor of 0.8. The length of the conductor is 100 km. The neutral point is earthed. The efficiency of the line is 90%. The resistance of the conductor 1 kg long and 1 sq cm is 0.040 ohm. The weight of 0.1 cm³ copper is 0.02 kg. Find the weight of the copper required.

Solution: Given,

Line voltage (V_L) = 360 kV = 360 $\times 10^3$

Length of the line (l) = 100 km = 100 $\times$ 10^3 m

Load power (P) = 150 MW, p.f. = 0.8

Efficiency of the line (η_L) = 90%

$$\text{Resistivity of the conductor } (\rho) = \frac{R \times A}{l}$$

$$= \frac{0.040 \times 1 \times 10^{-4}}{10^{-3}}\ \Omega\text{-m} = 4.0 \times 10^{-5}\ \text{m}$$

$$\text{Load current } (I_L) = \frac{P}{\sqrt{3} \times V_L \times p.f}$$

$$= \frac{150 \times 10^6}{\sqrt{3} \times 360 \times 10^3} = 240.56\ \text{A}$$

$\therefore$ Power loss in the line (P_L) = (100 – 90)%

$$= 10\% \text{ of Load} = \frac{10}{100} \times 150 = 15\ \text{MW}$$

$\therefore$ I^2R loss in the conductors are 15×10^6 W

$$\therefore\ I^2R \text{ loss perconductor} = \frac{15\times10^6}{3} = 5 \times 10^6 \text{ W}$$

$$\therefore\ \text{Resistance per conductor} = \frac{5\times10^6}{(I^2)}\ \Omega = \frac{5\times10^6}{(240.56)^2}$$

$$= 86.40\ \Omega$$

$$\text{or Resistance per conductor per km} = \frac{86.40}{100} = 0.864\ \Omega/\text{km}$$

$$\text{Conductor area (A)} = \frac{0.040}{0.864} = 0.0463 \text{ sq.cm}$$

Volume of the copper per metre = 0.0463×100 cm^3 = 4.63 cm^3

Weight of the copper per km run (w) = $4.63 \times 0.02 = 0.0926$ kg

Weight of copper for three conductor (w) = $W \times l \times 3$

$= 0.0926 \times 100 \times 10^3 \times 3 = 27{,}780$ kg **Ans.**

Problem 17.6. Find the most economical cross-section of a 3-core distributor cable 300m long supplying a load of 100 kW at 440 volts and 0.8 p.f. lagging for 4200 hours per annum and open circuited for the remainder of the year. The cost of cable including installation is Rs (20 a + 30) per meter where a is the area of each conductor in sq cm. Interest and depreciation total 10% and cost of energy wasted is 10 p per unit. The resistance per km of conductor of 1 cm^2 crosssection is 0.150 ohm.

Solution: Given,

Load power (p) = 100 kW

Operating voltage (V_2) = 440V, p.f. = 0.8

$$\text{Line current } (I_L) = \frac{P}{\sqrt{3}\times V_L \times \cos\theta}$$

$$= \frac{100\times10^3}{\sqrt{3}\times440\times0.8} = 164.02 \text{ A}$$

Energy wasted per annum

$$= [3I^2\ \frac{R}{a} \times l \times t] \times 10^{-3} \text{ kWh}$$

$$= 3 \times (164.02)^2 \times \frac{0.150}{a} \times 300 \times 10^{-3} \times 4200$$

$$\times 10^{-3} \text{ kWh}$$

$$= \frac{15254}{a} \text{ kWh}$$

Annual cost of energy wasted as losses

$$= \text{Cost per unit} \times \text{Total unit}$$

$$= \left(\frac{10}{100} Rs\right) \times \frac{15254}{a} = Rs\ \frac{1525.4}{a} = \frac{P_3}{a}$$

Capital cost of cable

$$= \text{Rs}\ (20\ a + 30) \times l$$

$$= \text{Rs}\ (20\ a + 30) \times 300$$

Annual cost of the cable

$$= (20\ a + 30) \times 300 \times \frac{10}{100}\ \text{Rs}$$

$$= (20\ a + 30) \times 30\ \text{Rs}$$

$$= \frac{600a}{P_2} + \frac{900}{P_1}$$

For most economical cost,

$$P_1 a = \frac{P_3}{a} = \frac{1525.4}{a} = 600\ a$$

$$\therefore \quad a = \sqrt{\frac{1525.4}{600}}\ \text{sq. cm} = 1.59\ \text{sq. cm}$$

Thus most economical area of the conductor = 1.59 sq.cm **Ans.**

Problem 17.7. The cost per km for each of the copper conductor of a section a sq. cm for a transmission line is Rs (25000 a + 1000). The load factor of the load current is 90 % and load factor for the losses is 70%. The rate of interest and depreciation is 12% and the cost of energy is 20 paisa per kWh. Find the most economical current density for the transmission line by the use of Kelvin's law.

Given $\rho = 1.76 \times 10^{-8}\ \Omega$-m

Solution: Given,

P_2 = Rs 2500 sq. cm, P_1 = 1000, r = 12%

Load factor = 90%

Load factor for losses = 70%, $\rho = 1.76 \times 10^{-8}\ \Omega$-m

Cost of energy per unit (c) = 20 paise/kWh.

Let l be the length of the conductor in km

For most economical area of the conductor

$$P_2 \times a = \frac{P_3}{a} \text{ or } a^2 = \frac{P_3}{P_2}$$

$$\therefore \quad a = \sqrt{\frac{P_3}{P_L}} \qquad \text{............ (1)}$$

Since current density (J)

$$= \frac{\text{Current (I)}}{\text{Area of cros-section}} \text{ or } J = \frac{I}{a}$$

$$\therefore \quad J = I\sqrt{\frac{P_2}{P_3}} \qquad \text{............ (2)}$$

Here $P_2 = 2500 \times l \times r = 2500 \times l \times \frac{12}{100}$

$= 300\ l$

and $\frac{P_3}{a} = C \times I^2 \times \frac{S \times l}{a} \times \frac{8760}{1000} \times$ loss factor for losses

$$= \text{Rs } \frac{20}{100} \times I^2 \times \frac{1.76 \times 10^{-8} \times l \times 1000}{a \times 10^{-4}} \times \frac{8760}{1000} \times 0.70$$

$\therefore \quad P_3 = 21.581 \times 10^{-2} \times I^2 l$

Now, from equation (2)

$$J = I\sqrt{\frac{300 \times l}{21.584 \times 10^{-2} \times I^2 l}} \text{ amp/sq. cm}$$

$= 37.28$ amp/sq.cm **Ans.**

Problem 17.8. A three phase cable supplies an industry whose maximum demand is 2 MW and whose total annual energy consumption is 4×10^6 kWh at 0.8 p.f. The capital cost of cable is Rs (70,000 a + 50,000) per km where a is its cross-sectioned area in sq-cm of each core, the energy cost 10 paise/kWh. The rate of interest and depreciation is 15%. The specific resistance of copper is 1.72×10^{-6} Ω/cm^2. The operating voltage is 15 kV. Loss load factor = 0.20 load factor + 0.80 (load factor)^2.

Solution: Given,

$$P = 2 \text{ MW} = 2 \times 10^6 \text{ W}$$

Operating voltage (V) = 15 kV, p.f. = 0.8, r = 15%

Total annual energy consumption = 4×10^6 kWh

Capital cost = Rs (70,000 a + 50,000)

This yeilds,

$$P_2 = \text{Rs. } 70{,}000 \text{ /sq.cm} \times l \times r$$

$$= \text{Rs } 70{,}000 \times l \times \frac{15}{100} = \text{Rs } 10500\ l$$

$$P_1 = \text{Rs } 50{,}000$$

Now, Load current

$$= \frac{P}{\sqrt{3} \times V_2 \times p.f} = \frac{2 \times 10^6}{\sqrt{3} \times 15 \times 10^3 \times 0.8}$$

$$= 96.23 \text{ A}$$

As we know

$$\text{Load factor} = \frac{\text{Annual energy consumption}}{\text{Maximun demand}}$$

$$= \frac{4 \times 10^6}{2 \times 10^3 \times 8760} = 0.228$$

∴ Loss load factor

$$= 0.20 \times \text{Load factor} + 0.80 \text{ (load factor)}^2$$

$$= 0.020 \times 0.228 + 0.80 \times (0.228)^2$$

$$= 0.0872$$

Hence,

$$\frac{P_3}{a} = \frac{3I^2 Rt}{1000} \times \text{loss load factor} \times \text{Energy cost/kWh}$$

$$= 3 \times \frac{(96.23)^2}{1000} \times \frac{\rho l}{A} \times 8760 \times 0.0872 \times \frac{10}{100} \times 8760 \text{ Rs}$$

$$= 3 \times \frac{(96.23)^2}{1000} \times \frac{1.72 \times 10^{-6} \times l \times 1000}{a \times 10^{-4}} \times 0.0872 \times \frac{10}{100} \times 8760 \text{ Rs}$$

$$\therefore \quad P_3 = \text{Rs } 36499.87\ l$$

Now, The most economical conductor size is given by

$$a = \sqrt{\frac{P_3}{P_2}} = \sqrt{\frac{36499.87l}{10500l}} = 1.864 \text{ sq.cm}$$

Size of the conductor = 1.864 sq. cm. **Ans.**

Table A.1 Electrical characteristics of bare aluminum conductors steel-reinforced †

Code word	Aluminium area, cmil	Stranding Al/St	Layers of aluminium	Outside diameter, in	Resistance: Dc, 20°C, Ω/1,000 ft	Resistance: Ac, 60 Hz, 20°C, Ω/mi	Resistance: Ac, 60 Hz, 50°C, Ω/mi	GMR D_s, ft	Reactance per conductor 1-ft spacing, 60 Hz: Inductive X_a, Ω/mi	Reactance per conductor 1-ft spacing, 60 Hz: Capacitive X'_a, MΩ-mi
Waxwing	266,800	18/1	2	0.609	0.0646	0.3488	0.3831	0.0198	0.476	0.1090
Partridge	266,800	26/7	2	0.642	0.0640	0.3452	0.3792	0.0217	0.465	0.1074
Ostrich	300,000	26/7	2	0.680	0.0569	0.3070	0.3372	0.0229	0.458	0.1057
Merlin	336,400	18/1	2	0.684	0.0512	0.2767	0.3037	0.0222	0.462	0.1055
Linnet	336,400	26/7	2	0.721	0.0507	0.2737	0.3006	0.0243	0.451	0.1040
Oriole	336,400	30/7	2	0.741	0.0504	0.2719	0.2987	0.0255	0.445	0.1032
Chickadee	397,500	18/1	2	0.743	0.0433	0.2342	0.2572	0.0241	0.452	0.1031
Ibis	397,500	26/7	2	0.783	0.0430	0.2323	0.2551	0.0264	0.441	0.1015
Pelican	477,000	18/1	2	0.814	0.0361	0.1957	0.2148	0.0264	0.441	0.1004
Flicker	477,000	24/7	2	0.846	0.0359	0.1943	0.2134	0.0284	0.432	0.0992
Hawk	477,000	26/7	2	0.858	0.0357	0.1931	0.2120	0.0289	0.430	0.0988
Hen	477,000	30/7	2	0.883	0.0355	0.1919	0.2107	0.0304	0.424	0.0980
Osprey	556,500	18/1	2	0.879	0.0309	0.1679	0.1843	0.0284	0.432	0.0981
Parakeet	556,500	24/7	2	0.914	0.0308	0.1669	0.1832	0.0306	0.423	0.0969
Dove	556,500	26/7	2	0.927	0.0307	0.1663	0.1826	0.0314	0.420	0.0965
Rook	636,000	24/7	2	0.977	0.0269	0.1461	0.1603	0.0327	0.415	0.0950
Grosbeak	636,000	26/7	2	0.990	0.0268	0.1454	0.1596	0.0335	0.412	0.0946
Drake	795,000	26/7	2	1.108	0.0215	0.1172	0.1284	0.0373	0.399	0.0912
Tern	795,000	45/7	3	1.063	0.0217	0.1188	0.1302	0.0352	0.406	0.0925
Rail	954,000	45/7	3	1.165	0.0181	0.0997	0.1092	0.0386	0.395	0.0897
Cardinal	954,000	54/7	3	1.196	0.0180	0.0988	0.1082	0.0402	0.390	0.0890
Ortolan	1,033,500	45/7	3	1.213	0.0167	0.0924	0.1011	0.0402	0.390	0.0885
Bluejay	1,113,000	45/7	3	1.259	0.0155	0.0861	0.0941	0.0415	0.386	0.0874
Finch	1,113,000	54/19	3	1.293	0.0155	0.0856	0.0937	0.0436	0.380	0.0866
Bittern	1,272,000	45/7	3	1.345	0.0136	0.0762	0.0832	0.0444	0.378	0.0855
Pheasant	1,272,000	54/19	3	1.382	0.0135	0.0751	0.0821	0.0466	0.372	0.0847
Bobolink	1,431,000	45/7	3	1.427	0.0121	0.0684	0.0746	0.0470	0.371	0.0837
Plover	1,431,000	54/19	3	1.465	0.0120	0.0673	0.0735	0.0494	0.365	0.0829
Lapwing	1,590,000	45/7	3	1.502	0.0109	0.0623	0.0678	0.0498	0.364	0.0822
Falcon	1,590,000	54/19	3	1.545	0.0108	0.0612	0.0667	0.0523	0.358	0.0814
Bluebird	2,156,000	84/19	4	1.762	0.0080	0.0476	0.0515	0.0586	0.344	0.0776

† Most used multilayer sizes.
Data, by permission, from Aluminum Association, "Aluminum Electrical Conductor Handbook," New York, September 1971.

Table A.2 ***ABCD*** **constants for various networks .**

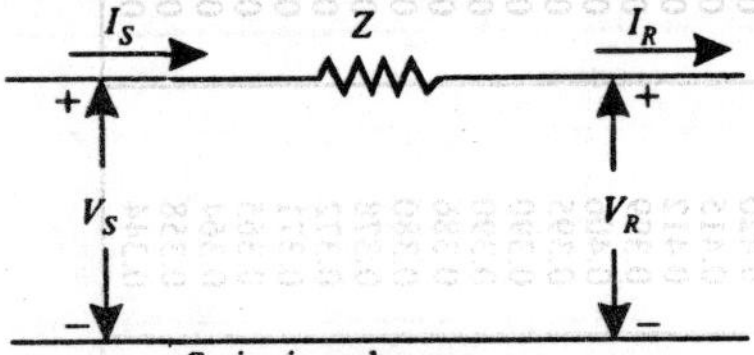

Series impedance

$A = 1$
$B = Z$
$C = 0$
$D = 1$

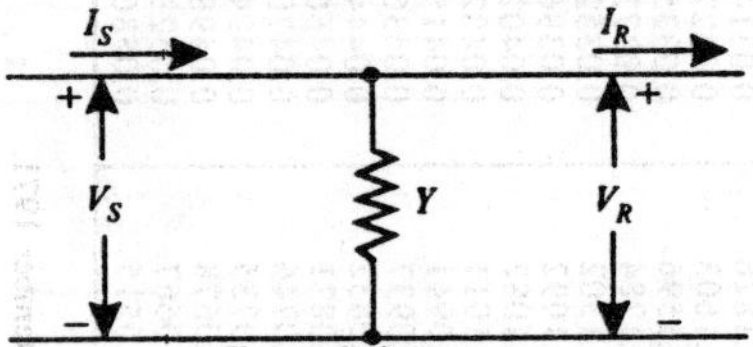

Shunt admittance

$A = 1$
$B = 0$
$C = Y$
$D = 1$

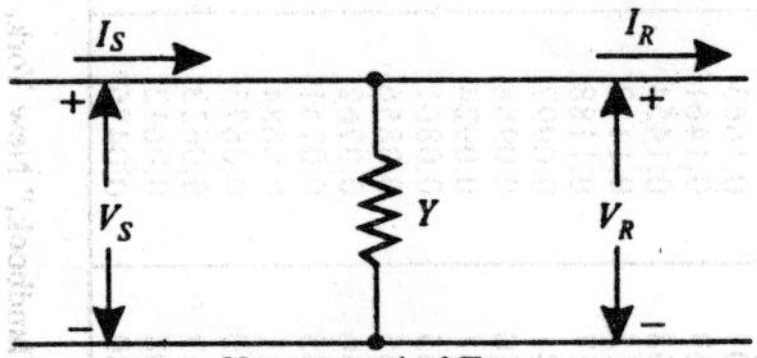

Unsymmetrical T

$A = 1 + YZ_1$
$B = Z_1 + Z_2 + YZ_1Z_2$
$C = Y$
$D = 1 + Yz_2$

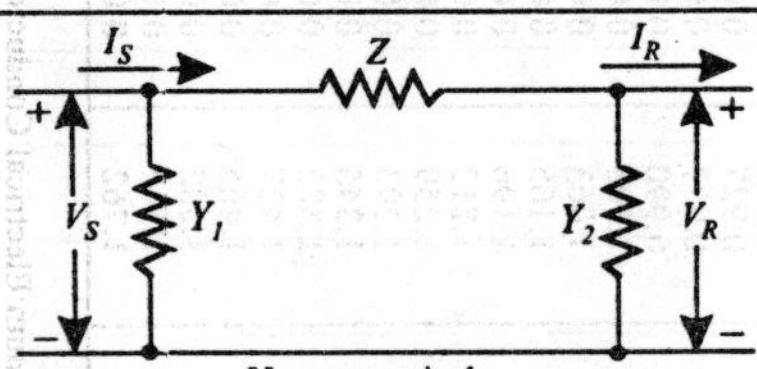

Unsymmetrical π

$A = 1 + Y_2 Z$
$B = Z$
$C = Y_1 + Y_2 + ZY_1Y_2$
$D = 1 + Y_1 Z$

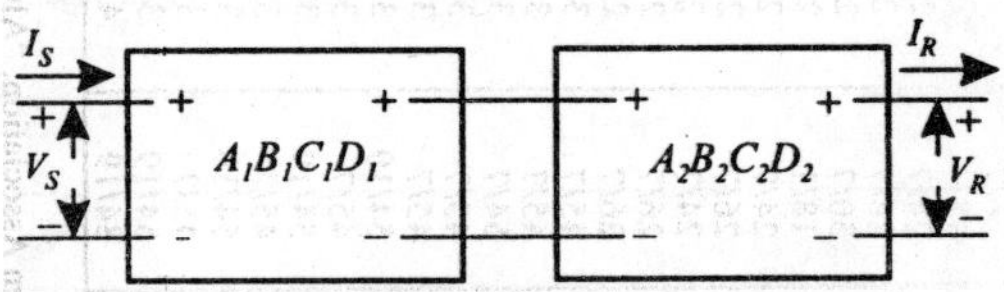

Network in series

$A = A_1A_2 + B_1C_2$
$B = A_1B_2 + B_1D_2$
$C = A_2C_1 + C_2D_1$
$D = B_2C_1 + D_1D_2$

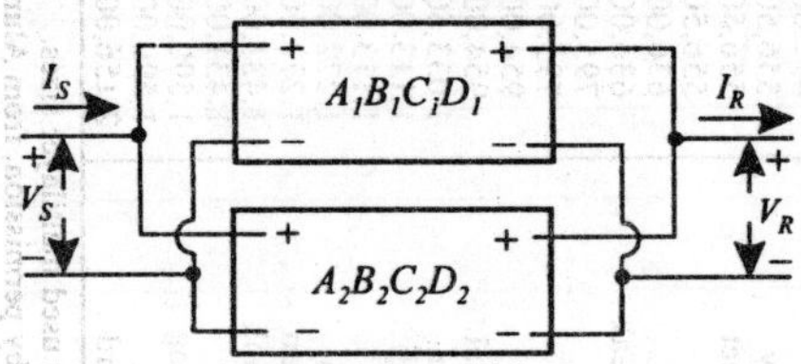

Network in parallel

$A = (A_1B_2 + A_2B_1) / (B_1 + B_2)$
$B = B_1B_2 / (B_1 + B_2)$
$C = C_1 + C_2 + (A_1 - A_2)(D_2 - D_1) / (B_1 + B_2)$
$D = (B_2D_1 + B_1D_2) / (B_1 + B_2)$

Reader's Notes

Reader's Notes

Reader's Notes